Thermal Effects of High Power Laser Energy on Materials

Bahman Zohuri

Thermal Effects of High Power Laser Energy on Materials

Springer

Bahman Zohuri
Galaxy Advanced Engineering Inc.
Albuquerque, NM, USA

ISBN 978-3-030-63063-8   ISBN 978-3-030-63064-5 (eBook)
https://doi.org/10.1007/978-3-030-63064-5

This Springer imprint is published by the registered company Springer Nature Switzerland AG
The registered company address is: Gewerbestrasse 11, 6330 Cham, Switzerland

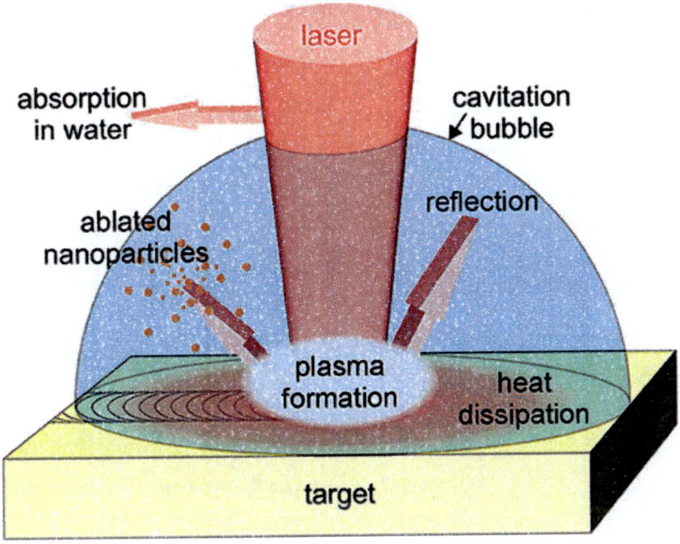
laser
absorption
in water
cavitation
bubble
reflection
ablated
nanoparticles
plasma
formation
heat
dissipation
target

This book is dedicated to my son Sasha, grandson Dariush, and my granddaughter Donya.

Preface

In this book, a tutorial discussion of the response of materials to laser radiation is presented, emphasizing simple, intuitive models. Topics discussed include optical reflectivity of metals at infrared (IR) wavelengths, laser-induced heat flow in materials, the effects of melting and vaporization, the impulse generated in materials by pulsed radiation, and the influence of the absorption of laser radiation in the blow-off region in front of the irradiated material.

The book aims to investigate the relations existing between the thermal parameters of simple metals and to check these relations based on available experimental data. This book also looks at the characteristics of materials and their prosperities, so one can investigate and perform theoretical analysis from heat conduction perspective mathematically.

The motivation of this investigation is to discover the constancy of a certain combination of thermal parameters, which is a mathematical condition for the linearization of one-dimensional, nonlinear, partial differential equation (PDE) of heat conduction. By solving these PDEs either numerically or to find a close solution by methods such as perturbation theory, Laplace transformation, or in some cases, asymptotic methods, and dimensional analysis. Using the Eulerian or Lagrangian approach, the solutions to boundary conditions and heat conduction problems in materials can be obtained.

The relationship of the above-mentioned combination of thermal parameters to results derived from the theory of solids was investigated, and applications of the resulting linearized or nonlinearized equations to problems in heat conduction are considered.

In this book we also have introduced commercially available and unclassified computer codes that deal with the responses of materials to laser radiation. Codes like PUFF-TFT, MOM, Asthma, ACE, and others are well described, and their capabilities are well introduced. Most of these codes were developed under some sort of government contracts during the era of Strategic Defense Initiative (SDI) in the past, and presently commercial and defense companies have them under development.

Chapter 1 of the book gives detailed descriptions of high-power laser energy. It covers the high-power laser and its energy definition. This chapter presents a tutorial discussion of the material responses to high-power laser energy radiations, with emphasis on simple, intuitive models. The topics discussed include optical reflectivity of metals at infrared (IR) wavelengths, laser-induced heat flow in materials, the effects of melting and vaporization, the impulse generated in materials by pulsed radiation, and the influence of the absorption of laser radiation in the blow-off region in front of the irradiated material.

Chapter 2 covers the material characteristics and their properties, where in modern society we are surrounded by an amazing variety of materials. Most of the materials discussed in this book are solids that have been modified from their natural states to make them more suitable for practical applications. The materials we use have such an impact on our lifestyles that historical eras have been named after them. Ancient artifacts found by archeologists have been dated and analyzed to reveal the increasing sophistication of the manufacturing methods. Historians have shown that technological advancements created new tools for agriculture and new weapons for armies. Explorers established trade routes to redistribute raw materials and finished products. Modern culture is also influenced by the availability of new materials.

Chapter 3 is the touching base of heat conduction in solids from holistic science of physics and mathematical theory point of view. One of the major goals of physics and so as ours in the case of laser in particular and its interaction with materials is to understand the nature of light. Due to the complex nature of light, this goal is very difficult to achieve, but this complication means that light offers many opportunities for different applications including optical interferences and the responses of materials to laser radiation and its interaction with matter and specifically metallic materials. This leads to dealing with the heat conduction equation in a very complex form; therefore, we need to deal with heat transport from the aspect of *conduction*, *convection*, and *radiation*.

Then comes Chap. 4. We have covered an introduction to electromagnetics, so our reader would refresh their background in physics and electrical engineering, when it comes to an understanding of the responses of materials to laser radiations, covered in Chap. 5. In order to comprehend laser interaction with matter, one needs to have some basic knowledge of the electromagnetic field in addition to conduction of heat in solids that was described in the previous chapter. Moreover, a general understanding of electromagnetic wave equation and consequently its behavior is necessary when it comes to laser irradiation. Thus, the Maxwell's equations and the electric field of the electromagnetic wave, and in this chapter, we try to introduce all these at a very holistic wave and leave the granule aspect of it to the readers so they can refer themselves to more detailed related books.

Chapter 5 covers the topic of the responses of materials to high-power laser energy radiation. The word laser is an acronym for light amplification by stimulated emission of radiation, although its common usage today is its noun form—laser—rather than as its acronym, LASER. A laser is a device that creates and amplifies a narrow, intense beam of coherent light. Atoms emit radiation. We see it every day when the "excited" neon atoms in a neon sign emit light. Normally, they radiate their light in random directions at random times. The result is incoherent light—a technical term for what you would consider a jumble of photons going in all directions. The trick in generating coherent light, of a single or just a few frequencies going in one precise direction, is to find the right atoms with the right internal storage mechanisms and create an environment in which they can all cooperate—to give up their light at the right time and all in the same direction.

Chapter 6 talks about the laser surface processing, where we cover high-power laser (HEL) and its application in the military for the purpose of precession as well as in industries as a cutting tool. High-power lasers (HPLs) have been used to process materials to improve their wear resistance since the late 1960s. Laser surface processing encompasses established techniques, such as laser transformation hardening, which has achieved production-level capabilities, and techniques that are still in the development stage, such as laser cladding, an emerging viable alternative to other hard-facing processes, such as plasma deposition. The laser beam is a chemically clean light source that delivers a precisely controlled quantity of energy to localized regions. Hence, only those areas exposed to wear are processed for improvement in wear resistance. The laser beam is easily maneuvered by optical elements and can be adapted to automation. High-power densities and low interaction times result in rapid heating and cooling, which produces a shallow heat-affected zone, low distortion of the workpiece, and minimal deterioration of bulk properties. Rapid solidification effects produce refined and novel microstructures that result in improved properties.

Finally, Chap. 7 would cover existing computer codes that this author has been involved with either to improve them to a customized level for a particular application or to use them both in the military and industries.

In this chapter, we will study and explain most of the computer codes that were developed during the administration of President Reagan and his idea of Strategic Defense Initiative (SDI) during the period 1980–1990. In this regard, scientists were looking at developing some means of computation capabilities that would deal with x-ray deposition into the matter and interaction of thermal and hydrodynamics of high-power laser energy with matter. These codes, in particular, were developed under the auspices of the Department of Energy (DOE) by tasking the Air Force Research Laboratory (AFRL) in support of ground-based laser (GBL), space-based laser (SBL), as well as airborne laser (ABL) with its supporting component known as Relay Mirror Experiment (RME) in order to overcome the effects and impacts of thermal blooming during the transport of the above laser system through the Earth's atmosphere and be able to shoot down any target around the globe.

Albuquerque, NM, USA, 2016 Bahman Zohuri

Acknowledgments

I am indebted to the many people who aided, encouraged, and supported me beyond my expectations. Some are not around to see the results of their encouragement in the production of this book, yet I hope they know of my deepest appreciations. I especially want to thank my true close friends, to whom I am deeply indebted, who have continuously given their support without hesitation. They all have always kept me going in the right direction.

Above all, I offer very special thanks to my late mother and father and to my children, in particular, my son Sasha, my daughters Natasha and Natalie, as well as my grandson Dariush and my granddaughter Donya. They have provided constant interest and encouragement, without which this book would not have been written. Their patience with my many absences from home and long hours in front of the computer to prepare the manuscript are especially appreciated.

Contents

About the Author

Bahman Zohuri currently works for Galaxy Advanced Engineering, Inc., a consulting firm that he started in 1991 when he left both the semiconductor and defense industries after many years of working as a chief scientist. After graduating from the University of Illinois in the field of physics and applied mathematics, he went to the University of New Mexico, where he studied nuclear engineering and mechanical engineering. He joined Westinghouse Electric Corporation, where he performed thermal hydraulic analysis and studied natural circulation in an inherent shutdown and heat removal system (ISHRS) in the core of a liquid metal fast breeder reactor (LMFBR) as a secondary fully inherent shutdown system for secondary loop heat exchange. All these designs were used in nuclear safety and reliability engineering for a self-actuated shutdown system. He designed a mercury heat pipe and electromagnetic pumps for large pool concepts of a LMFBR for heat rejection purposes for this reactor in 1978, for which he received a patent. He was subsequently transferred to the defense division of Westinghouse, where he oversaw dynamic analysis and methods of launching and controlling MX missiles from canisters. The results were applied to MX launch seal performance and muzzle blast phenomena analysis (i.e., missile vibration and hydrodynamic shock formation). Dr. Zohuri was also involved in analytical calculations and computations in the study of nonlinear ion waves in rarefying plasma. The results were applied to the propagation of so-called soliton waves and the resulting charge collector traces in the rarefaction characterization of the corona of laser-irradiated target pellets. As part of his graduate research work at Argonne National Laboratory, he performed computations and programming of multi-exchange integrals in surface physics and solid-state physics. He earned various patents in areas such as diffusion processes and diffusion furnace design while working as a senior process engineer at various semiconductor companies, such as Intel Corp., Varian Medical Systems, and National Semiconductor Corporation. He later joined Lockheed Martin Missile and Aerospace Corporation as a senior chief scientist and oversaw research and development (R&D) and the study of the vulnerability, survivability, and both radiation and laser hardening of different components of the Strategic Defense Initiative, known as Star Wars.

This included payloads (i.e., IR sensor) for the Defense Support Program, the Boost Surveillance and Tracking System, and Space Surveillance and Tracking Satellite against laser and nuclear threats. While at Lockheed Martin, he also performed analyses of laser beam characteristics and nuclear radiation interactions with materials, transient radiation effects in electronics, electromagnetic pulses, system-generated electromagnetic pulses, single-event upset, blast, thermo-mechanical, hardness assurance, maintenance, and device technology.

He spent several years as a consultant at Galaxy Advanced Engineering, Inc., serving Sandia National Laboratories, where he supported the development of operational hazard assessments for the Air Force Safety Center in collaboration with other researchers and third parties. Ultimately, the results were included in Air Force Instructions issued specifically for directed energy weapons operational safety. He completed the first version of a comprehensive library of detailed laser tools for airborne lasers, advanced tactical lasers, tactical high-energy lasers, and mobile/tactical high-energy lasers.

He also oversaw SDI computer programs, in connection with Battle Management C^3I and artificial intelligence, and autonomous systems. He is the author of several publications and holds several patents, such as for a laser-activated radioactive decay and results of a through-bulkhead initiator. He has published many books and articles with different publishers and publishing companies; they all can be found on amazon.com or researchgate.net as well as on the Internet if a search is performed under his name.

Chapter 1
High-Power Laser Energy

This chapter goes through high-power laser and its energy definition. This chapter presents a tutorial discussion of the material responses to high-power laser energy radiations, with emphasis on simple, intuitive models. Topics discussed include optical reflectivity of metals at infrared (IR) wavelengths, laser-induced heat flow in materials, the effects of melting and vaporization, the impulse generated in materials by pulsed radiation, and the influence of the absorption of laser radiation in the blow-off region in front of the irradiated material.

1.1 Introduction

Throughout this chapter, we will discuss high-power laser and its energy definition and all related applications from industrial and military applications. High-power lasers emit very high optical powers, as required for a number of applications such as laser material processing; material processing such as welding, cutting, drilling, soldering, marking, and surface modification; as well as weapon of directed energy dwelling with its target of interest as part of its military application. Other applications are also considered, such as large-scale laser displays (i.e., red, green, and blue (RGB) source), remote sensing (e.g., with Light Detection and Ranging, LiDAR), medical applications (e.g., surgery), military applications (e.g., antimissile weapons), fundamental science (e.g., particle acceleration), and laser-induced nuclear fusion (e.g., in the National Ignition Facility (NIF) project). Note that in principle LASER stands for Light Amplification by Stimulated Emission of Radiation.

B. Zohuri, *Thermal Effects of High Power Laser Energy on Materials*,
https://doi.org/10.1007/978-3-030-63064-5_1

Lasers with extreme high-power laser energies (HPLEs) are required for a number of applications such as:

- *Material processing* (i.e., welding, precision cutting, drilling, soldering, marking surface modification). Laser beam welding is dominant in the automotive industry. It is used in the area where large-volume production is required.

  The principle of laser beam welding equipment along with its advantages and disadvantages are described as below.

  Laser beam welding (LBW) is a fusion welding process in which two metal pieccs are joined together by the use of laser. The laser beams are focused to the cavity between the two metal pieces to be joined. The laser beams have enough energy and when it strikes the metal pieces produce heat that melts the material from the two metal pieces and fills the cavity. After cooling a strong weld is formed between the two pieces.

  It is a very efficient welding process and can be automated with robotics machinery easily. This welding technique is mostly used in the automotive industry.

  It works on the principle that electrons of an atom get excited by absorbing some energy. And then after some time when it returns back to its ground state, it emits a photon of light. The concentration of this emitted photon is increased by stimulated emission of radiation, and we get a high-energy concentrated laser beam.

  As Fig. 1.1 illustrates, the main parts of a laser welding equipment and beam welding are:

  1. **Laser machine:** It is a machine that is used to produce a laser for welding. The main components of the laser machine are shown in Fig. 1.1.
  2. **Power source:** A high-voltage power source is applied across the laser machine to produce a laser beam.
  3. **CAM:** It is a computer-aided machining in which the laser machine is integrated with the computers to perform the welding process. All the controlling action during the welding process by laser is done by CAM. It speeds up the welding process to a greater extent.
  4. **CAD:** It is called a computer-aided design. It is used to design the job for welding. Here computers are used to design the workpiece and how the welding is performed on it.
  5. **Shielding gas:** A shielding gas may be used during the welding process in order to prevent the welding processes from oxidation.

  The types of lasers that are used in welding industry mainly are as follows:

  1. **Gas lasers:** It uses mixtures of gases as a lasing medium to produce laser. Mixtures of gases such as nitrogen, helium, and CO_2 are used as the lasing medium.
  2. **Solid-state laser:** it uses several solid media such as synthetic ruby crystal (chromium in aluminum oxide), neodymium in glass (Nd:glass), and neodymium in yttrium aluminum garnet (Nd:YAG, most commonly used).

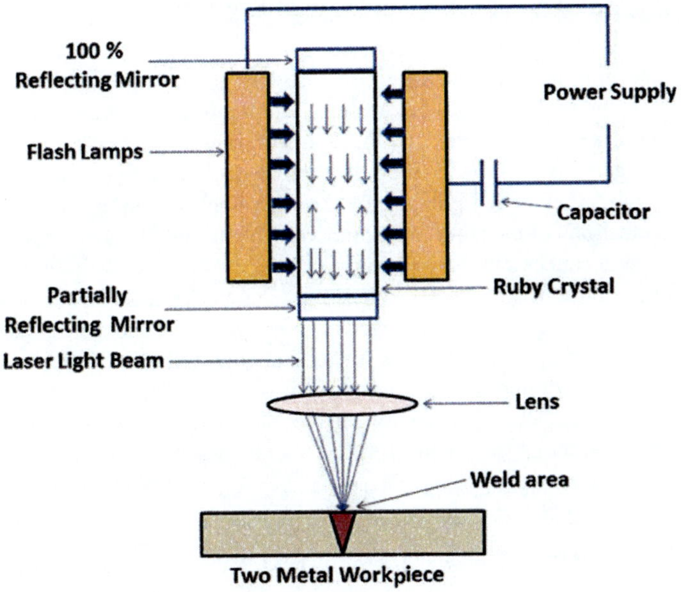

Fig. 1.1 Laser Beam Welding Equipment. *(Courtesy of* theweldingmaster.com)

3. **Fiber laser:** The lasing medium in this type of laser is optical fiber itself.

The characteristics of laser beam welding could be listed as follows:

1. The power density of laser beam welding is high. It is of the order 1 MW/cm^2. Because of this high energy density, it has small heat-affected zones. The rate of heating and cooling is high.
2. The laser beams produced are coherent (having the same phase) and monochromatic (i.e., having the same wavelength).
3. It is used to weld smaller sizes spot, but the spot sizes can vary from .2 mm to 13 mm.
4. The depth of penetration of the LBW depends upon the amount of power supply and location of the focal point. It is proportional to the amount of power supply. When the focal point is kept slightly below the surface of the workpiece, the depth of penetration is maximized.
5. Pulsed or continuous laser beams are used for welding. Thin materials are weld by using millisecond pulses, and continuous laser beams are used for deep welds.
6. It is a versatile process because it is capable of welding carbon steels, stainless steel, HSLA steel, aluminum, and titanium. Due to the high cooling rate, the problem of cracking is there when welding high-carbon steels.
7. It produces high-quality weld.
8. This welding process is the most popular in the automotive industry.

As per Fig. 1.1, laser welding typically works in the following steps:

- First, the setup of the welding machine at the desired location (in between the two metal pieces to be joined) is done.
- After setting it up, a high-voltage power supply is applied to the laser machine. This starts the flashlamps of the machine, and it emits light photons. The energy of the light photon is absorbed by the atoms of ruby crystal and electrons excited to their higher energy level. When they return back to their ground state (lower energy state), they emit a photon of light. This light photon again stimulates the excited electrons of the atom and produces two photons. This process continues, and we get a concentrated laser beam.
- This high concentrated laser beam is focused to the desired location for the welding of the multiple pieces together. Lens is used to focus the laser to the area where welding is needed. CAM is used to control the motion of the laser and workpiece table during the welding process.
- As the laser beam strikes the cavity between the two metal pieces to be joined, it melts the base metal from both the pieces and fuses them together. After solidification, we get a strong weld.
- This is how a laser beam welding works.

The advantages and disadvantages of the laser welding tool are listed below.

Advantages

- It produces high weld quality.
- LBW can be easily automated with robotic machinery for large-volume production.
- No electrode is required.
- No tool wears because it is a non-contact process.
- The time taken for welding thick section is reduced.
- It is capable of welding areas which are not easily accessible.
- It has the ability to weld metals with dissimilar physical properties.
- It can be weld through air and no vacuum is required.
- X-ray shielding is not required as it does not produce any X-rays.
- It can be focused on small areas for welding. This is because of its narrower beam of high energy.
- A wide variety of materials can be welded by using laser beam welding.
- It produces a weld of aspect ratio (i.e., depth-to-width ratio) of 10:1.

Disadvantages

- The initial cost is high. The equipment used in LBW has a high cost.
- High maintenance cost.
- Due to the rapid rate of cooling, cracks may be produced in some metals.
- High skilled labor is required to operate laser beam welding (LBW).
- The welding thickness is limited to 19 mm.
- The energy conversion efficiency in LBW is very low. It is usually below 10%.

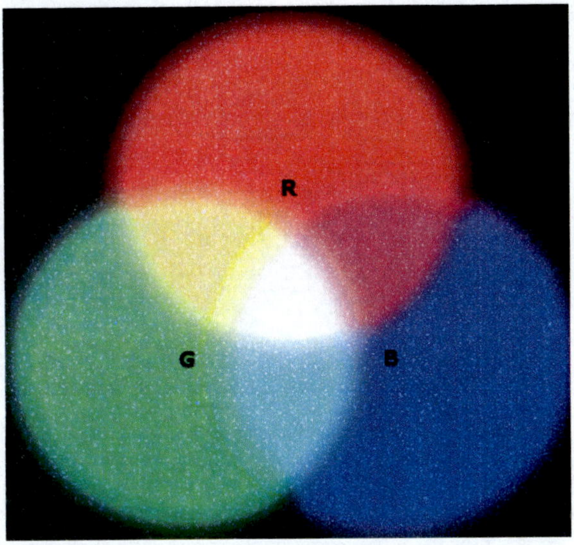

Fig. 1.2 Additive mixing of red, green, and blue light

Material processing with high-power lasers is the second largest segment of laser applications concerning global turnovers (after communications).

- **Large-scale laser displays** (i.e., red, green, blue (RBG) light sources, which are usually provided in the form of laser beams). An RGB source is a light source which emits at the same time red, green, and blue light. Such sources are required mainly for color display applications, for example, for large-screen video shows. A wide range of colors can be obtained by mixing different amounts of red, green, and blue light (*additive color mixing*; see Fig. 1.2).

A possible combination of wavelengths is 630 nm for red, 532 nm for green, and 465 nm for blue light. As Fig. 1.2, illustrates, superimposing inputs from three light sources can lead to white light, seen in the center. A wide range of other color tones could be achieved by mixing variable intensities of those contributions.

Many currently used projection displays ("beamers") are based on an *arc lamp*, combined with various color *filters*. While an arc lamp is much cheaper than a *laser* source of comparable output power, it does not allow for a high *wall-plug efficiency*, it has a limited lifetime, and the poor spatial *coherence* of the output imposes restrictions on the achievable image quality. Also, the available color space is not very large. Therefore, laser sources are under investigation that could offer a wider color space, much better spatial coherence (*beam quality*), and higher power efficiency. Such RGB laser sources emit light in the form of *laser beams*—either one beam for each color or all colors combined in one beam.

There are special RGB sources which can be used for stereoscopic displays, called 6P RGB sources. For each of the red, green, and blue colors, they have two spectral components which differ in wavelength, e.g., by 20 nm. For viewing, one

Fig. 1.3 Light Detection and Ranging (LiDAR)—GIS Lounge. (Source: USGS)

can then use special filter glasses which transmit only one set of wavelengths (R1, G1, B1) for the left eye and another set (R2, G2, B2) for the right eye. That way, one can produce independent images for the two eyes [1].

- **Remote sensing** (e.g., LiDAR). The acronym LiDAR stands for, basically, in principle, Light Detection and Ranging. LiDAR was originally understood as *laser radar* but is nowadays taken as an acronym meaning *Light Detection and Ranging*, which is somewhat more general (Fig. 1.3). It can be used for the creation of three-dimensional images based on *distance measurements*, e.g., with the *time-of-flight method*. In other cases, one measures other quantities such as the concentration of substances in air or wind velocities [2].

The capitalization of the term LiDAR is not consistent in the literature. Many authors simply write *lidar* (just like *radar*), although for acronyms full capitalization is more common.

LiDAR's first application was in the 1960s and came soon after the invention of lasers. Used in combination with radar, LiDAR was originally used in atmospheric studies to measure clouds by the National Center for Atmospheric Research. LiDAR later became more known by the public when the Apollo 15 mission used it to map the surface of the moon more accurately than ever before.

LiDAR has many other practical applications including mapping crop yields in agriculture, mapping terrestrial and extraterrestrial bodies, studying the atmosphere, and contributing to military technology. LiDAR is also used to detect objects underground and is vitally important for archeologists. LiDAR can find previously unmapped or unknown fault lines contributing to localized earthquakes as well as faults in building structures.

Also bear in mind that there exists equipment known as Laser Radar (LADAR) that stands for Laser Detection And Ranging as well. LADAR systems use light to determine the distance of an object, as illustrated in Fig. 1.4.

Fig. 1.4 LADAR Images. (Source: US Army provided by short-wave infrared (SWIR) camera of www.swirconops.com)

Note that in Fig. 1.4, on the left is a LADAR image from the front of a vehicle stopped at a crosswalk in Santa Barbara, CA. On the right is the same LADAR data "viewed" from an overhead location highlighting the 3-D nature of the data collected.

Many short-wave infrared (SWIR) camera capabilities are considered sensitive. If you would like to learn more about SWIR capabilities for the military, you can register at a special website, www.swirconops.com. You must be a US Government employee or a USG contractor and be a US Citizen to access this site.

Military application of SWIR presently is utilized in tracking of rocket and missile launches, flight paths, and intercept.

Visible and thermal tracking of rockets and missiles benefits from the addition of short-wave infrared (SWIR) cameras. Visible cameras suffer from atmospheric distortion and haze over long distances, while imaging with the short-wave infrared cuts through the interferences. Visible cameras only see the hottest portions of the rocket plume and have difficulty following missiles over long distances or bright ambient day lighting. The hot gases, however, are much more visible to SWIR and thermal cameras (also known as mid- and long-wave infrared cameras).

The thermal cameras easily see the thermal plume, but they lack the ability to see the hard body or debris because they aren't equally as hot. Nor can they "see" with the resolution of SWIR cameras. The InGaAs-based SWIR cameras share the visible camera's ability to use glass based optics, requiring less expensive lens solutions. SWIR cameras also operate with lower power and size than thermal cameras, allowing for more platform flexibility. Awaiting similar test confirmation is our new DR1 High Dynamic Range control algorithm, a hardware solution that will increase the ability of InGaAs cameras to extract additional imaging information from such high contrast scenes.

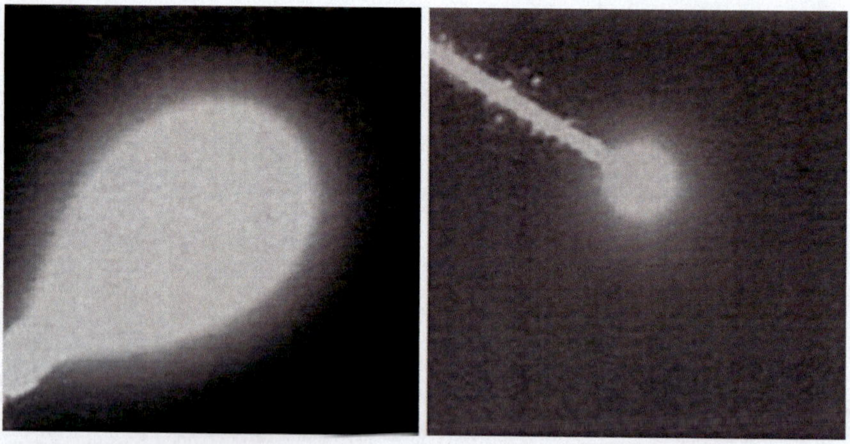

Fig. 1.5 Flying debris tracking. (Source: US Army provided by short-wave infrared (SWIR) camera of www.swirconops.com)

In the image below, on the left, a missile is launched into the air as a SWIR camera tracks its path. Hot debris is shown coming off of the plume. On the right, the descending missile is still visible to a SWIR camera even from many miles away. Flying debris can still be detected at this distance as demonstrated in Fig. 1.5.

Note that in Fig. 1.5, Sensors Unlimited offers video and high-frame rate InGaAs cameras with 320×256 or 640×512 formats that are ideal for integration on missile tracker systems or perimeter defense scanning for shoulder-fired rockets.

Founded in 1991 as an advanced research and development company, Sensors Unlimited quickly grew as a world leader in and innovator of indium gallium arsenide semiconductor technology. Tracking missile and rockets is only one of the many applications for our short-wave infrared technology; SUI products are also used for night vision, laser designation, and the tracking of lasers.

Many SWIR camera capabilities are considered sensitive. If you would like to learn more about SWIR capabilities for the military, you can register at a special website, www.swirconops.com. You must be a US Government employee or a USG contractor and be a US Citizen to access this site [3, 4].

Since the speed of light is well-known, LADAR can use a short-pulsed laser to illuminate a target and then time how long it takes the light to return. The advantage of LADAR over RADAR (radio detection and ranging) is that LADAR can also image the target at the same time determine the distance. This allows a 3-D view of the object in question. This provides long-range reconnaissance with greater fidelity and thus greater recognition range than other technologies.

Newer LADAR systems, with InGaAs detectors on board, can use eye-safe lasers (traditionally 1.55 or 1.57 microns) to minimize the users' eye damage, as well as eye damage to other combatants and non-combatants on the battlefield. In addition to eye safety concerns, these lasers are also used as night vision goggles and in older night vision technologies, allowing for the location of the detected object and the laser to be kept secret.

- **Medical applications** (e.g., surgery). There are various kinds of lasers which are applied for medical purposes. In most cases, they essentially do a kind of laser material processing, just on biological materials. In a few cases, laser light is used in other ways, for example, for triggering photochemical reactions.

Lasers are not only used for medical therapies but also for diagnostic purposes and for medical research. For example, there are methods of ocular imaging with optical coherence tomography (which may involve wavelength-swept lasers), laser microscopy, and laser spectroscopy, which all can be useful for planning treatments. Optogenetics is another important field in medical research which profits from laser technology.

Besides lasers, medical laser system may contain various other kinds of photonics and optics technology, e.g., frequency converters, photodetectors for monitoring purposes, cameras, and beam delivery systems.

Generally, medical laser systems need to fulfill relatively strict conditions. In particular, they need to be highly reliable, since a failure during a medical operation may have serious consequences. Their performance parameters must be well reproducible and precisely monitored during operation. Various legal conditions (e.g., fulfilling certain standards), which can substantially vary between countries, must be fulfilled before such lasers can be applied.

The following sections describe some typical applications of medical lasers. Note, however, that the article cannot completely cover the huge field of medical applications, with many new applications being under development [5].

1. **Lasers in Ophthalmology**

   Ophthalmology, i.e., the branch of medicine and surgery dealing with the diagnosis and treatment of problems of the eye, is a classical field of medical laser applications. This is essentially because the eye, including essential inner parts like the retina (the light-sensitive region), is well accessible by light, while other approaches, e.g., to the retina are difficult.
2. **Lasers in Dermatology**

   The skin is also a particularly accessible part for laser radiation, and various laser treatment methods have been developed for treatment of different skin conditions. For example, vascular and pigmented lesions can be treated, also scars and wrinkles. Further, lasers are frequently applied for the removal of tattoos and for hair removal.
3. **Lasers in Dentistry**

   A common application is the preparation of cavities in the context of curing caries. This can be done with short pulses from a Q-switched Er:YAG laser. Compared with conventional drilling, this procedure may incur less pain.

   There are various other laser applications in dentistry, which concern either the hard tissue (teeth) or soft tissues, e.g., for exposing unerupted or partially erupted teeth.
4. **Lasers in Urology**

   A popular method is the treatment of benign prostatic hyperplasia, which causes problems for many men. There are endoscopes containing optical fibers which allow minimally invasive operations, where high-power laser light, e.g., from laser diodes is used to ablate tissues. Compared with open operations, there are reduced risks of complications, and the recovery of the patient is quicker.

Laser lithotripsy has been developed for removing stones in the urinary tract. Different microscopic mechanisms can be used to break apart stones with intense laser pulses, such that the remaining small pieces can be easily washed out of the urinary tract. One mostly uses Q-switched holmium lasers emitting around 2.1 μm, a spectral region with good absorption in water. However, it is currently investigated whether the use of thulium fiber lasers could be advantageous due to advantages like a better beam quality and higher pulse repetition rates. The method appears to be a good alternative to extracorporeal shockwave lithotripsy with ultrasound waves.

Lasers are also used for treating urothelial tumors, prostate cancer, urolithiasis, and urinary tract strictures, for example.

5. **Lasers in General Surgery**

Lasers may also be used in other kinds of surgery for cutting (*laser scalpel*), ablating, or cauterizing tissues. A wide range of laser types can in principle be used, including CO_2 lasers, diode lasers, fiber lasers, gas lasers, and excimer lasers. Both continuous-wave operation and pulsed operation can be useful.

Typical advantages of laser surgery are that the affected tissue are automatically sterile and the tendency for bleeding is reduced through the effect of photocoagulation. However, there may also be disadvantages, such as a lack of haptic feedback and a low processing speed. In some cases, lasers need to be developed, e.g., with increased pulse repetition rates for faster processing.

- **Military applications** (e.g., directed-energy weapons). Directed-energy weapons are nothing new to mankind; historically the origin of such weapons began centuries ago when the famous Greek mathematician, physicist, engineer, inventor, and astronomer Archimedes of Syracuse used different mirrors to collect sunbeams and focused them on the Roman fleet in order to destroy enemy ships with fire. This is known as the Archimedes heat ray. Archimedes may have used mirrors acting collectively as a parabolic reflector to burn ships attacking Syracuse. The device was used to focus sunlight onto approaching ships, causing them to catch fire. Of course, the myth or reality of the Archimedes heat ray still is questionable, but with the help of a group of students from Massachusetts Institute of Technology, certain experiments were carried out with 127 1-foot (30 cm) square mirror tiles in October of 2005 that were focused on a mockup wooden ship at a range of about 100 feet (30 m). The flames broke out on a patch of the ship, but only after the sky had been cloudless and the ship had remained stationary for around 10 min. It was concluded that the device was a feasible weapon under these conditions.

The battles of tomorrow will be fought with different weapons that have more lethal effects and faster delivery systems. One of mankind's greatest achievements in the twentieth century is the ability to destroy the entire human race several times over. At this time of intensive arms, more money is invested in the next generation of

Fig. 1.6 Overview of the National Ignition (NIF) Facility

weapons. It is in the best interest of every citizen to be aware and able to make an informed judgment on the best possible direction for the arms race. Offensive or defensive weapons are a cruel reality that nevertheless must be addressed [6].

- **Laser-induced nuclear fusion** (e.g., National Ignition Facility (NIF)) project is taking advantages and process of the extreme conditions of matter that are encountered both in nature and in the laboratory, for example, in the center of stars and in relativistic collisions of heavy nuclei, now inertial confinement fusion (ICF), where a temperature of 10 °K and a pressure exceeding a billion atmospheres can be achieved (Fig. 1.6). A sound knowledge of the equation of state is a prerequisite for understanding processes of very high temperature and pressures, as noted in some recent developments [7, 8].

The National Ignition Fusion (NIF) experiment in Lawrence Livermore National Laboratory in California is a large laser-based inertial confinement fusion research device using a laser to confine fuel of D-T, in a pellet, as illustrated in Fig. 1.6.

Fundamental science (e.g., particle acceleration). At present the highest energy accelerators are all circular colliders, but both hadron accelerators and electron accelerators are running into limits. Higher-energy hadron and ion cyclic accelerators will require accelerator tunnels of larger physical size due to the increased beam rigidity.

For cyclic electron accelerators, a limit on practical bend radius is placed by synchrotron radiation losses, and the next generation will probably be linear accelerators 10 times the current length. An example of such a next-generation electron accelerator is the proposed 40-km-long International Linear Collider.

It is believed that plasma wakefield acceleration in the form of electron beam "afterburners" and standalone laser pulsers might be able to provide dramatic increases in efficiency over RF accelerators within two to three decades. In plasma wakefield accelerators, the beam cavity is filled with a plasma (rather than vacuum). A short pulse of electrons or laser light either constitutes or immediately precedes the particles that are being accelerated. The pulse disrupts the plasma, causing the charged particles in the plasma to integrate into and move toward the rear of the bunch of particles that are being accelerated. This process transfers energy to the particle bunch, accelerating it further, and continues as long as the pulse is coherent [9].

Energy gradients as steep as 200 GeV/m have been achieved over millimeter-scale distances using laser pulsers [10], and gradients approaching 1 GeV/m are being produced on the multi-centimeter scale with electron beam systems, in contrast to a limit of about 0.1 GeV/m for radio-frequency acceleration alone. Existing electron accelerators such as that of the Stanford Linear Accelerator Center (SLAC) could use electron beam afterburners to greatly increase the energy of their particle beams, at the cost of beam intensity. Electron systems in general can provide tightly collimated, reliable beams; laser systems may offer more power and compactness. Thus, plasma wakefield accelerators could be used—if technical issues can be resolved—to both increase the maximum energy of the largest accelerators and to bring high energies into university laboratories and medical centers.

Higher than 0.25 GeV/m gradients have been achieved by a dielectric laser accelerator, [11] which may present another viable approach to building compact high-energy accelerators [12]. Using femtosecond duration laser pulses, an electron accelerating gradient 0.69 Gev/m was recorded for dielectric laser accelerators [13]. Higher gradients of the order of 1 to 6 GeV/m are anticipated after further optimizations [14].

1.2 Heat Conduction in Simple Metals

Material processing with high-power lasers is the second largest segment of laser applications concerning global turnovers (after communications).

There is no commonly accepted definition of the property "high power"; in the context of laser material processing, it usually means multiple kilowatts or at least a few hundred watts, whereas for laser displays, some tens of watts may already be considered high. In some areas, this label is assigned simply for generating a significantly higher output power than other lasers based on the same technology; for example, some "high-powered" laser pointers emit a few hundred milliwatts, whereas ordinary laser pointers are limited to a few milliwatts.

Additional aspects come into play for pulsed lasers. For example, the peak power may be as important as the average output power for a Q-switched laser. Depending on the pulse repetition rate and pulse duration, the peak power may be very high even for a laser with a moderate average output power. Usually, a high average power and not only a high peak power is expected from a high-power laser.

Theoretically the relations that exist between the thermal parameters of simple metals can be investigated, and we can check these relations in more detail in the following chapters of this book. The motivation behind such investigation is the discovery of the constancy of a certain combination for the linearization of the one-dimensional, non-linear, partial differential equation of heat conduction. The relationship of the abovementioned combination of thermal parameters to result derivable from the theory of solid will also be investigated in further chapters, and applications of the resulting linearized equation to problems in heat conduction will be considered.

1.3 Technical Challenges

One of the technical challenges about producing a laser beam as a high-power energy is its generation. The generation of high optical power in lasers involves the following number of technical issues:

1. One requires one or several powerful pump sources. While lamp pumping was originally the only viable approach for most solid-state lasers, pumping with high-power laser diodes (diode bars or diode stacks) has become more and more widespread. Diode-pumped lasers now offer the highest output powers in continuous-wave operation. For very high pulse energies (e.g., tens of joules), lamp pumping is still more practical.
2. At least for long-term continuous-wave operation, a high wall-plug efficiency is an important economic factor. Unfortunately, various technical challenges (e.g., thermal effects; see below) tend to make it more difficult at very high power levels to achieve a good efficiency.
3. Even in a fairly efficient gain medium, a substantial fraction of the pump power is converted into heat, which can have a number of detrimental side effects. In the worst case, thermally induced stress leads to fracture of the laser crystal. High-power solid-state lasers also exhibit strong thermal lensing, making it substantially more difficult to achieve a high beam quality. In lasers with polarized output, depolarization loss often compromises the efficiency. Efficient heat removal and thermal management are therefore important issues, and additional measures (e.g., in the context of resonator design) are often required for coping with various kinds of thermal effects.
4. Particularly in Q-switched lasers, very high optical intensities can occur, which may lead to laser-induced damage of optics (such as laser mirrors), e.g., via laser-induced breakdown. Even if the optical intensities remain well below the damage threshold of all optical elements, tiny dust particles can provoke damage phenomena. It can therefore be essential to keep the laser setup very clean, e.g., by operating it in a sealed case which may be opened only in a clean room. In addition, it can be imperative to use precision optics with a high optical damage threshold.

5. Laser resonators with large effective mode areas tend to be sensitive to misalignment and vibrations of optical components. It can therefore be more challenging to achieve robust maintenance-free operation and a good beam pointing stability.

As it can be observed from the above points, to produce a high-quality high-power laser energy, we need to technically overcome these challenges [5].

1.4 Types of High-Power Lasers

There are several different types of high-power lasers.

- High-power *diode bars* and *diode stacks* have already been mentioned above as possible pump sources for *solid-state lasers*. They allow the generation of kilowatts of output power, but with a poor *beam quality*. For some applications, where beam quality is not essential, the direct use of high-power laser diodes (→*direct diode lasers*), e.g., for laser welding, soldering and brazing, cladding, and heat treatment, is an interesting option, offering a comparatively simple, compact, cost-effective, and energy-efficient solution.
- There are various types of lamp-pumped or diode-pumped solid-state bulk lasers. Rod lasers can be optimized for several kilowatts of output power, but diffraction-limited beam quality is possible only up to a few hundred watts (with significant efforts). Slab lasers can be developed for tens of kilowatts or more with relatively high beam quality. Thin-disk lasers easily generate hundreds of watts with diffraction-limited beam quality and have the potential to reach that even at power levels well above 10 kW. The power efficiency is usually fairly good.
- High-power fiber lasers and amplifiers can generate up to a few kilowatts with close to diffraction-limited beams and high power efficiency. With relaxed beam quality requirements, even significantly higher powers are possible. Strictly, such fiber devices are often not lasers, but master oscillator power amplifier (MOPA) configurations.
- Some gas lasers, e.g., CO_2 lasers and excimer lasers, are also suitable for hundreds or thousands of watts of output power. They typically operate in different other regions than solid-state lasers, e.g., in the mid-infrared or ultraviolet region.
- There are chemical lasers with multi-kilowatt or even megawatt output powers, explored e.g., in the context of antimissile weapons.

A perhaps not very practical but theoretically very interesting high-power laser concept is that of the radiation-balanced laser. Here, heat generation in the gain medium is essentially eliminated by optical refrigeration.

An aspect of great importance for further laser development is that of *power scaling*, based on certain power-scalable laser architectures. Even for non-power-scalable laser types, it can be very helpful to understand the scaling properties of various parts or techniques.

1.5 Laser-Related Safety Issues

Laser has become a common tool of civilians and soldiers all over the world. Many lasers, perhaps most of them, are in some way dangerous to people. For several reasons, it is our eyesight that is most threatened, but there are many other dangers to deal with as well. Laser safety is a very complex problem.

The use of high-power lasers raises important issues on laser safety:

- The output powers are far higher than what any eye can tolerate, so that even tiny parasitic reflections must be safely prevented from reaching an eye. Even the use of strongly absorbing laser goggles may not be sufficient, as such glasses may not be able to stand such high optical intensities for more than a brief moment of time.
- The skin and clothes of workers are also at risk in environments where optical powers and intensities are sufficient, e.g., for laser cutting of metals.
- High-power laser beams may incinerate materials such as plastics or wood. That happens easily already for laser powers of the order of 1 W. Fire protection is therefore an important issue. Also, the formation of poisonous fumes needs to be avoided, or such fumes have to be efficiently removed.
- There are various types of risks which are not related to laser beams themselves. In particular, high-power electric power supplies often involve high electric voltages, which can cause electric shocks. Power cables, which can be damaged in a harsh industrial environment, can also create hazards.

An important safety principle in the area of high-power lasers is to enclose the laser setup with a solid housing and ideally also the whole area where dangerous laser beams can be present. Interlocks can prevent the operation of a laser at times where persons are in a hazardous area.

The use of lasers almost always carries with it some kind of danger, either at the laser site itself or wherever there is a direct, reflected, or scattered laser beam. At the laser site, it is not only the actual laser beam which can be dangerous, but electrical, chemical, and other hazards exist as well. Most laser power supplies can cause server electrical shocks, possibly even electrocution. Furthermore, many highly explosive and toxic substances are used in solid, liquid, or gas form to power laser cooling systems. For many reasons, it is useful to divide the hazards from laser beam into two main groups: those to the eye and those to the skin. The eyes may be severely damaged and even permanently blinded by rather low-energy laser beams, while the skin is not nearly as sensitive. To get severe skin burns in the visible and infrared part of the spectrum, it is normally necessary to use a very high-energy laser beam, which delivers, at least, several watts per second square centimeter (W/cm^2) to the target. Safety threshold limits both for the skin and eyes are well-defined and have resulted in very strict safety regulation [6].

There are a variety of laser safety standards including federal and state regulations and nonregulatory standards. The most important and most often quoted is the American National Standards Institute's Z136 series of laser safety standards.

These standards are the foundation of laser safety programs in industry, medicine, research, and government. The ANSI Z136 series of laser safety standards are referenced by the Occupational Safety and Health Administration (OSHA) and many US states as the basis of evaluating laser-related occupational safety issues.

ANSI Z136.1 Safe Use of Lasers, the parent document in the Z136 series, provides information on how to classify lasers for safety, laser safety calculations and measurements, laser hazard control measures, and recommendations for Laser Safety Officers and Laser Safety Committees in all types of laser facilities. It is designed to provide the laser user with the information needed to properly develop a comprehensive laser safety program [6].

For manufacturers of laser products, the standard of principal importance is the regulation of the Food and Drug Administration's (FDA) Center for Devices and Radiological Health (CDRH), which regulates product performance. All laser products sold in the United States since August 1976 must be certified by the manufacturer as meeting certain product performance (safety) standards, and each laser must bear a label indicating compliance with the standard and denoting the laser hazard classification [6].

The establishment of the threshold levels for different laser injuries is basic to the whole question of laser safety. The threshold level is an exposure value below which adverse changes have a low probability of occurrence and no significant risk exists. There is always some question about what the actual value of the threshold is, because it varies with the wavelength and exposure duration but also with the individual.

The value of the threshold may be set by using a statistical analysis to determine a certain damage probability (usually 50%) and then setting the safety level at a selected level of probability below this, usually at the 0.01 or 0.001% level. This energy/power level is often a factor of 10 below that for the 50% damage point. In order to calculate a correct threshold level, it is also necessary to try to simulate some kind of worst-case scenario—when the eye is hit in its most sensitive part and takes in as much laser light from the laser in question as is possible under the circumstances [6].

1.6 High-Power Laser Suppliers

Among few suppliers, two of them are named here that can manufacture high-power laser energy at commercial level.

1. **Kapteyn-Murnane Laboratories (KMLabs)**

   Beyond serving the multitude of applications for which Ti:sapphire lasers have become ubiquitous, RAEA is especially designed as an unparalleled source for producing short-wavelength light via high harmonic generation (Fig. 1.7). The strength of RAEA for this purpose is realized in its contribution as one of the critical components in the Pantheon EUV/soft X-ray source. With patented

Fig. 1.7 KMLABS commercial high-power laser energy

Permacool technology, the RAEA enables repetition rate flexibility (up to 200 kHz) for applications that benefit from high average power.

The award-winning RAEA has been engineered for the most challenging applications rather than the most common and was specifically designed with the XUUS system in mind for high harmonic generation. From better beam quality and stability to remote alignment and lower maintenance needs, RAEA serves as a repeatable and reliable source for any of your high harmonic generation needs. In addition, with the advancements we've made in both pulse quality and energy, you no longer need to fully commit your amplifier to just high harmonic generation. RAEA provides enough energy to drive a XUUS system and then some, allowing you to expand your lab capability without needing another system.

Notably, KMLabs is a pioneer in cryo-cooling laser as illustrated in Fig. 1.8.

The enabling technology behind RAEA's unmatched performance is the award-winning Permacool™ cryo-cooling technology, patented by KMLabs.

Permacool is at the heart of the RAEA, allowing you to push your system harder without having to worry about the thermal lensing that can distort your beam in both time and space in other systems. KMLabs' 20+ years of experience in cryogenically cooling Ti:sapphire crystals have allowed us to fine-tune this technology to make the best use of it in your system [15].

2. **Lumibird**

Lumibird manufactures a wide range of high-power lasers, thanks to its expertise in three key technologies: pulsed solid-state lasers (nanosecond range), CW and pulsed fiber lasers and fiber amplifiers, and laser diodes. Various application areas are addressed, in industry (manufacturing, LiDAR sensors), science (laboratories and universities), medical (ophthalmology), and defense. Fig. 1.9.

The Laser division of Lumibird, incorporating Keopsys, Quantel, and SensUp, offers state-of-the-art lasers for industrial, defense, scientific, and medical applications. Its products are based on three key technologies: fiber lasers, solid-state lasers, and laser diodes.

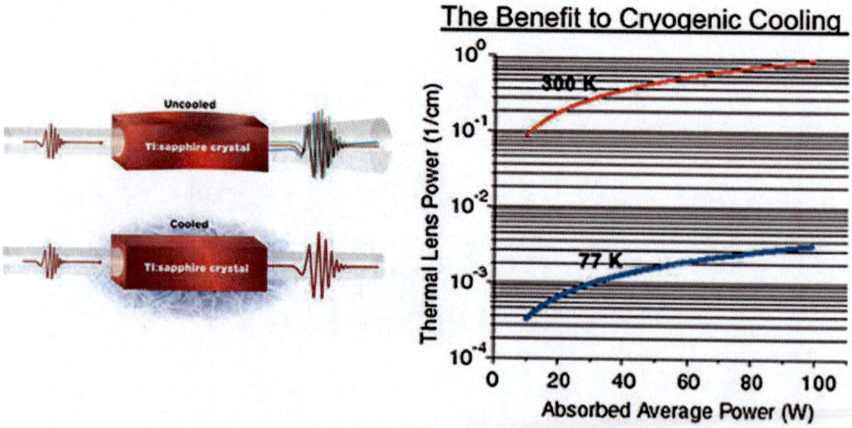

Fig. 1.8 KMLabs cryo-cooling laser

LUMIBIRD
MORE THAN LASER

Fig. 1.9 Lumibird high-power laser

The Quantel Medical division of Lumibird mainly offers diagnostic and treatment equipment for ophthalmology based on lasers and ultrasonic sensors [16].

3. **Frankfurt Laser Company**

Frankfurt Laser Company offers fiber-coupled high-power diode laser systems with up to 300 W output power. Separate diode bars are also available. See Fig. 1.10.

The new-generation laser diode system is a customer-friendly design and delivers 2 mW up to 300 W at the fiber output. Possible wavelengths are in the range of 450 nm to 1550 nm. The operation conditions such as power, current, temperature, as well as pulse width and frequency are clearly visible on the LCD display and can be adjusted easily. System parameters can be controlled manually as well as via RS232 port with the software provided. The system is mounted in a 19″ case which makes it convenient to install it into a standard rack or cabinet [17].

Fig. 1.10 Frankfurt laser diode system

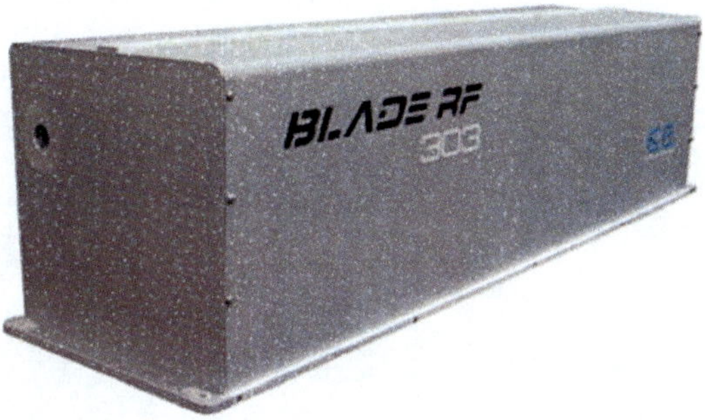

Fig. 1.11 El.En. high-power laser system

4. **El.En.**

   Blade RF 177 and 303 are two RF CO_2 laser sources that combine great performance and ease of integration. Their power output ranges from 150 W for the RF 177 to 300 W for the RF 303. If operated in the pulsed mode, this laser source can deliver up to 1100 W of output peak power! (Fig. 1.11)

   Blade RF 177 and *303* are two RF *CO2 laser sources* that combine great performance and ease of integration. Their power output ranges from 150 W for the RF 177 to 300 W for the RF 303. If operated in the pulsed mode, this laser source can deliver up to 1100 W of output peak power! Both models share the same compact size, which makes them easy to integrate with your existing laser systems. Like every El.En. device, the *Blade RF 177* and *303* are built with care: their sealed-off technology will ensure long-time power stability and thus optimal laser beam parameters. This device is perfect for applications such as plastic and leather cutting, thin plastic film *processing*, display and LCD processing, galvo-scanner applications, and many others [18].

5. **Laser Quantum**

   Laser Quantum manufactures both continuous wave and ultrafast lasers with high output power, such as the finesse laser family with output powers of up to 16 W. See Fig. 1.12.

   The axiom 532 is a high-powered continuous wave (CW) laser in a new architecture, incorporating all the active optics within the laser head. Offering

Fig. 1.12 Laser Quantum high-power laser system

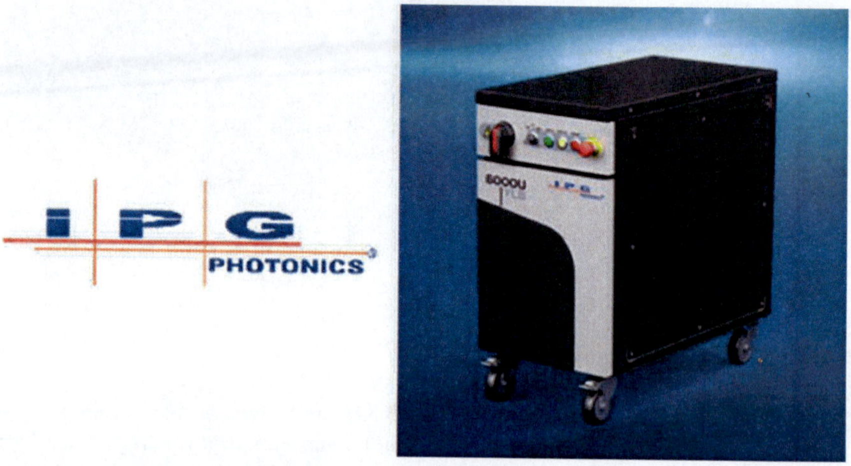

Fig. 1.13 IPG photonics high-power laser system

power levels up to 12 W and RMS noise below 0.02%, the axiom 532 is highly suitable for integration in an ultrafast laser or semiconductor industries. The axiom 532 requires minimal cooling due to its high efficiency, making it also highly suited to Ti:sapphire pumping [19].

6. **IPG Photonics**

   High-power kilowatt-class fiber laser sources are available from one to hundreds of kilowatts average power. Housed in rugged cabinets, these systems are designed to operate in industrial manufacturing environments. A variety of beam delivery and process head options provide the ultimate flexibility in optimizing one laser for many different applications. IPG high-power industrial fiber lasers are recognized as reliable and productive tools in demanding industries such as automotive, aerospace, and oil and gas. See Fig. 1.13.

   IPG's high-power CW fiber lasers cover output power range from 1 kW to over 100 kW and feature a wide range of operating wavelengths, single-mode and multimode options, high stability, and extremely long pump diode lifetime. These

Fig. 1.14 Aerodiode high-power laser system

lasers are water-cooled and can be supplied with a built-in or standalone chiller. The lasers are available with a wide variety of fiber terminations, collimation optics, and processing heads.

IPG manufactures high-power CW ytterbium lasers in 1 to >100 kW range and erbium, thulium, and raman fiber lasers in 1 to 5 kW range [20].

7. **Aerodiode**

Aerodiode offers high-power laser diodes with up to 150 W output at 808 nm, 915 nm, 976 nm, or 1064 nm—available as stock items or assembled with a turn-key laser diode driver with air-cooled temperature regulation. Versions with a high-power optical fiber connector or with a 3 mm output collimator are available. See Fig. 1.14.

These fiber-coupled 808 nm laser diodes are offered as stock items or associated with a CW or pulsed laser diode driver. They are compatible with our high-speed nanosecond pulsed drivers or high-power CW drivers with air cooling for the multimode laser diode versions.

These single mode 808 nm laser diodes can reach high power in nanosecond pulse regime up to 600 mW. Most turn-key diode + driver solutions are optimized for single-shot to CW performances with pulse width lengths down to 1 nanosecond. The 808 nm laser diode precision pulses are generated internally by an on-board pulse generator, or on-demand from an external TTL signal. Three multimode versions are offered for CW emission up to 100 W in a 200 μm core multimode fiber [20].

1.7 Summary

In summary, this chapter did cover very basic and elementary aspects of high-power laser energy (HiPLE) both from commercial and to some degree its military applications point of view. It also covered some research facilities that make these high-power lasers.

The High-Power laser Energy Research (HiPER) facility is a proposed experimental laser-driven inertial confinement fusion (ICF) device undergoing preliminary design for possible construction in the European Union. As of 2019, the effort appears to be inactive.

HiPER was designed to study the "fast ignition" approach to generating nuclear fusion, which uses much smaller lasers than conventional ICF designs, yet produces fusion power outputs of about the same magnitude. This offers a total "fusion gain" that is much higher than devices like the National Ignition Facility (NIF) and a reduction in construction costs of about ten times. This opened a window for a small machine to be rapidly built that would reach ignition before NIF. HiPER and the Japanese FIREX designs intended to explore this approach.

However, research into the fast ignition approach on smaller machines like the Omega laser in the United States demonstrated a number of problems with the concept. Another alternative approach, *shock ignition*, began to take over future development starting around 2012. HiPER and FIREX both appear to have seen no additional development since that time.

HiPER should not be confused with an earlier ICF device in Japan known as "HIPER", which has not been operational for some time.

In the case of HiPER, the driver laser system is similar to existing systems like NIF, but considerably smaller and less powerful.

The driver consists of a number of "beamlines" containing Nd:glass laser amplifiers at one end of the building. Just prior to firing, the glass is "pumped" to a high-energy state with a series of xenon flash tubes, causing a population inversion of the neodymium (Nd) atoms in the glass. This readies them for amplification via stimulated emission when a small amount of laser light, generated externally in a fiber-optic, is fed into the beamlines. The glass is not particularly effective at transferring power into the beam, so to get as much power as possible back out, the beam is reflected through the glass four times in a mirrored cavity, each time gaining more power. When this process is complete, a Pockels cell "switches" the light out of the cavity. One problem for the HiPER project is that Nd:glass is no longer being produced commercially, so a number of options need to be studied to ensure supply of the estimated 1300 disks.

From there, the laser light is fed into a very long spatial filter to clean up the resulting pulse. The filter is essentially a telescope that focuses the beam into a spot some distance away, where a small pinhole located at the focal point cuts off any "stray" light caused by inhomogeneities in the laser beam. The beam then widens out until a second lens returns it to a straight beam again. It is the use of spatial filters that lead to the long beamlines seen in ICF laser devices. In the case of HiPER, the filters take up about 50% of the overall length. The beam width at exit of the driver system is about 40 cm $\times$ 40 cm.

One of the problems encountered in previous experiments, notably the Shiva laser, was that the infrared light provided by the Nd:glass lasers (at ~1054 nm in *vaco*) couples strongly with the electrons around the target, losing a considerable amount of energy that would otherwise heat the target itself. This is typically addressed through the use of an optical frequency multiplier, which can double or triple the frequency of the light into green or ultraviolet, respectively. These higher frequencies interact less strongly with the electrons, putting more power into the target. HiPER will use frequency tripling on the drivers [21].

The military aspect of High-Energy Laser (HEL) looks at the tactical aspect of it. The Tactical High-Energy Laser, or THEL, was a laser developed for military use, also known as the Nautilus laser system. The mobile version is the Mobile Tactical High-Energy Laser, or MTHEL. In 1996, the United States and Israel entered into an agreement to produce a cooperative THEL called the Demonstrator, which would utilize deuterium fluoride chemical laser technologies. In 2000 and 2001, THEL shot down 28 Katyusha artillery rockets and 5 artillery shells. On November 4, 2002, THEL shot down an incoming artillery shell. The prototype weapon was roughly the size of six city buses, made up of modules that held a command center, radar, and a telescope for tracking targets, the chemical laser itself, fuel and reagent tanks, and a rotating mirror to reflect its beam toward speeding targets. It was discontinued in 2005 [22].

The start of President Regan's Strategic Defense Initiative (SDI) program otherwise known as Star Wars was a new era for opening the directed-energy weapon (DEW), a ranged weapon that damages its target with highly focused energy, including laser, microwaves, and particle beams. Potential applications of this technology include weapons that target personnel, missiles, vehicles, and optical devices [23].

In the United States, the Pentagon, DARPA, the Air Force Research Laboratory (AFRL), United States Army Armament Research Development and Engineering Center, and the Naval Research Laboratory (NRL) are researching directed-energy weapons and railguns to counter ballistic missiles, hypersonic cruise missiles [24], and hypersonic glide vehicles. These systems of missile defense are expected to come online no sooner than the mid- to late 2020s [25].

Russia [26–31], China [32–35], India [36], and the United Kingdom [37, 38] are also developing directed-energy weapons, while Iran [39–42] and Turkey [43] have directed-energy weapons in active service [43–45], and the first usage of directed-energy weapons in a combat was claimed to have been happened in Libya in August 2019 by Turkey which claimed to use the ALKA directed-energy weapon [46, 47].

After decades of research and development, directed-energy weapons are still at the experimental stage, and it remains to be seen if or when they will be deployed as practical, high-performance military weapons [48, 49].

References

1. https://www.rp-photonics.com/rgb_sources.html, Last Time Was Accessed on Internet about July 3, 2020
2. https://www.rp-photonics.com/lidar.html, Last Time Was Accessed on Internet about July 3, 2020
3. http://www.sensorsinc.com/applications/military/rocket-missile-tracking/. Last time was accessed on Internet about July 3, 2020

4. http://www.sensorsinc.com/applications/military/night-vision-systems, Last time was accessed on Internet about July 3, 2020
5. https://www.rp-photonics.com/medical_lasers.html. Last time was accessed on Internet about July 3, 2020
6. Zohuri, B.: Directed Energy Weapons, Physics of High Energy Lasers (HEL), 1st edn. Springer (2016)
7. Zohuri, B.: Interrail Confinement Fusion Drive Thermonuclear Energy, 1st edn. Springer (2017, 2016)
8. Zohuri, B.: Magnetic Confinement Fusion Driven Thermonuclear Energy, 1st edn. Springer (2017)
9. Wright, M.E.: Riding the plasma wave of the future. Symmetry Magazine. **2**(3), 12 (2005) Archived from thc Original on 2006-10-02. Retrieved 2005-11-10
10. Briezman, B.N., et al.: Self-focused particle beam drivers for plasma wakefield accelerators. (PDF). AIP Conference Proc. **396**, 75–88 (1997) Bibcode:1997AIPC..396...75B. 10.1063/1.52975. Archived from the original (PDF) on 2005-05-23. Retrieved 2005-05-13
11. Peralta, E.A., et al.: Demonstration of electron acceleration in a laser-driven dielectric microstructure. Nature. **503**(7474), 91–94 (2013) Bibcode:2013. Naturc 503. 91P. 10.1038/nature12664. PMID 24077116
12. England, R.J., Noble, R.J., Fahimian, B., Loo, B., Abel, E., Hanuka, A., Schachter, L.: Conceptual layout for a wafer-scale dielectric laser accelerator. AIP Conference Proc. **1777**, 060002 (2016). https://doi.org/10.1063/1.4965631
13. England, R.J., Byer, R.L., Soong, K., Peralta, E.A., Makasyuk, I.V., Hanuka, A., Cowan, B.M., Wu, Z., Wootton, K.P.: Demonstration of acceleration of relativistic electrons at a dielectric microstructure using femtosecond laser pulses. Opt. Lett. **41**(12), 2696–2699 (2016) Bibcode:2016OptL. 41.2696W. 10.1364/OL.41.002696. ISSN 1539-4794. PMID 27304266
14. Hanuka, A., Schächter, L.: Operation regimes of a dielectric laser accelerator. Nuclear Instruments and Methods in Physics Research Section A: Accelerators, Spectrometers, Detectors and Associated Equipment. **888**, 147–152 (2018) Bibcode:2018NIMPA.888..147H. 10.1016/j.nima.2018.01.060. ISSN 0168-9002
15. https://www.kmlabs.com/raea-ultrafast-amplifier. Last time was accessed on Internet about July 4, 2020
16. https://www.lumibird.com/en/. Last time was accessed on Internet about July 4, 2020
17. http://www.frlaserco.com/. Last time was accessed on Internet about July 4, 2020
18. https://elenlaser.com/. Last time was accessed on Internet about July 4, 2020
19. https://www.laserquantum.com/. Last time was accessed on Internet about July 4, 2020
20. https://www.aerodiode.com/?v=7516fd43adaa. Last time was accessed on Internet about July 4, 2020
21. https://en.wikipedia.org/wiki/HiPER. Last time was accessed on Internet about July 4, 2020
22. https://en.wikipedia.org/wiki/Tactical_High_Energy_Laser. Last time was accessed on Internet about July 4, 2020
23. Zohuri, B.: Directed Energy Beam Weapons, 1st edn. Springer (2019)
24. Zohuri, B., McDaniel, P.J., Lee, J., Rodgers, J.C.: New weapon of tomorrow's battlefield driven by hypersonic velocity. J. Energy Power Eng. **13**, 177–196 (2019)
25. Thaad-ER In Search Of A Mission – Aviationweek.com, 20 January 2015
26. https://www.rbth.com/defence/2017/07/12/a-farewell-to-traditional-arms-russia-develops-weapons-for-the-future_801080
27. https://www.mirror.co.uk/news/world-news/russian-battle-laser-hypersonic-missiles-9658436
28. http://www.pravdareport.com/news/russia/kremlin/27-01-2017/136730-mig_35_laser-0/
29. http://www.pravdareport.com/russia/economics/03-08-2016/135198-russia_laser_weapons-0/
30. https://www.mirror.co.uk/tech/russia-unveils-worlds-first-death-9043360
31. https://sputniknews.com/russia/201610141046325336-russia-energy-weapon/
32. http://nextshark.com/the-chinese-military-now-have-laser-weapons/
33. http://www.popsci.com/china-new-weapons-lasers-drones-tanks

34. http://www.popularmechanics.com/military/research/news/a24973/chinese-microwave-weapon-short-out-ieds-tanks/
35. http://thediplomat.com/2017/01/how-china-is-weaponizing-outer-space/
36. http://indiatoday.intoday.in/education/story/indias-top-secret-weapon/1/479199.html/
37. Dragonfire laser turret unveiled at DSEI 2017 – MBDA. MBDA. Retrieved 2017-10-21
38. Dragonfire, a guide to the new British laser weapon. UK Defence J. 2017-09-18. Retrieved 2017-10-21
39. Iran starts producing laser weapons for air defense
40. Iran achieves Laser Weapon System's deflection technology
41. Iran is Producing Laser Air Defense System - Islam Times
42. IRGC Navy develops anti-laser weapon - Tehran Times
43. Turkey's laser weapon ARMOL passes acceptance tests. DailySabah. Retrieved 2020-02-08
44. Janes
45. Turkey's laser gun passes acceptance tests. www.aa.com.tr. Retrieved 2020-02-08
46. Army Recognition. www.armyrecognition.com. Retrieved 2020-02-08
47. Is Turkey the first country to shoot down a drone with a laser?. Ahval. Retrieved 2020-02-08
48. https://thebulletin.org/navys-new-laser-weapon-hype-or-reality8326
49. https://spectrum.ieee.org/tech-talk/aerospace/military/no-quick-laser-missile-defense

Chapter 2
Materials Characteristics and Materials Properties

In modern society, we are surrounded by an amazing variety of materials. Most of the materials discussed in this book are solids that have been modified from their natural states to make them more suitable for practical applications. The materials people use have such an impact on their lifestyles that historical eras have been named for them. Ancient artifacts found by archaeologists have been dated and analyzed to reveal the increasing sophistication of their manufacturing methods. Historians have shown that technological advancements created new tools for agriculture and new weapons for armies. Explorers established trade routes to redistribute raw materials and finished products. Modern culture is also influenced by the availability of new materials.

2.1 Introduction

Our today's modern culture, regardless how much plastic we use on a daily basis, such as our simple bottle of water or cup of coke or any other drinks in a fast-food restaurant, is still influenced by availability of new materials. Plastics, starting from the 1960s era started being used to make colorful toys and house wares at such a low cost at high production level that they were frequently disposed of and replaced with latest style. Influence of plastics todays can even be seen in our household and automobile industries by far out, regardless of their life cycle being so short, and more often we have to replace plastic parts in our car due to continuous exposure to heat if cars are not covered and protected from direct or indirect sun radiation or cold weather of any harsh winter. Inside homes, more than 80% are made of plastics, from our standard dish plate to any electronic gadget that we use in our kitchen. See Fig. 2.1.

In the 1980s, silicon-based electronics started spreading through businesses and homes. Since the 1960s, homes (and landfills) in the United States have become bigger and more crowded with objects for applications never dreamed of by Stone

B. Zohuri, *Thermal Effects of High Power Laser Energy on Materials*,
https://doi.org/10.1007/978-3-030-63064-5_2

Fig. 2.1 A standard place setting includes metal cutlery, a polymer napkin, and a ceramic dish

Age humans. In the twenty-first century, there is much discussion of "globalization." Materials definitely follow a global cycle. Raw materials are collected, processed into useful materials, sold to consumers, and eventually discarded as waste. Each stage may occur on a different continent. Supply and demand of materials can affect international relations.

Traditionally the three major classes of materials are:

1. **Metals**
2. **Polymers**
3. **Ceramics**

Examples of these are steel, cloth, and pottery. These classes usually have quite different sources, characteristics, and applications.

In this chapter, we are going to study macroscopic (visible) and microscopic (invisible, physical and chemical) properties of matter or materials in relation to each other as well as when it comes to interaction with high-power laser energy (HPLE). This particular interest, since the Star Wars program of Strategic Defense Initiative (SDI), has been escalating among defense communities globally.

The atoms and molecules that compose matter are too small to see. But if we know what something is made of, and how it's held together, then we can break it apart and rearrange it. The result can be something with totally different properties. Nylon is made from oil!

One of way comparing the materials with each other is compresence methodology.

The above four categories are useful ways to sort different materials. Metals, polymers, and ceramics tend to have great differences in these categories. Each category will be briefly discussed here and then used in later chapters to highlight the special qualities of each material.

Again, in case of materials' response to HPLE, communities that are working on directed-energy weapons (DEWs) (Zohuri [1, 2]) are very much interested in certain materials both visible and invisible properties, and references such as by Tebo ([3]) provide wealth of information about selected values of the physical properties of

Table 2.1 Comparing materials

Performance composition and structure	*Versus*	*Physical and chemical properties processing and synthesis*

various materials. In cases, reference [3], presents values for thermal conductivity, specific heat, density, viscosity and Prandtl numbers certain liquid are given as functions of temperature. A table of room temperature values of some properties have been complied in reference [3] for which data as a function of temperature were not found.

At the end a short list of references to additional physical property, data on many different materials is provided as result of these compilations after reviewing the values presented in classified and unclassified literatures, and no guarantee is given that they are the best existing information [4–10].

Now getting back to Table 2.1, we can briefly explain each of these categories as they are presented here.

I. **Performance**

The performance of a material is discussed in the context of an application. For example, many materials are used for building houses. However, in an earthquake, the brick house would be the worst place to be!

The vibrations can rattle the bricks apart. California's building codes favor wood or steel reinforced concrete.

What should a good house material do? Protect the things inside from weather—wind, cold, heat, rain, snow, and hail—and from fire. A medieval castle was designed to withstand cannonballs, but since that is not usually a concern in modern construction, most people would decide that its superior performance is not worth the extra cost.

II. **Properties**

To choose a material with the best performance for a particular application, we will need to consider the properties of the available materials. Properties are the observed characteristics of a sample. The properties of matter can be divided into subcategories listed below.

1. **Physical Property**

Some physical properties describe how an object responds to mechanical forces. Hardness is one example of a mechanical property. If you drag a steel knife blade across a hard object, such as a plate, the hard surface is unchanged; if you drag the blade across a soft object, such as a piece of chalk, the soft surface will be scratched.

An object is tough if force is unable to break or tear it. The response to force depends on the material's structure and also on its shape and size. A piece of notebook paper can be torn easily, but a telephone book requires much greater force.

We can easily bend a flexible object such as a nylon jacket, but more force is required to bend a stiff object like a polyethylene milk jug. If an object returns to its original shape and size when the force is removed, we call it elastic. If the deformation remains, it is plastic. An object that breaks rather than bending is brittle.

An object is strong if an applied force is unable to deform or break it. A nylon windbreaker is strong, since pulling on it does not change its length. Sometimes the manner of applying a force makes a difference to the strength of an object. Ceramics can bear a lot of weight but will break if stretched or bent. Nylon survives compression, pulling, and twisting.

Color, texture, and reflectivity can be observed by shining light on a sample. Mirrors are colorless, smooth, and shiny. Electrical conductivity is detected by applying a voltage across an object. Applying heat to a sample reveals its ability to conduct heat, its melting point (temperature at which a solid changes to liquid), and its boiling point (temperature at which a liquid changes to gas).

Some properties are independent of the amount of sample. Melting point does not change if a sample is divided in half. Other properties, including mass and volume, increase with the amount of sample being studied.

When observing properties in a laboratory, scientists use the metric system of measurements. A ruler can be used to find the volume of a regularly shaped object.

2. **Chemical Property**

This property would describe what chemical reactions are likely to occur to the material of interest that is under study and investigation. We can observe how a sample reacts when mixed with other chemicals (water, acid). A material that can burn is described as flammable. Some materials rust (a type of oxidation reaction). Some materials dissolve in water or other liquids.

Usually a chemical reaction involves a transformation of the sample into a different substance, and it may be difficult to reverse the process. For example, wood is flammable. When it burns, it combines with oxygen from the air. The reaction produces ashes, smoke, and water; it cannot be reversed to make wood. The products of the reaction have quite different properties from the original wood. Table 2.2 is a presentation of properties of major classes.

Table 2.2 Characteristic properties of major classes

Metals	Polymers	Ceramics
Hard bur malleable	Stiff of flexible	Hard but brittle
Shiny	Dull	Shiny if glazed
Little color	Colorless	Many colors
Intermediate	Nonconductive	Nonconductive
High density	Low density	Intermediate density
Difficult to burn	Flammable	Not flammable

III. **Composition**

Composition tells what chemicals are in a sample. The most specific description will reveal the chemical elements that are present in the sample.

Chemists determined that matter is composed of combinations of about 100 elements. The simplest pieces chemists can make are atoms. Atoms of the same element are identical (have the same properties), and atoms of different elements are different. Some physical properties of elements are listed in Appendix II. Chemists can change oil into nylon only because they are composed of the same elements.

The periodic table of elements lists all known elements. Each element's square has its atomic number, name, and one- or two-letter chemical symbol (Table 2.3).

As shown in Fig. 2.2, which is a chemical periodic table of elements, different sets of elements are found in metal, polymer, and ceramic samples. The elements for metallic and polymeric materials do not overlap at all. This is one of the reasons that the properties of those two classes are so different. The elements found in ceramics are also found in metals or polymers.

Table 2.3 Chemical property of hydrogen and oxygen

1	**Hydrogen**	H
8	**Oxygen**	O

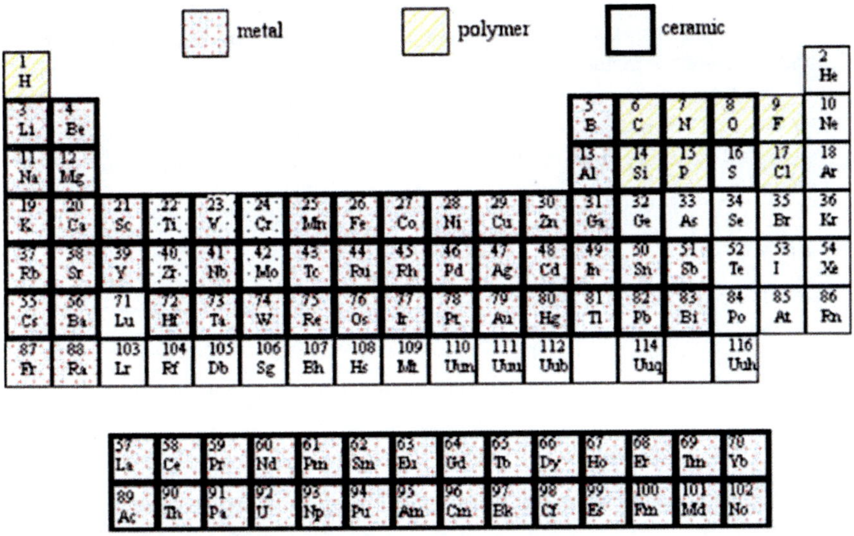

Fig. 2.2 Elements used in different classes

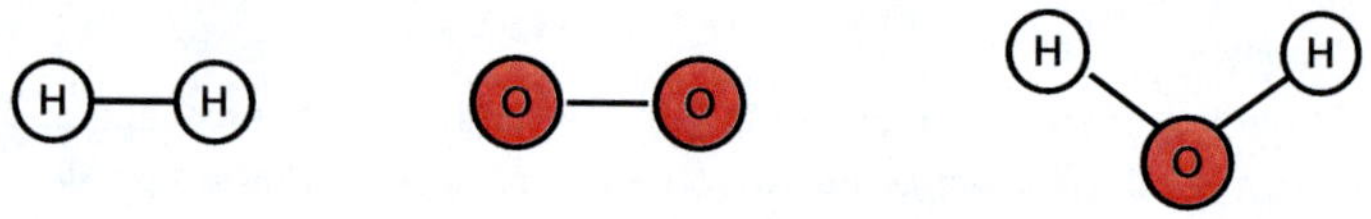

Fig. 2.3 The elements hydrogen and oxygen and the compound water

IV. **Compounds**

A compound has a uniform composition: every sample removed and tested, no matter how small, contains the same elements in the same proportions. The elements in compounds are held together tightly and can only be separated from each other by chemical reactions.

A chemical formula relays the chemical composition of a compound in a compact manner. It includes the chemical symbol for each element, listed from left to right as they appear on the periodic table (Fig. 2.2). The symbol for each element is immediately followed with the number of atoms of that element as a subscript. Fe_2O_3 contains two iron atoms and three oxygen atoms. If there is only one atom of the element, we do not bother to write the number 1. Water has two hydrogens and one oxygen, so its chemical formula is written as H_2O. See Fig. 2.3.

Table salt is NaCl. Sand always contains one silicon for every two oxygen, so its formula is
SiO_2.

1. **Mixture**

If you mix sand and salt, any proportion of elements is possible. In a mixture, the different elements are not coordinated with each other. This means that samples from a mixture won't always have identical proportions. Bricks, steel, and concrete are materials that are mixtures.

The elements in a mixture are not tightly bound, so they often can be separated by physical processes. To separate a mixture of sand and salt, you can add water since salt dissolves, but sand does not. After straining out the sand, boil the water away from the salt (Fig. 2.4).

V. **Structure**

The three-dimensional arrangement of atoms in a sample creates its structure. When sand changes into glass, its silicon and oxygen atoms shift positions to make a continuous sheet.

Even though the chemical composition is the same, some of the properties have changed.

VI. **Processing and Synthesis**

Various methods can be used to create materials from existing substances. Processing a material could be as simple as hammering a piece of copper or flaking arrowheads from a piece of flint. When sand is melted and formed into

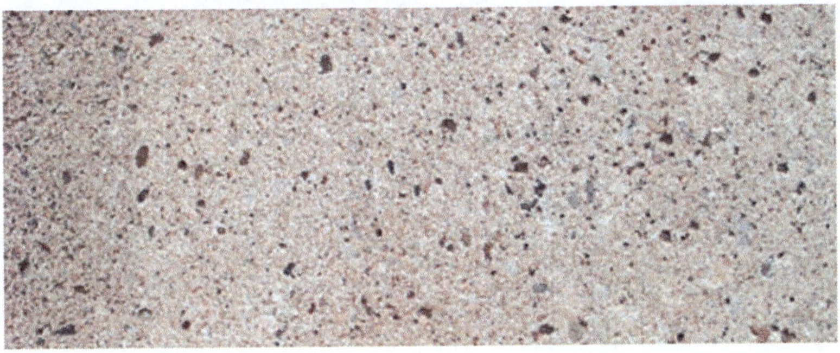

Fig. 2.4 Concrete is composed of cement, sand, and pebbles

glass, the primary change occurring is in the arrangement of the silicon and oxygen atoms. There are several ways to change iron ore, a compound, into iron metal. The different processes used to produce cast and wrought iron result in different ratios of iron and carbon which create different properties in the final objects and make them suitable for different applications.

Synthesis implies a major change in chemical composition; for example, polymers are synthesized by cooking mixtures of chemicals. New molecular structures result.

2.2 Atomic Structure and Chemical Properties Relationship

We can ask ourselves: How does atomic structure relate to chemical properties? The answer can be found below.

1. **The Components of an Atom**

It has been mentioned that elements are the purest forms of chemicals. Chemical reactions cannot make them simpler. Each atom in a sample of an element has the same chemical properties.

In 1897, it was discovered that atoms can be separated into negatively charged and positively charged parts. Electrons are the pieces with negative charge. Chemists usually say that one electron has a charge of −1. Electrons can be rearranged by chemical reactions.

If all the electrons are removed from an atom, what remains is the nucleus. It is positively charged and contains most of the mass of an atom. The positive charge of the nucleus attracts the negative charge of the electrons.

Further experiments showed that a nucleus is composed of two types of particles. Each proton in the nucleus has a charge of +1. A neutron has no charge. Both types of particle have much larger masses than electrons. Neutrons and protons are held together so tightly that an atomic nucleus will not change during a chemical reaction. See Fig. 2.5.

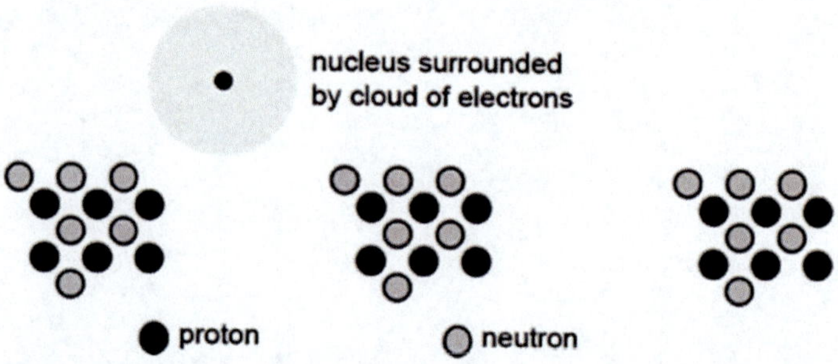

Fig. 2.5 Nucleus surrounded by a cloud of electrons

As Fig. 2.5 illustrates, carbon (C) always has six protons in its nucleus, but it can have six, seven, or eight neutrons.

Since an atom has no overall charge, its number of protons is equal to its number of electrons. A carbon atom always has six protons and six electrons.

Atoms are so small that 6.02x1023 carbon atoms are needed to make a 12-gram sample! The mass of an atom may be reported in atomic mass units (amu). 1 amu = $1.66 \mathrm{x} 10^{-24}$ g.

2. **Atomic Number and the Periodic Table of Chemistry**

The identity of an atom is determined by its number of protons. The atomic number found on the periodic table in Fig. 2.6 is the number of protons. Any atom with one proton is a hydrogen atom (with atomic number 1). Scientists have found atoms with as many as 116 protons in their nuclei, but the largest nuclei are unstable and not used for materials.

The order of elements on the table is the number of protons. Elements are similar to their left- and right-hand neighbors in mass and size. Elements farther down the periodic table show an increase in mass and size.

Notice that the rows labelled 1 through 7 (called "periods") are not all the same length. The pattern provides information about the chemical properties of elements. Elements in the same column (called a "group") undergo similar reactions. For example, group I contains the elements Li, Na, and K which all react violently with water.

The tallest groups, labelled with Roman numerals, contain the main group elements. The central section of the table, starting with element #21, contains transition elements. The two rows at the bottom of the table contain inner transition elements.

3. **Classes of Elements**

Left/bottom corner: metals. These are almost all solids at room temperature. They are shiny and conduct electricity well.

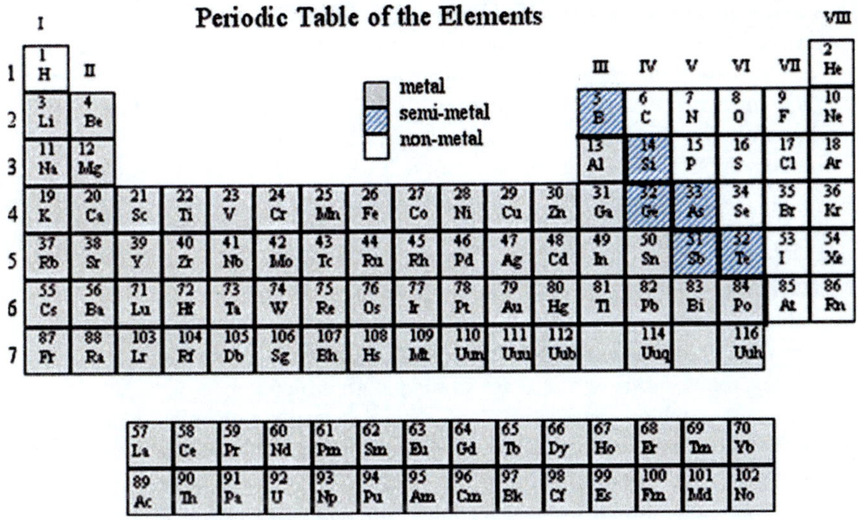

Fig. 2.6 Chemical periodic table of elements

Right/top corner: non-metals. These are more often gas or liquid than solid and are nonconductors. The last group on the right, labelled VIII, contains the noble gases. These almost never react and do not form useful materials.

Along the diagonal division: semi-metals. These have properties that are intermediate between those of metals and nonmetals. They are shiny, but too brittle to make wires. Their most useful property is that they are semiconductors. This means that they conduct electricity under controlled conditions.

4. **Valence Electrons**

The chemical properties of an element are determined by the number of electrons available for reactions. The theory called quantum mechanics explains that electrons are arranged in shells 1–12 around each atomic nucleus. If a shell is completely filled, the electrons are very stable. It is hard to remove or add any. Noble gases have filled shells and so rarely undergo chemical reactions. Elements from other groups will undergo reactions to gain or lose electrons.

The reactive electrons are called valence electrons. The number of valence electrons for a main group element is the same as its group number, I through VIII. If we know an element's valence, we can predict how it is likely to react to fill the shell: lose, gain, or share electrons. We will see later that the valence electron concept does not work as well for transition metals.

5. **Ions**

If energy is available, an atom can lose or gain electrons until it matches the closest noble gas (achieve a filled shell). The name of the atom is unchanged; since it has the same number of protons as before, it is still the same element. No chemical process can change the number of protons in a nucleus. However, the positive charge of the nucleus is no longer exactly cancelled by negative electrons. An ion is a particle for which the number of protons is not equal to the number of electrons.

We write the chemical symbol with a superscript showing the ion's charge.

Example. Na has one valence electron; remove it to match Ne. Na + 11 p+, 10 e-
Na could add seven electrons to match Ar, but that large a change would require too much energy. In general, elements on the left of the periodic table (metals) form positive ions.

Elements on the right side of the periodic table (nonmetals) usually form negative ions.

Example. O has six valence electrons; we can add 2 to match Ne. O2–8 p+, 10 e-.

Losing six electrons to match He would be too difficult.

The most common ionic compound, table salt, has the chemical name sodium chloride. It contains equal numbers of sodium ions (Na+) and chloride ions (Cl-) and is known by the formula NaCl. No superscripts for charges appear in the compound's chemical formula because the total compound is neutral. The charges on the cations cancel out the charges on the anions.

When we read the formula, we must remember that a compound composed of a metal and a nonmetal is always ionic.

2.3 Thermal Conductivity

The thermal conductivity integrals for metallic solids are presented in this section as a function of temperature from 4 to 300 °K defined from the Fourier equation for steady unidirectional conduction which is:

$$Q = \lambda A \frac{dT}{dL} \tag{2.1}$$

where:

Q = rate of heat conduction, negative in the direction of increasing length.
λ = thermal conductivity.
A = cross-sectional area of the heat conduction path, normal to the direction of heat flow.
$\frac{dT}{dL}$ = temperature gradient along the path at the section under consideration.
T = temperature.
L = length.

For a constant cross-sectional area, this may be reduced to:

$$Q\frac{L}{A} = \int_{T_1}^{T_2} \lambda dT \tag{2.2}$$

where T_1 and T_2 are the temperature at any two points along the path of the flow.

The thermal conducting integrals tabulated in this section are values of:

$$\int_{T_0}^{T_L} \lambda dT \tag{2.3}$$

where T_0 and T_L are temperatures along a heat flow path communicating between heat reservoirs at $T_0 = 4^0K$ and T_L at length L from T_0. The heat flow along a conductor of constant cross-sectional area A, through length L, may then be determined from the difference of the thermal conductivity integrals, i.e.,

$$Q\frac{L}{A} = \int_{T_1}^{T_2} \lambda dT = \int_{T_0}^{T_2} \lambda dT - \int_{T_0}^{T_1} \lambda dT \tag{2.4}$$

(In calculating the values of the thermal conductivity integrals presented in this section, a linear interpolation was assumed between the temperature intervals tabulated on the data sheets.)

Of the metals included, four become superconducting above 4 °K. These metals and their transition temperatures are shown in Table 2.4.

The values of the thermal conductivity for these metals below the superconducting transition temperatures are for the normal state rather than the superconducting state.

Several of the data sheets include extrapolated values and are so noted on the individual data sheets. based on the characteristics of the thermal conductivity curves of metals in the same series classification. The estimated deviation of the thermal conductivity integrals over the extrapolated range is not more than 10% from the probable values.

Now, if we define general definitions and units for heat conduction, then thermal conductivity can be treated as a non-equilibrium property usually determined in a

Table 2.4 Four metals and their transition temperatures

Metals	Index	Transition temperature [15]
Lead	13.142	7.22 °K
Niobium	13.151	8.7–8.9 °K
Tantalum	13.151	4.38 °K
Vanadium	13.151	4.89 °K

steady-state experiment utilizing the Fourier Law for linear flow in a homogeneous, isotropic substance very similar to Eq. 2.1 and can be written as:

$$\dot{Q} = -\lambda A \frac{dT}{dx} \tag{2.5}$$

where $\dot{Q}$ is the thermal energy current, A is the cross-sectional area, dT/dx is the temperature gradient, and λ is the thermal conductivity coefficient as before.

Now we can define the following symbols and units as they can be found in most heat transfer textbooks or literatures:

A = cross-sectional area, meters^2.
c_p = specific heat at constant pressure, joules/kilogram, kelvin.
c_v = specific heat at constant volume, joules/kilogram, kelvin.
J = heat current density, watts/meter^2.
L = Lorenz ratio = $\lambda/\sigma T$, volts^2/kelvin^2.
l = mean free path, meters.
M = molecular weight, kilograms/mole.
$\dot{Q}$ = heat current, watts.
R = gas constant per mole, 8.3143 joule/kelvin, mole.
T = temperature, kelvin.
t = time, second.
v = velocity, meter/second.
x = space coordinate, meters.
α = thermal diffusivity, meters^2/second.
λ = thermal conductivity, watts/meter, kelvin.
μ = dynamic viscosity, Newton second/meter^2, (= 10 poise).
ρ = density, kilograms/meter^3.
σ = electrical conductivity, 1/ohm, meter.

Now we can state that commonly used units and their conversion factors are given in series of plots as illustrated in Fig. 2.7.

For non-isotropic bodies such as some dielectric crystals, the basic differential equation of Eq. 2.5 can be modified to:

$$\vec{J} = -\lambda \, grad \, T = -\lambda \frac{\partial T}{\partial t} \tag{2.6}$$

where $\vec{J}$ is the vector thermal current density and λ is a symmetric tensor of second order. Heat-conduction equations and their solutions for nonhomogeneous and nonlinear systems are discussed at length by McAdams [16], Schneider [17], and Carslaw and Jaeger [18].

In general, the total heat *transport* is affected by radiation, convection, and conduction mechanisms and may depend on temperature, pressure, density,

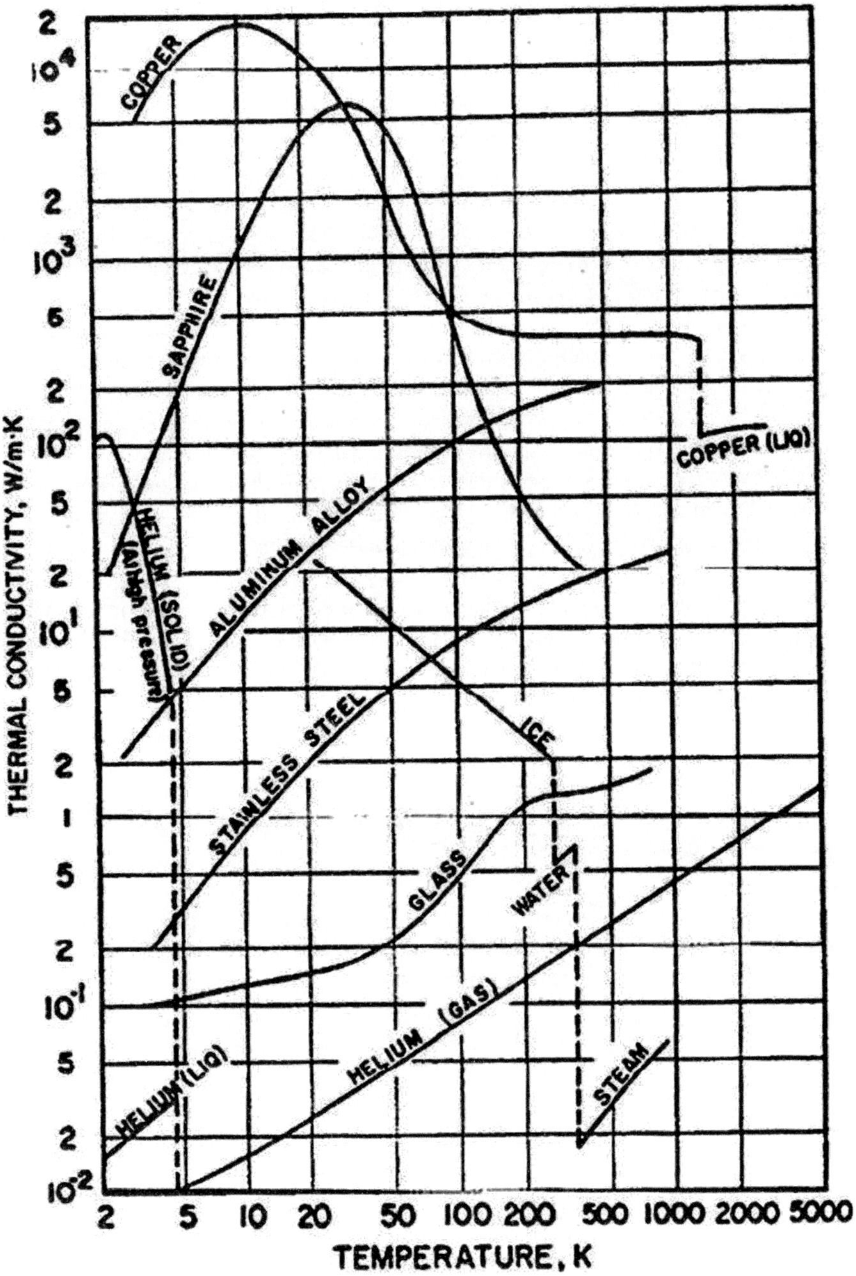

Fig. 2.7 Typical curves showing temperature dependence of thermal conductivity

materials, temperature gradient, etc. However, the coefficient λ, as defined by the Eqs. 2.5 and 2.6, refers only to heat transport by conduction mechanism, and that is where our concentration in writing this book is.

It is usually assumed that the thermal conductivity is not a function of the temperature gradient but is a function of the temperature, composition, purity, perfection, and other similar intensive parameters of the system. It is also assumed that the conductivity is not size- or shape-dependent, though this is not always true. Size and shape effects become significant whenever the size of the conductor is comparable to the mean free path for motion of the particles or quasi-particles that transport the thermal energy. These effects have been observed for conduction by molecules in rarefied gases and for conduction by phonons (quantized normal modes of lattice vibration) in small, high-purity dielectric crystal at low temperatures.

Reprehensive values for the temperature dependence of several substances are given in Fig. 2.7. They are typical curves for a high purity metal (copper), high-purity crystalline dielectric (sapphire), nonferrous alloy (aluminum alloy), ferrous alloy (stainless steel), disordered dielectric (glass), fluid (helium), and water.

The Thermophysical Properties Research Center at Purdue University has published a large compilation of thermal conductivity data and graphs over large temperature ranges for many solids and fluids [19].

A general survey of the experimental and theoretical aspects of thermal conductivity is given in the book edited by Tye [20]. Proceedings of the annual conferences on thermal conductivity [21] are usually available from the sponsoring agency; sometimes they are formally published.

A similar coefficient useful in *transient* heat-flow problems is the thermal diffusivity α defined by:

$$\alpha \equiv \frac{\lambda}{\rho c_p} \tag{2.7}$$

where ρ is the density and c_p is the specific heat at constant pressure. For an isotropic, homogeneous body without local heat sources or sinks, the basic partial differential equation for transient heat conduction is given by:

$$\frac{\partial T}{\partial t} = \alpha \nabla^2 T \tag{2.8}$$

The more complicated equations and their solutions are discussed by Jakob [22], Schneider [17], and Carslaw and Jaeger [18].

Values of the diffusivity α are not tabulated but may be calculated using conductivity values from this section, densities from Sec. 2b, and specific heats from Sec. 4e of Powell [4].

2.4 Heat Transport in Solids

Two principal mechanisms are responsible for the transport of heat energy in a solid. The first is the drift motion of conduction electrons; the second is the directional cooperative vibration of interacting lattice ions, represented by the quasi-particle concept, phonons. Other mechanisms such as internal radiation or excitons may be important in some materials. Electron conduction is predominant in metals and alloys and phonon conduction in dielectrics and some highly disordered alloys. Both electron and phonon conductions are limited at low temperatures by impurity and imperfection scattering. Therefore, the thermal conductivity at low temperature is critically dependent on the exact amount and types of impurities and imperfections.

At high temperatures, electron conduction is limited by phonon or lattice scattering primarily and therefore is not critically dependent on the impurities. The conductivities of various metals or alloys of approximately the same composition tend to converge at high temperatures. Phonon conduction in crystalline dielectrics is limited at high temperatures by phonon-phonon scattering. Therefore, the thermal conductivities of crystalline dielectrics also tend to converge to common values at high temperatures. Phonon conduction in disordered dielectrics is highly limited at most temperatures by imperfection scattering.

Therefore, the thermal conductivity of disordered dielectrics is very low compared with that of metals or crystalline dielectrics. A review of the phenomena and various mechanisms of metal at low temperatures was given by Powell [23] and a more detailed study of the theories and concepts for solids by Rosenberg [24]. The book edited by Tye [20] also contains reviews of both the theoretical and experimental aspects for solids.

The thermal conductivity of a metal or alloy can be estimated by using the Wiedemann–Franz–Lorenz Law as illustrated in Eq. 2.9:

$$\lambda = L\sigma T \tag{2.9}$$

where σ is the electric conductivity, λ is the thermal conductivity, T is temperature, and L is the Lorenz ratio whose Sommerfeld classical value is given by:

$$L \approx 2.45 \times 10^{-8}\ (\text{volt/kelvin})^2 \tag{2.10}$$

For both pure metals and alloys, the Lorenz ratio generally approaches the Sommerfeld value at high and very low temperatures. For pure metal, L is lower than the above number at temperatures below the ice point; for alloys, it is higher, as much as 10 times greater (near 20 K) for very disordered, multicomponent alloys.

Data for the conductivity of metals are given in Table 2.5. It should be noted that the values quoted at 4.2 K, and often at 20 K, are for the purest specimen that has

Table 2.5 Thermal conductivity of solid elements (in watts/meter, kelvin)

Element*	4.2 K†	20 K†	77 K	194 K	273 K	373 K	573 K	973 K‡
Ag	14,500	5,100	481	430	428	422	407	376
Al	17,000	11,500	440	238	235	234	233	
Au	2,190	1,570	354	328	318	313	306	279
B	40	350	270	54	32	22	17	
Be	....	3,500	1,800	318	220	172	134	96
Bi (⊥c)	1,590	100	27	13	11	9		
Cd	9,500	226	107	99	98	95	89	
Ce	0.5	1.9	5.0	......	11			
Co	95	450	205	130	100	85		
Cr	165	575	192	112	95	87	81	66
Cs	110	61						
Cu	11,800	10,500	610	419	401	393	384	359
Dy	2.8	15	12	9	11			
Er	5	12	11	14	14			
Fe	8,500	15,000	218	95	83.5	71.9	56.3	33.9
Ga (‖b)	16,000	630	105	90	85			
Gd	16	32	16	13	14			
Ge	900	1,500	330	98	67	47	29	18
Hf	3.6	18	26	......	22	22	21	
Hg	180	....	41.5	34				
Ho	5.8	18	7.7	9	10.5			
In	850	183	....	100	87			
Ir	550	1,800	230	162	160	157	152	
K	650	150	114	112	109			
La	9							
Li	256	720	152	90	82	80		
Lu (‖c)	20	40	28	25	23			
Mg	510	1,390	200	156	153	150	146	
Mn	0.94	2.4	5.5	7.0	7.7			
Mo	66	277	215	138	135	132	130	113
Na	4,750	590	137	128	125			
Nb	27	85	56	51	51	53	55	60
Nd	....	....	....	......	16			
Ni	300	800	195	110	91	81	67	71
Os	125	540	135	91	88	85	85	
P	0.6	27	35	18	13			
Pb	2,100	59	41	37	35	34	32	
Pd	800	600	82	76	76	76		
Pr	....	....	6.8	11	12.5			
Pt	910	490	87	75	73	72	72	74
Pu	....	....	2.8	4.6	6.2	8.8		
Rb	185	70	61	60	60			
Re	700	840	64	51	49	47	44	44
Rh	1,110	3,800	250	155	151	147	137	
Ru	660	2,300	190	120	117	117	117	
S	11	2.4	0.67	0.37	0.29	0.15		
Sb	200	220	58	31	20	21	19	
Sc	2.8	12	14	18	22			
Se (‖c)	150	59	14					
Si	260	4,900	1,400	270	170	108	65	32
Sm	4.8	6.9	7.1	11	13			
Sn	7,500	230	80	73	67	60		
Ta	48	147	60	57	57	57	59	61
Tb	6.0	20	12	10	13			
Tc	....	....	....	......	51	50	50	
Te (⊥c)	850	90	15	6.2	4.4	3.2	2.4	
Th	17	54	50	39	37	37	37	33
Ti	5.8	28	33	25	22	21	19	20
Tl	1,800	80	58	49	47			
Tm	9.0	18	12	16	17			
U	4.6	15	21	25	27	28	30	40
V	2.3	12	23	27	30	32	34	
W	4,000	5,400	264	180	170	166	141	122
Y	2.2	10	....	......	15	15	15	15
Zn	970	690	138	124	119	113	103	
Zr	40	105	37	25	22	20	19	21

Note: *Solid A, H_2, and He are in Table 2.8a, b. Symbols in parentheses indicate heat flow parallel or perpendicular to b or c axes. For other elements, the results are for isotropic or polycrystalline elements

†The low-temperature conductivity depends critically on the exact amount and types of impurities and imperfections. The values quoted are for the purest specimen that has been measured at the present

‡For temperatures greater than 973 K, see reviews by reference [25, 26]

Values quoted are from the reviews and compilations by R. L. Powell and Blanpied ([25] and R. W. Powell et al. [26]. The values are generally for the highest-purity metals tested. Disagreements among different modern authors are caused primarily by difference in sample purity and preparation been measured at the present. Further measurements on more pure metals may give substantially higher conductivities. Above 20 K, the quoted values should not change substantially (more than 5%) as more pure metals are measured with more refined techniques. Date for the conductivity of some commercial alloys are given in Table 2.6; semiconductor in Table 2.7; crystalline dielectrics and optical materials in Table 2.8a, b; and disordered materials in Table 2.9.

A few selected values for solids below 1 K are given in Table 2.10. The conductivity of normal metals varies linearly with temperature at temperatures below their conductivity maximum. In superconductors the electrons do not contribute significantly to the transport of heat.

Therefore, their conductivity of crystalline solids is usually size-dependent below 1 K and therefore depends on the exact specimen configuration.

A recent review of the literature by Childs et al. [30] includes data, tables, and graphs on most solids for temperatures at and below 300 K.

2.5 Conclusions

A lot of techniques and methodologies have been developed in the industry that allow to measure materials properties. For example, a technique was developed around 1971 by the Physics International Company (PIC) under the supervision of Stefansky et al. [11]. This technique was developed for measuring in a direct and straightforward manner the elevated-temperature tensile properties of materials following "instantaneous" in-depth heating.

High-energy pulsed (HEP) electron beams are used *to* heat specimens to the required temperature in approximately 0.08 μsec, and a Hopkinson bar-type device, in a tensile configuration, is used to perform the mechanical tests at any desired time after irradiation.

Data on the temperature-induced degradation of Young's modulus and the yield strength of 2014-T6 aluminum and commercially pure alpha titanium have been obtained. The data extend up to approximately 700 F for three different times: 0.2 msec, 1.0 msec, and 10 msec. The strain rates were approximately 500 in./in./s. The results indicate that the dynamic tensile properties of the materials studied are degraded at these temperatures and times at temperature, following instantaneous in-depth heating. More specifically, Young's modulus at ambient pressure appears to be a function of temperature only. The degradation of the yield strength, on the other hand, is time- and temperature-dependent but not as severe as that shown by the long-heating-time handbook data.

Results are also presented on direct measurements of the Gruneisen coefficient of 2014-T6 aluminum and alpha titanium at energy densities up to melt. The measurements were made using the transmitted stress technique: a pressure

Table 2.6 Thermal conductivity of selected commercial alloys (in watts/meter, kelvin)

Alloy*†	Ref.	4.2 K	20 K	77 K	194 K	273 K	373 K	573 K	973 K
Aluminum:									
1100	45	50	240	270	220	220			
2024	45	3.2	17	56	95	130			
3003	45	11	58	140	150	160			
5052	45	4.8	25	77	120	140			
5083, 5086	45	3.0	17	55	95	120			
Duralumin	72	5.5	30	91	140	160	180		
Bismuth:									
Rose metal	46	......	5.5	8.3	14	16			
Wood's metal	72	4	17	23					
Copper:									
Electrolytic tough pitch	45	330	1,300	550	400	390	380	370	350
Free cutting, leaded	45	200	800	460	380	380			
Phosphorus deoxidized	45	7.5	42	120	190	220			
Brass, leaded	45	2.3	12	39	70	120			
Beryllium	72	2.0	17	36	70	90	113	172	
German silver	46, 72	0.75	7.5	17	20	23	25	30	40
Silicon bronze, A.	45	......	3.4	11	23	39			
Manganin	46	0.48	3.2	14	17	22			
Constantan	46	0.9	8.6	17	19	22			
Ferrous:									
Commercial pure iron	46, 48	15	72	106	82	76	66	54	34
SAE 1020	46	13	20	58	65	65			
SAE 1095	46	......	8.5	31	41	45			
3 Ni, 0.7 Cr, 0.6 Mo	48	......	6	22	.....	33	35	36	30
4 Si	48	......	......	......	.....	20	24	28	26
Stainless	11	0.3	2	8	13	14	16	19	25
27 Ni, 15 Cr	48	......	1.7	55	.....	11	12	16	21
Gold:									
Gold-cobalt thermocouple	45	1.2	8.6	20					
Lead:									
60 Pb, 40 Sn	72	......	28	44					
Nickel:									
80 Ni, 20 Cr	48	......	......	......	.....	12	14	17	23
Contracid	46	0.2	2	7.3	9.5	13			
Inconel	27, 48	0.5	4.2	12.5	13	15	16	19	26
Monel	46	0.9	7.1	15	20	21	24	30	43
Platinum:									
10 Ir	46	......	......	......	.....	31	31.4		
10 Rh	46	......	......	......	.....	30.1	30.5		
Silver:									
Silver solder	46	......	12	34	58				
Normal Ag thermocouple	46	48	230	310					
Tin (60 Sn, 40 Pb)	72	16	55	51					
Titanium:									
5.5 Al, 2.5 Sn, 0.2 Fe	27	......	1.8	4.3	6.4	7.8	8.4	10.8	
4.7 Mn, 3.99 Al, 0.14 C	46	......	1.7	4.5	6.5	8.5			

Note: *Commercial alloys of the same nominal composition may vary in conductivity from 5% to 25% because of differences in heat treatment and uncontrolled impurities. Contracid, Inconel, and Monel are registered trade names for nickel alloys. See reference [25] for additional data
†When composition is given, it is by weight percent

Table 2.7 Thermal conductivity of semiconductors * (in watts/meter, kelvin)

Substance	Ref.	Dopant	Electrical conductivity at room temp., ohm-m	Type	Carriers at room temp., no./m³	Temperature					
						4K	20K	77K	194K	273K	373K
Graphite	50	....	9.8×10^{-6}	⊥ C	....	0.25	15	190	300	250	
	50	....	4.1×10^{-5}	∥ C	....	0.2	7.4	68	84	80	
	50, 72	....	2.4×10^{-5}	Ext.	....	0.02	0.7	15			
	50, 72	....	5.7×10^{-5}	Molded	....	....	0.08	1.2	4	5	
	70	....	1.2×10^{-6}	Natural	....	6	150	610	270	160	
	50, 72	....	2.4×10^{-4}	Deposit	....	....	....	....	....	580	540
	50, 72	....	2.1×10^{-3}	Coke	....	....	....	....	....	28	28
Ge	70	Cu	....	*p,n*	10^{20}	1000	1500	300	90	70	50
	9	In	3×10^{-2}	*p*	10^{21}	290	1300	310			
	9	Ga	3×10^{-5}	*p*	10^{25}	8.5	270	200			
Si	56	O_2	5.5×10^{-7}	*n*	7.8×10^{20}	310	1900	900	270	150	
	66	P	4.6×10^{-1}	*p*	4×10^{20}	200	4200	1500	270	150	
	9	Au	2.2×10^{-1}	*p*	10^{26}	87	1300	840	250	150	
	66	B	4.5×10^{-2}	*p*	4×10^{21}	140	3500	1500	270	150	
Bi_2Te_3	72	....	....	*n*	3×10^{23}	....	....	6	3	3	4.9
GaAs	26	....	....	....	7×10^{15}	1000	1900	300	80	50	
InAs	72	....	....	*n*	10^{25}	....	....	....	....	6.7	5.0
				n	3×10^{22}	....	....	....	....	6.7	5.2
InSb	26	....	....	*p*	10^{22}	....	1000	110	....	17	12
				n	10^{23}	....	....	....	....	18	14
PbS	20	....	....	*p*	2×10^{24}	17	45	8			
				n	8×10^{23}	0.7	2.8	3.9			
PbSe	20, 56	....	....	*p*	5×10^{24}	24	33	5.5			
PbTe	20	....	....	*p*	2×10^{24}	11	31	7	4	2.4	1.7
				n	6×10^{24}	....	....	8	1.3	2.3	2.4
SiC	66	N	10×10^{-2}	*n*	10^{23}	30	2000	3000	1000	500	
	66	Al	6×10^{-3}	*p*	4×10^{25}	0.08	1.5	50	150	140	

Note: *Pure Ge and Si are given in Table 2.8 (a, b)
Various authors differ from 5% to 10% on the experimental results for the conductivity of semiconductors
See also the review by F. Steigmeier vol. 2 pp 251 of ref. [20]

Table 2.8 Thermal conductivity of crystalline dielectrics and optical materials (in watts/meter, kelvin)

Material	Ref.	*T*, K	Conductivity	Material	Ref.	*T*, K	Conductivity
Ar	52	8	6.0	Glass (plastic perspex)	46	4.2	0.058
		10	3.7			20	0.074
		20	1.4	Glass (Pyrex)	46	77	0.44
		77	0.31			194	0.88
AgCl	39	223	1.3			273	1.0
		273	1.2	H_2 (para + 0.5 % ortho)	46	2.5	100
		323	1.1			3.0	150
		373	1.1			4.0	200
Al_2O_3 (sapphire) 36 deg to *c* axis	46	4.2	110			6.0	30
		20	3,500			10	3
		35	6,000	H_2O (ice)	72	173	3.5
		77	1,100			223	2.8
Al_2O_3 (sapphire) ⊥ to *c* axis	48, 72	373	2.6			273	2.2
		523	3.9	He^3	52	0.6	25
		773	5.8			1.0	2
Al_2O_3 (sintered)	46, 48	4.2	0.5			1.5	0.57
		20	23			2.0	0.21
		77	150	He^4	52	0.5	42
		194	48			0.8	120
		273	35			1.0	24
		373	26			2.0	0.18
		973	8	I	52	300	0.45
As_2S_3 (glass)	72	283	0.16			325	0.42
		323	0.21			350	0.40
		373	0.27	KBr	39, 75	2	150
BaF_2	39	225	20			4.2	360
		260	13.4			100	12
		305	10.9			273	5.0
		370	10.5			323	4.8
BeO	46, 48	4.2	0.3			373	4.8
		20	16	KCl	39, 75	4.2	500
		77	270			25	140
		373	210			80	35
		573	120			194	10
		1,273	29			273	7.0
C (diamond)	46	4.2	75			323	6.5
		20	1,600			373	6.3
		77	3,400	KI	75	4.2	700
		194	870			80	13
		273	660			194	4.6
$CaCO_3$ ∥ to *c* axis	72	83	25			273	3.1
		273	5.5	Kr	52	4.2	0.48
$CaCO_3$ ⊥ to *c* axis	72	83	17			10	1.7
		194	6.5			20	1.2
		273	4.6			77	0.36
		373	3.6	LiF	5, 72	4.2	620
CaF_2	39	83	39			20	1,800
		223	18			77	150
		273	10	$MgO{\cdot}Al_2O_3$ (spinel)	48, 68	373	13
		323	9.2			773	8.5
		373	9.0	MnO	72	4.2	0.25
CsBr	39	223	1.2			40	55
		273	0.94			120	8.0
		323	0.81			573	3.5
		373	0.77	NaCl	39, 77	4.2	440
CsI	39	223	1.4			20	300
		273	1.2			77	30
		323	1.0			273	6.4
		373	0.95			323	5.6
Glass (phoenix)	46	4.2	0.095			373	5.4
		20	0.13	NaF	69, 75	5	1,100
		77	0.37			50	250
						100	90

(continued)

Table 2.8 (continued)

Material	Ref.	T, K	Conductivity	Material	Ref.	T, K	Conductivity
Ne	52	2	3.0	SiO_2 (quartz) ⊥ to c axis	72	20	370
		3	4.6			194	10
		1.2	1.2			273	6.8
		10	0.8	SiO_2 (fused)	72	4.2	0.25
		20	0.3			20	0.7
NH_4Cl	46	77	17			77	0.8
		194	23			194	1.2
		230	38			273	1.4
		273	27			373	1.6
$NH_4H_2PO_4$ ∥ to optic axis	72	315	0.71			673	1.8
		339	0.71	TlBr	72	316	0.59
$NH_4H_2PO_4$ ⊥ to optic				TlCl	72	311	0.75
axis	72	313	1.26	TiO_2 (rutile) to optic			
		342	1.34	axis	72	4.2	200
NiO	72	4.2	5.9			20	1,000
		40	100			273	13
		194	82	TiO_2 (rutile) ⊥ to optic			
SiO_2 (quartz) ∥ to c axis	72	20	720	axis	72	4.2	160
		194	20			20	690
		273	12			273	9

Note: Various authors differ from 5% to 25% on the experimental results for the conductivities of crystalline dielectrics and optical materials. See reference [25]

transducer located just beyond the region of energy deposition measures the pressures generated by the deposition of electron energy at constant volume. The results agree well with those calculated from equilibrium thermodynamic data, and they suggest, therefore, a temperature-independent Gruneisen coefficient at internal energies at least up to melt.

The chosen method was based on property degradation of materials due to "instantaneous" in-depth heating. In this case, the temperature dependence of the mechanical properties of metals and alloys is well documented (Reference [12]) for times at a temperature where properties have reached steady-state values.

Typical long soak times yield stress data versus temperature are shown on Fig. 2.8. As a general rule, for long soak times, the material no longer has any significant strength at temperatures equal to half the melting point.

In the region where strength properties vary with time and temperature, the data that are available (Reference [13]) are rather fragmentary and become increasingly so for high heating rates and short soak times. Although some evidence exists that strength properties can undergo significant temperature-induced degradation for times-to-heat plus times-to-test as short as 1 s (Reference [14], where these data were obtained using a resistive heating technique coupled with a conventional Hopkinson bar apparatus), the data are not in the range of direct interest to a

Table 2.9 Thermal conductivity of disordered dielectric, ceramic, refractory oxides, and insulating materials (in watts/meter, kelvin)

Material	Density g/cc	t, °C	Conductivity
Alumina	3.8	100	30
		400	13
		1,300	6
		1,800	7.4
	3.5	100	17
		800	7.6
Alumina + MgO	....	100	15
		400	10
		1,000	5.6
Asbestos	0.4	−100	0.07
		0	0.09
		100	0.1
Asbestos + 85% MgO	0.3	30	0.08
Barium titanate	....	50	3
		100	2.8
		150	2.6
		200	2.4
Beryllia	2.8	100	210
		400	90
		1,000	20
		1,800	15
	1.85	50	64
		200	40
		600	23
Brick, dry	1.54	0	0.04
Brick refractory:			
Aloxite	....	1,000	1.3
Aluminous	1.99	400	1.2
		1,000	1.3
Diatomaceous	0.77	100	0.20
		500	0.24
	0.40	100	0.08
		500	0.10
Fireclay	2.0	400	1
		1,000	1.2
Silicon carbide	2	200	2
		600	2.4
Vermiculite	0.77	200	0.26
		600	0.31
Calcium oxide	....	100	16
		400	9
		1,000	7.5
Cement mortar	2.0	90	0.55
Charcoal	0.2	20	0.055
Concrete	1.6	0	0.8
Cork	0.05	0	0.03
		100	0.04
	0.35	0	0.06
		100	0.08
Cotton wool	0.08	30	0.04
Diatomite	0.2	0	0.05
		400	0.09
	0.5	0	0.09
		400	0.16
Ebonite	1.2	0	0.16
Felt, flax	0.2	30	0.05
	0.3	30	0.04
	0.3	30	0.05
Fuller's earth	0.53	30	0.10
Glass wool	0.2	−200 to 20	0.005
		50	0.04
		100	0.05
		300	0.08
Graphite, 100 mesh	0.48	40	0.18
20–40 mesh	0.70	40	1.29
Linoleum, cork	0.54	20	0.08

Material	Density g/cc	t, °C	Conductivity
Magnesium oxide	....	100	36
		400	18
		1,200	5.8
		1,700	9.2
Magnesium + SiO_2	....	100	5.3
		400	3.5
		1,500	2.3
Mica, muscovite	....	100	0.72
		300	0.65
		600	0.69
Phlogopite	....	100	0.66
Canadian	....	300	0.19
		600	0.20
Micanite	....	30	0.3
Mineral wool	0.15	30	0.04
Paper, fiber glass and Al foil layers	0.1	−200 to 20	0.0001
Perlite, expanded	0.1	−200 to 20	0.002
Plastic:			
Celluloid	1.4	30	0.02
Polystyrene foam	0.05	−200 to 20	0.033
Aluminized Mylar foil	0.05	−200 to 20	0.0001
Porcelain	....	90	1
Rock, basalt	....	20	2
Chalk	....	20	0.92
Granite	2.8	20	2.2
Limestone	2.0	20	1
Sandstone	2.2	20	1.3
Slate, $\perp$	....	95	1.4
Slate, $\parallel$	....	95	2.5
Rubber, sponge	0.2	20	0.05
Rubber, 92%	....	25	0.16
Sawdust	0.2	30	0.06
Shellac	....	20	0.23
Silica aerogel	0.1	−200 to 20	0.003
Silica aerogel + 50% Al	0.1	−200 to 20	0.003
Snow	0.25	0	0.16
Steel wool	0.1	55	0.09
Thoria	....	100	10
		400	5.8
		1,500	2.4
Titanium dioxide	....	100	6.5
		400	3.8
		1,200	3.3
Uranium dioxide	....	100	9.8
		400	5.5
		1,000	3.4
Wood, balsa, $\perp$	0.11	30	0.04
Fir, $\perp$	0.54	20	0.14
Fir, $\parallel$	0.54	20	0.35
Pine, $\perp$	0.45	60	0.11
Pine, $\parallel$	0.45	60	0.26
Walnut, $\perp$	0.65	20	0.14
Wool	0.09	30	0.04
Zinc oxide	....	200	17
		800	5.3
Zirconia	....	100	2
		400	2
		1,500	2.5
Zirconia + SiO_2	....	200	5.6
		600	4.6
		1,500	3.7

Note: Values quoted are from W. D. Kingery et al. ref. [27], R. W. Powell ref. [28] and "International Critical Tables" ref. [29]
Various authors differ from 10% to 50% on the experimental results for the conductivities of disordered dielectrics, ceramics, refractory oxides, and insulating materials

Table 2.10 Thermal conductivity of some materials below 1 K ‡ **(in watts/meter, kelvin)**

Material	Ref.	Temperature				
		0.2 K	0.4 K	0.6 K	0.8 K	1.0 K
Cr K Alum	72	0.05	0.3*	0.8*	2*	3.2
Epibond 104‡	2	0.00018	0.001	0.0025	0.0048	0.006*
In (superconducting)	18, 72	0.1*	1.2	7	15	23
KCl	4, 65	0.2*	1.6	5	16	30
KI	65	1.5*	6	22	50	100
Kel-F‡	2	0.00029	0.0012	0.0027	0.0045	0.006*
LiF	23	0.09	0.8	2	6.5	10
Nb (normal, superconduct-	13, 72	4.5*	9*	12*	20*	25
ing)	13	0.1*	0.5	1.5	3.5	6
Nylon‡	2	0.00018	0.0006	0.0011	0.0015*	0.002*
Pyrex	2	0.001	0.003	0.0065	0.011	0.016*
Rubber (hard)	2	0.0025	0.007	0.015	0.02*	0.03*
Sn (normal, superconduct-	52, 78	100*	2000*	4500	6000	7000
ing)	52, 35	0.4	2.5	15	80	150
Ta (normal, superconduct-	13, 72	3*	6*	8*	10*	11
ing)	13	0.15*	0.9	3	9	10
Teflon‡	2	0.00008	0.0004	0.0013	0.0022*	0.0035*
Tl (normal, superconduct-	52, 78	2000*	4000*	6000*	7000*	8500*
ing)	52, 78	0.06	5	120	600	1200

Note: The conduction of materials below 1 K is strongly dependent upon their chemical purity and physical perfection and, for small specimens, depend on the actual dimension surfaces. The values quoted are typical for these materials. See ref. [25] for additional data

†The thermal conductivity of normal metals in the impurity range varies as $\lambda = \alpha T$

‡Epibond is a diglycidyl ether of bisphenol, which is a fluid mixed in a ratio of 4:1 with powdered Bentonite clay. Kel-F is a polychlorotrifluorethylene. Nylon is polyamide. Teflon is a polychlorotrifluorethylene

nuclear radiation effects analyst concerned with structural response. From this standpoint, a more reasonable and desirable set of data would be generated under the following conditions:

- The specimen is heated in a small fraction of a microsecond.
- The specimen is held at temperature, and its high strain rate mechanical properties (yield stress, Young's modulus, etc.) are measured for times radiating from a fraction of a millisecond to many milliseconds after heating.

- The testing time is very short compared with the time at temperature.
- The specimen must be sufficiently large so that macroscopic properties can be measured.

The Physics International Company report shows a technique that is described for measuring the dynamic yield stress and Young's modulus in tension under the above noted condition, and readers who are interested in further details should refer to this report [11].

Fig. 2.8 Long-heating-time temperature-induced tensile property degradation

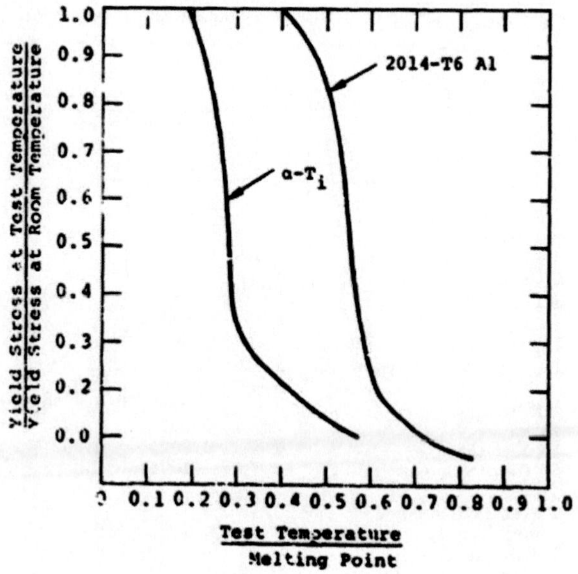

References

1. Zohuri, B.: Directed Energy Weapons: Physics of High Energy Lasers (HEL), 1st edn. Springer (2016)
2. Zohuri, B.: Directed-Energy Beam Weapons, 1st edn. Springer (2019)
3. Tebo, F.J.: Selected Values of the Physical Properties of Various Materials. Argonne National Laboratory, ANL-5914
4. Powell, R.L., Blanpied, W.A.: Thermal Conductivity of Metals and Alloys at Low Temperatures NBS Circular 556. National Bureau of Standards (1954)
5. Pagerey, P.R The Thermal Conductivity of Organic Liquids. Presented of paper contributed by Hear Transfer Division of the American Society of Mechanical Engineers for presentation at the semi-annual meeting. Paper no. 54-SA-55, June 20, 1954
6. Sakiadis, B.C., Coates, J.: A Literature Survey of the Thermal Conductivity of Liquids Engineering, Experimental Station, Bulletin No. 34. College of Engineering, Louisiana State University (1954)
7. Liquid Metals Handbook, edited by R. N. Lyon, Department of the Navy and U.S. Atomic Energy Commission, NAVEXOS P-733, Chapter 2, June 1952
8. Evans, J. E., "Thermal Conductivity of Fourteen Metals and Alloys Up to 1100 °F". Report NACA-RM-E50L07 (1951)
9. The Reactor Handbook, Vol. 3, Materials, 1953
10. Metal Handbook, Complied by the American Society for Metals
11. Stefansky, T., Mazzella, A., Triebes, K., Shea, J.: Material Properties Measurements us Ing Pulsed Electron Beams Report AD729364. Physics International Company, San Leandro (1971)
12. ASM Metals Handbook, American Society for Metals, Metals Park, Ohio, 1964
13. Aerospace Materials Handbook, 1968
14. Minutes for the Sixth Technical Direction Meeting of the DASA PREDIX Program, November 5, 6, 1969

15. American Institute of Physics Handbook, McGraw-Hill Book Co., Inc., New York (1957) sec. 4, p. 49
16. McAdams, W.H.: Heat Transmission, 3rd edn. McGraw-Hill Book Company, New York (1954)
17. Schneider, P.J.: Conduction Heat Transfer. Addison-Wesley Press Inc, Cambridge, MA (1955)
18. Carslaw, H.S., Jaeger, J.C.: Conduction of Heats in Solids, 2nd edn. Oxford University Press, London (1959)
19. Touloukian, Y. S., director, Thermophysical Properties research Center Data Book, Vols I, II, and III, Purdue University, Lafayette, Indiana, 1964. Note that, this reference is no longer available; is being superseded by "TPRC Series on Thermophysical Properties of Matter". Covered the following properties in each volume: thermal conductivity, viscosity, radiative properties, thermal diffusivity, and specific heat. Vol. I: Metallic Elements and Their Alloys (in Solid, Liquid, or Gaseous State). Vol. II: Nonmetallic Elements, Compounds and Mixture (In Liquid or Gaseous State at Normal Temperature and Pressure). Vol. III: Nonmetallic Elements, Compounds and Mixtures (in Solid State at Normal Temperature and Pressure). TPRC *Reports in the Standard References Data Series*
20. Tye, R.P. (ed.): Thermal Conductivity, 2 Vols. Academic Press, Inc., London (1969)
21. Thermal Conductivity Conferences, unpublished proceeding: e.g. 6th – Dayton, Ohio, 1966; 7th – National Bureau of Standards, Washington, D. C., 1967 (published as NBS Spec. Publ. 302); 8th – Lafayette, Inc. 1968
22. Jacob, M.: Heat Transfer, vol. I and II, p. 1957. Wiley, New York (1949)
23. Powell, R.L.: ASTM Spec. Tech Publication. **387** (1966)
24. Rosenberg, H.M.: Low Temperature Solid State Physics. Oxford University Press, London (1963)
25. Powell, R.L., Blanpied, W.A.: Thermal conductivity of metals and alloys at low temperatures. NBS Circ. **556** (1954) Revised and Expanded by G. E. Childs, L. J. Ericks, and R. L. Powell: Thermal Conductivity of Solids at Low Temperatures, NBS Monograph, to be published in 1971
26. Powell, R.W., Ho, C.Y.: In: Liley, P.E. (ed.) Thermal Conductivity of Selected Materials National Standard Reference Data Series. National Bureau of Standards: no. 8 Part 1 (1966) C. Y. Ho, et al.: no 16, Part 2, 1968; no. 18, Part 3 (to be published)
27. Kingery, W.D., Franel, R.L.C., Vasilos, T.: J. Am. Ceram. Soc. **37**, 107 (1954)
28. Powell, R.W.: Research. **7**, 492 (1954)
29. National Academy of Sciences and National Research Council: International Critical Tables. McGraw-Hill Book Company, New York (1920–1990)
30. Childs, G.E., Hanley, H.J.M.: NBS Tech. Note 350, October, 1966. Cryogenics. **8**, 94 (1968)

Chapter 3
Heat Conduction in Solids

One of the major goals of physics and so as ours in case of laser in particular and its interaction with materials is to understand the nature of light. Due to the complex nature of light, this goal is very difficult to fully achieve, but this complication means that light offers many opportunities for different applications including optical interferences and in our case response of materials to laser radiation and its interaction with matter and specifically metallic materials. This leads to dealing with heat conduction equation in a very complex form; therefore we need to deal with heat transport from aspects of *conduction*, *convection*, and *radiation*.

3.1 Introduction

When there exists a body or matter with different parts exposed to a different temperature, then heat flows from the hotter to the colder part of that body or matter. Consider the metal rod *ABCD*, as depicted in Fig. 3.1, and suppose it is heated at the end *A* by some means of external heat source (i.e. laser radiation).

For some time, the temperature of the rod gradually rises, the parts near *A* being heated first, but no change takes place as *CD* until *BC* has had its temperature raised. Ultimately, if the end *A* is heated long enough, it is found that a steady state of temperature is reached, in which, while the temperature may vary from point to point, it remains the same at each point as the time changes.

This transference of heat from the hotter portions of the rod to the colder is called *conduction of heat*. However, it should be distinguished from *convection of heat*, on the one hand, and *radiation of heat*, on the other hand.

In a typical, heat convection scenario, the transference of heat is due to the motion of the heated body itself, as, for example, when the different parts of a liquid are at different temperature is reached. However, in heat radiation, the hotter body loses heat, and the colder body gains it by means of a process in some intervening medium.

B. Zohuri, *Thermal Effects of High Power Laser Energy on Materials*,
https://doi.org/10.1007/978-3-030-63064-5_3

Fig. 3.1 Single rod

Mathematically, *conductivity* of *heat* may be defined by hypothesis suggested by the French physicist and scientist Joseph Fourier known as Fourier's Law as defined by Eq. 3.1.

$$\vec{q} = -k\vec{\nabla}T \tag{3.1}$$

Equation 3.1 simply indicates that when there exists a temperature gradient within a body, heat energy will flow from the region of high temperature to the region of low temperature, and as we stated, this phenomenon is known as conduction of heat transfer.

This equation determines the heat flux vector $\vec{q}$ for a given temperature profile T and thermal conductivity k. The minus sign ensures that heat flows down the temperature gradient. In the following section, we expound on this heat equation to determine temperature within a body, where the heat conduction takes place.

3.2 Heat Equation-Driven Temperature Determination

The temperature profile within a body depends upon the rate of its internally generated heat, its capacity to store some of this heat, and its rate of thermal conduction to its boundaries (where the heat is transferred to the surrounding environment). Mathematically this is stated by the heat equation, as it is shown in Eq. 3.2

$$\vec{\nabla}^2 T - \frac{1}{\alpha}\frac{\partial T}{\partial t} = \frac{1}{k} q_{generated} \tag{3.2}$$

along with its boundary conditions, equations that prescribe either the temperature T on, or the heat flux $q_{generated}$ through all of the body boundaries (Eq. 3.3).

$$\begin{cases} T(\Omega_a) = T_{prescribed} \\ q(\Omega_b) = q_{prescribed} \\ \Omega_a \cup \Omega_b = \Omega \end{cases} \tag{3.3}$$

In the heat equation, the power generated per unit volume is expressed by $q_{generated}$. The thermal diffusivity α is related to the thermal conductivity k, the specific heat c, and the density ρ by:

$$\alpha = \frac{k}{\rho c} \tag{3.4}$$

For steady-state problems, the heat equation simplifies to:

$$\vec{\nabla}^2 T = -\frac{1}{k} q_{generated} \tag{3.5}$$

3.3 Derivation of Heat Equation

The heat equation follows from the *conservation of energy* for a small element within the body:

Heat conducted in	+	Heat generated within	=	Heat conducted out	+	Change in energy stored within

We can combine the heats conducted in and out into one "net heat conducted out" term to give:

Net heat conducted out.	=	Heat generated within	–	Change in energy stored within

Mathematically, this equation is expressed as:

$$\vec{\nabla} \cdot \vec{q} = q_{generated} - \frac{de}{dt} \tag{3.6}$$

The change in internal energy e is related to the body's ability to store heat by raising its temperature, given by:

$$\frac{de}{dt} = \rho c \frac{dT}{dt} \tag{3.7}$$

One can substitute for **q** using Fourier's law of heat conduction from above to arrive at the heat equation:

$$\begin{cases} \vec{\nabla} \cdot \left(k \vec{\nabla} T\right) = q_{generated} - \rho c \dfrac{dT}{dt} \\ -k \vec{\nabla}^2 T + \rho c \dfrac{\partial T}{\partial t} = q_{generated} \\ \vec{\nabla}^2 T - \dfrac{1}{\alpha} \dfrac{\partial T}{\partial t} = -\dfrac{1}{k} q_{generated} \end{cases}$$

Blackbody	A body with a surface emissivity of 1. Such a body will emit all of the thermal radiation it can (as described by theory) and will absorb 100% of the thermal radiation striking it. Most physical objects have surface emissivities less than 1 and hence do not have blackbody surface properties
Density, ρ	The amount of mass per unit volume. In heat transfer problems, the density works with the specific heat to determine how much energy a body can store per unit increase in temperature. Its units are kg/m^3
Emissive power	The heat per unit time (and per unit area) emitted by an object. For a blackbody, this is given by the Stefan–Boltzmann relation σT^4
Graybody	A body that emits only a fraction of the thermal energy emitted by an equivalent blackbody. By definition, a graybody has a surface emissivity less than 1 and a surface reflectivity greater than 0
Heat flux, q	The rate of heat flowing past a reference datum. Its units are W/m^2
Internal energy, e	A measure of the internal energy stored within a material per unit volume. For most heat transfer problems, this energy consists just of thermal energy. The amount of thermal energy stored in a body is manifested by its temperature
Radiation viewfactor, F_{12}	The fraction of thermal energy leaving the surface of object 1 and reaching the surface of object 2, determined entirely from geometrical considerations. Stated in other words, F_{12} is the fraction of object 2 visible from the surface of object 1 and ranges from 0 to 1. This quantity is also known as the radiation shape factor. Its units are dimensionless
Rate of heatgeneration, q_{gen}	A function of position that describes the rate of heat generation within a body. Typically, this new heat must be conducted to the body boundaries and removed via convection and/or radiation heat transfer. Its units are W/m^3
Specific heat, c	A material property that indicates the amount of energy a body stores for each degree increase in temperature, on a per unit mass basis. Its units are J/kg-K
Stefan–Boltzmannconstant, σ	Constant of proportionality used in radiation heat transfer, whose value is 5.669×10^{-8} W/m^2-K^4. For a blackbody, the heat flux emitted is given by the product of ε and the absolute temperature to the fourth power
Surface emissivity	The relative emissive power of a body compared to that of an ideal blackbody. In other words, the fraction of thermal

radiation emitted compared to the amount emitted if the body were a blackbody. By definition, a blackbody has a surface emissivity of 1. The emissivity is also equal to the absorption coefficient, or the fraction of any thermal energy incident on a body that is absorbed

Thermal conductivity, *k* A material property that describes the rate at which heat flows within a body for a given temperature difference. Its units are W/m-k

Thermal diffusivity, *α* A material property that describes the rate at which heat diffuses through a body. It is a function of the body's thermal conductivity and its specific heat. A high thermal conductivity will increase the body's thermal diffusivity, as heat will be able to conduct across the body quickly. Conversely, a high specific heat will lower the body's thermal diffusivity, since heat is preferentially stored as internal energy within the body instead of being conducted through it. Its units are m^2/s

3.4 Heat Transport: Conduction, Convection, Radiation

Here we will define and explain each of heat transport conditions such as conduction, convection, and radiation briefly in below sub-sections.

3.4.1 Conduction

If different parts of an isolated solid are at different temperatures, heat will flow from the hot places to the cold ones until eventually all are at the same temperature. By "isolated" here we mean that the solid is not able to exchange heat with the outside world.

Experimentally, it is found that for most substances, the rate of heat flow at any point is proportional to the temperature gradient—how fast the temperature is changing with position. To give an example, consider heat flowing down a thin rod, heated at one end, and assume the rod is wrapped in insulation so all the heat flows *down* the rod, none escapes from the surface. The natural unit of heat flow down the rod is how many joules per second pass a fixed point in the rod. It is found that:

$$\frac{dQ}{dt} \propto \frac{dT}{dx}$$

where Q is in joules, T in degrees Kelvin, and x in meters down the rod. The heat flow rate is then in joules per second, or watts. It is evident from this equation that if heat is supplied at a steady rate to one end of the rod, and drains from the other end, the temperature distribution will ultimately settle down to $dT/dx =$ constant, a linear drop along the rod from one end to the other.

It is also found experimentally that a rod of double the cross section carries twice the heat current at the same temperature difference. (This is also true for electric current, but remember it is not true for water in a pipe—the "caloric" fluid evidently doesn't act like a viscous liquid.)

This makes it possible to define a *coefficient of κ thermal conductivity* κ for a particular material by

$$\frac{dQ}{dt} = \kappa A \frac{dT}{dx}$$

for heat flow across an area A (in square meters) for a given temperature gradient dT/dx.

The units of κ are watts/K·meters. Some values: copper 390, stainless steel 13, glass around 0.8, white pine 0.11, air 0.026.

3.4.2 Microscopic Picture of Conduction

When a solid is warmed, the atoms jiggle around more—the heat energy is partially their kinetic energy of motion, partly the extra energy stores in the springy bonds between them as a result of their motion. If one end of a solid is heated, the more vigorously moving atoms there bounce against their neighbors, which then begin to move more vigorously, and the motion diffuses down the line. Obviously, this cannot be the whole story, because if we hit one end of the rod, a sound wave travels down by neighbor hitting neighbor and moves far faster than heat. A more accurate picture (for a nonmetal) is that when one end is heated, tiny sound waves (called phonons) are generated by the fast moving atoms near the surface. These phonons travel into the solid at the speed of sound, but, unlike the massive compression wave when the end of the rod is hit, these phonons bounce off impurities or imperfections in the solid and follow random paths, only a few tens of atomic spaces between hits, typically. This, then, is very like the diffusion of a molecule in a gas we studied earlier, and it takes several minutes for heat to make its way through, say, half a centimeter of glass. The picture is different for metals: the electrons which conduct electricity so efficiently do the same for heat. However, heat transport by electrons cannot be understood without quantum mechanics: Pauli's exclusion principle means only about 1% of the electrons take part in the heat conduction, but it also means that they travel far faster. The thing to remember at this stage is that the electrons carry the heat, phonons do too, but make a negligible contribution. As you might expect, good conductors of electricity are also good conductors of heat. This is why copper is used in saucepans (also, it doesn't corrode too badly).

Conductivity in liquids and gases can be measured—but usually heat transport in fluids is dominated by convection (see below). Exceptions are, for example, a fluid heated from above, or a pond below 4 degrees Celsius being cooled from above on a winter night.

3.4.3 *British Units*

In the real world out there, the units are different. Heat flow (e.g., in construction jobs) is measured in BTU per hour, temperature gradients in degrees Fahrenheit per inch thickness, and cross-sectional area in square feet. The R-value of "thermal resistance" is the inverse of the thermal conductivity. For 1 square foot of material, 1 inch thick, R relates the heat current to the temperature drop by an Ohm's law equation $\Delta T = IR$. Different areas and thicknesses scale in the obvious way. For a wall made of layers of different materials, the R-values just add.

3.4.4 *Convection*

Convection is gravitationally induced heat transport, driven by the expansion of a fluid on heating. The hot expanded fluid has lower density and so will rise to the top of colder, and therefore denser, fluid. The simplest example is water in a kettle heated from below: hot water will rise in a central column, spread through the top layer, cool, and then flow back down around the outside. The pattern becomes more complicated if a fluid is being heated over a large area, with no obvious center. Convection cells can arise, each having a pattern like that in the kettle, the cells in a hexagonal pattern. This can happen in weather: a storm can be such a cell. However, many patterns are possible: the fluid mechanics is extremely complex. One important example of convection currents is inside the earth. Such currents deep inside are believed drive the surface movement of plates, causing earthquakes, tsunamis, etc.

3.4.5 *Radiation*

Heat from the sun reaches us as radiation; much of it is visible light, the rest similar electromagnetic waves but at wavelengths our eyes are not sensitive to. All bodies not at absolute zero temperature radiate; at room temperature, the radiation is in the infrared, wavelengths longer than those of the visible spectrum. Microscopically, the radiation comes about because the oscillating ions and electrons in a warm solid are accelerating electric charges, and as you will find next semester, such charges radiate. Different substances radiate with different efficiencies, and those that radiate better also absorb incoming radiation better. A perfect absorber is called a blackbody (such perfection is not found in nature, but some things are close). This, then, is also a perfect radiator. It was found experimentally that for a perfect blackbody at an even temperature, the radiant energy output in watts per square meter of surface went as the fourth power of the absolute temperature:

$$P = \sigma T^4$$

P being power per square meter, σ is Stefan's constant, 5.67×10^{-8} Watts/sq.m./K^4.

For a given T, the radiant power peaks at a certain wavelength, $\lambda_{max}T =$ constant. This was well established: on heating a piece of metal, say, such as turning a dimmer on an ordinary light bulb, the first visible radiation is in the red, and an extremely hot object becomes white or even bluish. However, this was theoretically incomprehensible without quantum mechanics, and in fact the mystery of blackbody radiation led Planck to the first formulation of the idea of the quantum.

3.5 Heat Equation with Conduction and Convection

We consider the heat equation on the (0,1) with two extra terms that correspond to heat conduction and convection.

$$u_t(x,t) = k\left[\left(u_{xx}(x,t) - 2au(x,t)_x + bu(x,t)\right]0 < x < L, t > 0 \qquad (3.8)$$

$$u(0,t) = 0 \qquad (3.9)$$

$$u(a,t) \qquad (3.10)$$

$$u(x,0) = \varphi(x) \qquad (3.11)$$

There are many different ways to approach this problem given the boundary and initial condition, and one would be able to apply separation of variable directly. The disadvantage to this is that one gets a more complicated ODE for $X(x)$ and there is a more difficult analysis of the eigenvalue and eigenvectors.

We will take a different approach which allows us to use our earlier work after a change of dependent variable. So, to this end, let us define another variable such as $\vartheta(x,t)$ via:

$$u(x,t) = e^{ax+\beta t}\vartheta(x,t), \beta = k\left(b - a^2\right) \qquad (3.12)$$

Thus, we have:

$$\vartheta(x,t) = e^{-(ax+\beta t)}u(x,t)$$

and we can compute:

$$\vartheta_t - k\vartheta_x x = e^{-(ax+\beta t)}(-\beta u + u_t) - k\left[\left(e^{-(ax+\beta t)}(-au + u_x)\right]_x\right.$$

$$= e^{-(ax+\beta t)}\{(-\beta u + u_t) - k[-a(-au + u_x) + (-au_x + u_{xx})]\}$$

$$= e^{-(ax+\beta t)}\left[u_t - k\left(u_{xx} - 2au_x + a^2u\right) + \beta u\right]$$

$$= e^{-(ax+\beta t)}\left[u_t - k\left(u_{xx} - 2au_x + a^2u + a^2u\left(b - a^2\right)u\right)\right]$$

$$= e^{-(ax+\beta t)}\left[u_t - k\left(u_{xx} - au_x + a^2u + bu\right)\right] = 0$$

Furthermore:

$$\vartheta(0,t) = e^{-\beta t}u(0,t) = 0, \vartheta(1,t) = e^{-a-\beta t}u(1,t) = 0$$

and:

$$\vartheta(x,0) = e^{-ax}u(x,0) = e^{-ax}\varphi(x)$$

Therefore, $\vartheta(x,t)$ is the solution of the following:

$$\vartheta_t = k\vartheta_x x$$
$$\vartheta(0,t) = 0 \text{and} \vartheta(1,t) = 0$$
$$\vartheta(x,0) = e^{-ax}\varphi(x)$$

With $\lambda_n = -(n\pi)^2$ the solution to this problem is:

$$\vartheta(x,t) = \sum_{n=1}^{\infty} b_n e^{\lambda_n t} \sin(n\pi x) \text{with} b_n = 2\int_0^1 e^{-ax}\varphi(x)\sin(n\pi x)dx$$

Finally, our solution to (1)–(4) can be written as:

$$u(x,t) = e^{ax+\beta t}\sum_{n=1}^{\infty} b_n e^{\lambda_n t}\sin(n\pi x)$$

3.6 Solving Heat Equation Using Different Methods

For simplicity of this approach, we deal with solving heat conduction in the Cartesian coordinate system, and we do it for a one-dimentional case and then extended to more variables (3-D) and other coordinate systems. In this section, we examine the solution of boundary-value problems of heat conduction in the Cartesian coordinate system for the 1-D, 2-D, and 3-D finite, semi-finite, and infinite regions. Separation of variables method will be utilized, so long as the problem falls within homogeneous-type boundary-value problems. The integral transform technique such as Fourier, Laplace, and other transformation will be applied for the solution of non-homogeneous problems.

As we said, the method of separation of variables has been used in the solution of homogeneous heat conduction problem. The multidimensional steady-state heat conduction problems with no heat flux or generation can also be solved with this method if only one of the boundary conditions is non-homogeneous; problem involving more than one non-homogeneous boundary conditions can be split up into simpler problems each containing only one non-homogeneous boundary condition.

3.6.1 *Separation of Variables Method*

The method of separation of variables cannot always be used, and even when it can be used, it will not always be possible to get much past the first step in the method. However, it can be used to easily solve the 1-D heat equation with no sources, the 1-D wave equation, and the 2-D version of Laplace's equation, $\nabla^2 T = 0$.

In order to use the method of separation of variables, we must be working with a linear homogeneous partial differential equations with linear homogeneous boundary conditions. At this point we're not going to worry about the initial condition (s) because the solution that we initially get will rarely satisfy the initial condition(s). As we'll see however, there are ways to generate a solution that will satisfy initial condition(s) provided they meet some fairly simple requirements.

The method of separation of variables relies upon the assumption that a function of the form

$$T(x,t) = X(x)Y(t) \tag{3.13}$$

will be a solution to a linear homogeneous partial differential equation in x and t. This is called a product solution, and provided the boundary conditions are also linear and homogeneous, this will also satisfy the boundary conditions. However, as noted above, this will only rarely satisfy the initial condition, but that is something for us to worry about in the next section.

Now, before we get started on some examples, there is probably a question that we should ask at this point, and that is: Why? Why did we choose this solution, and how do we know that it will work? This seems like a very strange assumption to make. After all there really isn't any reason to believe that a solution to a partial differential equation will in fact be a product of a function of only x's and a function of only t's. This seems more like a hope than a good assumption/guess.

Unfortunately, the best answer is that we chose it because it will work. As we'll see it works because it will reduce our partial differential equation down to two ordinary differential equations, and provided we can solve those, then we're in business, and the method will allow us to get a solution to the partial differential Equ 3.8.

So, let's do a couple of examples to see how this method will reduce a partial differential equation down to two ordinary differential equations.

Example 3.1: **Use Separation of Variables on the Following Partial Differential Equation**

$$\frac{\partial T(x,t)}{\partial t} = k\frac{\partial^2 T(x,t)}{\partial x^2}$$

Initial Condition.I.C. $\quad T(x,0) = f(x)$

Boundary Condition. B.C $\quad T(0,t) = 0$

$\qquad T(L,t) = 0$

Solution:

So, we have the heat equation with no sources, fixed temperature boundary conditions (that are also homogeneous), and an initial condition. The initial condition is only here because it belongs here, but we will be ignoring it until we get to the next section.

The method of separation of variables tells us to assume that the solution will take the form of the product:

$$T(x,t) = X(x)Y(t)$$

So, all we really need to do here is plug this into the differential equation and see what we get.

$$\frac{\partial}{\partial t}(X(x)Y(t)) = k\frac{\partial^2}{\partial x^2}(X(x)Y(t))$$

$$X(x)\frac{\partial Y(t)}{\partial t} = kY(t)\frac{\partial^2 X(x)}{\partial x^2}$$

(continued)

Example 3.1 (continued)

As shown above, we can factor the *X*(*x*) out of the time derivative, and we can factor the *Y*(*t*) out of the spatial derivative. Also notice that after we've factored these out, we no longer have a partial derivative left in the problem. In the time derivative, we are now differentiating only *Y*(*t*) with respect to t, and this is now an ordinary derivative. Likewise, in the spatial derivative, we are now only differentiating *X*(*x*) with respect to x, and so we again have an ordinary derivative.

At this point it probably doesn't seem like we've done much to simplify the problem. However, just the fact that we've gotten the partial derivatives down to ordinary derivatives is liable to be a good thing even if it still looks like we've got a mess to deal with.

Speaking of that apparent (and yes I said apparent) mess, is it really the mess that it looks like? The idea is to eventually get all the *t*'s on one side of the equation and all the *x*'s on the other side. In other words, we want to "separate the variables" and hence the name of the method. In this case let's notice that if we divide both sides by *X*(*x*)*Y*(*t*), we get want we want, and we should point out that it won't always be as easy as just dividing by the product solution. So, dividing out gives us:

$$\frac{1}{Y}\frac{dY}{dt} = k\frac{1}{X}\frac{d^2X}{dx^2} \Rightarrow \frac{1}{kY}\frac{dY}{dt} = \frac{1}{X}\frac{d^2X}{dx^2}$$

Notice that we also divided both sides by *k*. This was done only for convenience down the road. It doesn't have to be done, and nicely enough if it turns out to be a bad idea, we can always come back to this step and put it back on the right side. Likewise, if we don't do it and it turns out to maybe not such a bad thing, we can always come back and divide it out. For the time being, however, please accept our word that this was a good thing to do for this problem. We will discuss the reasoning for this after we're done with this example.

Now, while we said that this is what we wanted, it still seems like we've got a mess. Notice however that the left side is a function of only *t* and the right side is a function only of *x* as we wanted. Also notice these two functions must be equal.

Let's think about this for a minute. How is it possible that a function of only *t*'s can be equal to a function of only *x*'s regardless of the choice of *t* and/or *x* that we have? This may seem like impossibility until you realize that there is one way that this can be true. If both functions (i.e., both sides of the equation) were in fact constant and not only a constant, but the same constant, then they can in fact be equal.

So, we must have:

(continued)

Example 3.1 (continued)

$$\frac{1}{kY}\frac{dY}{dt}=\frac{1}{X}\frac{d^2X}{dx^2}=-\lambda$$

where $-\lambda$ is called the *separation constant* and is arbitrary.

The next question that we should now address is why the minus sign? Again, much like dividing out the k above, the answer is because it will be convenient down the road to have chosen this. The minus sign doesn't have to be there, and in fact there are times when we don't want it there.

So how do we know it should be there or not? The answer to that is to proceed to the next step in the process (which we'll see in the next section), and at that point we'll know if it would be convenient to have it or not, and we can come back to this step and add it in or take it out depending what we choose to do here.

Okay, let's proceed with the process. The next step is to acknowledge that we can take the equation above and split it into the following two ordinary differential equations:

$$\frac{dY}{dt}=-k\lambda Y\frac{d^2X}{dx^2}=-\lambda X$$

Both of these are very simple differential equations; however because we don't know what λ is, we actually can't solve the spatial one yet. The time equation however could be solved at this point if we wanted to, although that won't always be the case. At this point we don't want to actually think about solving either of these yet however.

The last step in the process that we'll be doing in this section is to also make sure that our product solution, $T(x, t) = X(x)Y(t)$, satisfies the boundary conditions, so let's plug it into both of those.

$$T(0,t)=X(0)Y(t)=0T(L,t)=X(L)Y(t)=0$$

Let's consider the first one for a second. We have two options here: either $X(0) = 0$ or $Y(t) = 0$ for every t. However, if we have $Y(t) = 0$ for every t, then we'll also have $T(x, t) = 0$, i.e., the trivial solution, and as we discussed in the previous section, this is definitely a solution to any linear homogeneous equation we would really like, a non-trivial solution.

Therefore, we will assume that in fact we must have $X(0) = 0$. Likewise, from the second boundary condition, we will get $Y(t) = 0$ to avoid the trivial solution. Note as well that we were only able to reduce the boundary conditions down like this because they were homogeneous. Had they not been homogeneous, we could not have done this.

(continued)

Example 3.1 (continued)

So, after applying separation of variables to the given partial differential equation, we arrive at a 1st order differential equation that we'll need to solve for and a 2nd order boundary value problem that we'll need to solve for $X(x)$. The point of this section however is just to get to this point, and we'll hold off solving these until the next section.

Let's summarize everything up that we've determined here.

$$\frac{dY}{dt} = -k\lambda Y \frac{d^2X}{dx^2} + \lambda X = 0$$

$$X(0) = 0X(L) = 0$$

And note that we don't have a condition for the time differential equation and is not a problem. Also note that we rewrote the second one a little.

The time-dependent equation can really be solved at any time, but since we do not know what λ is yet, let's hold off on that one. Also note that in many problems only the boundary value problem can be solved at this point, so do not always expect to be able to solve either one at this point.

The spatial equation is a boundary value problem (BVP), and we know from our work in Sect. 1.10 Appendix C that it will only have a non-trivial solution (which we want) for certain values of λ, which we will recall are called eigenvalue. Once we have those, we can determine the non-trivial solutions for each λ, i.e., eigenfunctions.

Now, we actually solved the spatial problem

$$\frac{d^2X}{dx^2} + \lambda X = 0$$
$$X(0) = 0X(L) = 0$$

in Example 1 of Sect. 3.0 of Appendix C for $L = 2\pi$. So, because we have solved this once for a specific L and the work is not all that much different for a general L, we are not going to put a lot of effort here, and if you need to review, just get Example 1 from Sect. 3.0 of Appendix C.

We have three cases to deal with:

$$\underline{\lambda > 0}$$

In this case, we know the solution to the differential equation is:

$$X(x) = c_1 \cos\left(\sqrt{\lambda}x\right) + c_2 \sin\left(\sqrt{\lambda}x\right)$$

Applying again the first boundary condition gives:

(continued)

Example 3.1 (continued)

$$0 = X(0) = c_1$$

Now applying the second boundary condition, and using the above result of course, gives:

$$0 = X(L) = c_2 \sin\left(L\sqrt{\lambda}\right)$$

Now, we are after non-trivial solutions, and this means we must have:

$$\sin\left(L\sqrt{\lambda}\right) = 0 \Rightarrow X_n(x) = \sin\left(\frac{n\pi x}{L}\right) n = 1, 2, 3, \cdots$$

Note that we did not need the c_2 in the eigenfunction as it will just get absorbed into another constant that we will be picking up later on.

$\underline{\lambda = 0}$

The solution to the differential equation in this case is:

$$X(x) = c_1 + c_2 x$$

Applying the boundary conditions gives:

$$0 = X(0) = c_1 0 = X(L) = c_2 L c_2 = 0$$

So, in this case, the only solution is the trivial solution, and so $\lambda = 0$ is not an eigenvalue for this boundary value problem.

$\underline{\lambda < 0}$

Here the solution to the differential equation is:

$$X(x) = c_1 \cosh\left(\sqrt{-\lambda x}\right) + c_2 \sinh\left(\sqrt{-\lambda x}\right)$$

Applying the first boundary condition gives:

$$0 = X(0) = c_1$$

And applying the second gives:

$$0 = X(L) = c_2 \sinh\left(L\sqrt{-\lambda}\right)$$

(continued)

Example 3.1 (continued)

So, we are assuming $\lambda < 0$, and so $L\sqrt{-\lambda} \neq 0$, and this means $\sinh\left(L\sqrt{-\lambda}\right) \neq 0$. We therefore we must have and so we can only get the trivial solution in this case.

Therefore, there will be no negative eigenvalues for this boundary value problem. The complete list of eigenvalues and eigenfunctions for this problem is then:

$$\lambda_n = \left(\frac{n\pi}{L}\right)^2 X_n(x) = \sin\left(\frac{n\pi x}{L}\right) n = 1, 2, 3, \ldots$$

Now let us solve the time differential equation:

$$\frac{dY}{dt} = k\lambda_n Y$$

And note that even though we know λ, we are not going to plug it in quite yet to keep the mess to a minimum. We will however now use λ_n to remind us that we actually have an infinite number of possible values here. This is a simple linear (and separable for that matter) 1st order differential equation, and so we will let you verify that the solution is:

$$Y(t) = ce^{-k\lambda_n t} = ce^{-k\left(\frac{n\pi}{L}\right)^2 t}$$

Now that we have both ODEs solved and we can finally write down a solution. Note however that we have in fact found infinitely many solutions since there are infinitely many solutions (i.e., eigenfunctions) to the spatial problem. Our solution is a product of both solutions are then

$$u_n(x,t) = B_n \sin\left(\frac{n\pi x}{L}\right) e^{-k\left(\frac{n\pi}{L}\right)^2 t} n = 1, 2, 3, \ldots$$

We've denoted the product solution u_n to acknowledge that each value of n will yield a different solution. Also note that we've changed the c in the solution to the time problem to B_n and to denote the fact that it will probably be different for each value of n as well and because had we kept the c_2 with the eigenfunction, we'd have absorbed it into the c to get a single constant in our solution. The function above will satisfy the heat equation and the boundary condition of zero temperature on the ends of the bar.

Okay, so just what have we learned here? By using separation of variables, we were able to reduce our linear homogeneous partial differential equation with linear homogeneous boundary conditions down to an ordinary differential equation for one of the functions in our product solution (Eq. E-1), *Y*(*t*) in this case, and a boundary value problem that we can solve for the other function, *X*(*x*) in this case.

Note as well that the boundary value problem is in fact an eigenvalue/ eigenfunction problem. When we solve the boundary value problem, we will be identifying the eigenvalues, λ, that will generate non-trivial solutions the their corresponding eigenfunctions. Again, we'll look into this more in the next section. At this point all we want to do is identify the two ordinary differential equations that we need to solve to get a solution.

Before we do a couple of other examples, we should take a second to address the fact that we made two very arbitrary seeming decisions in the above work. We divided both sides of the equation by k at one point and chose to use $-\lambda$ instead of λ as the separation constant.

Both of these decisions were made to simplify the solution to the boundary value problem we got from our work. The addition of the k in the boundary value problem would just have complicated the solution process with another letter we'd have to keep track of, so we moved it into the time problem where it won't cause as many problems in the solution process. Likewise, we chose $-\lambda$ because we've already solved that particular boundary value problem (albeit with a specific L, but the work will be nearly identical) when we first looked at finding eigenvalues and eigenfunctions. This by the way was the reason we rewrote the boundary value problem to make it a little clearer that we have in fact solved this one already.

We can now at least partially answer the question of how we know to make these decisions. We wait until we get the ordinary differential equations and then look at them and decide of moving things like the k or which separation constant to use based on how it will affect the solution of the ordinary differential equations. There is also, of course, a fair amount of experience that comes into play at this stage. The more experience you have in solving these easier it often is to make these decisions.

Again, we need to make clear here that we're not going to go any farther in this section than getting things down to the two ordinary differential equations. Of course, we will need to solve them in order to get a solution to the partial differential equation, but that is the topic of the remaining sections in this chapter. All we'll say about it here is that we will need first to solve the boundary value problem, which will tell us what λ must be, and then we can solve the other differential equation. Once that is done, we can then turn our attention to the initial condition.

Okay, we need to work a couple of other examples, and these will go a lot quicker because we won't need to put in all the explanations. After the first example, this process always seems like a very long process, but it really isn't. It just looked that way because of all the explanation that we had to put into it.

Example 3.2: Use Separation of Variable on the Following Partial Differential Equation

$$\frac{\partial u}{\partial t} = k\frac{\partial^2 u}{\partial t^2}$$

$$u(x,0) = f(x)\frac{\partial u(0,t)}{\partial x} = 0\frac{\partial u(L,t)}{\partial x} = 0$$

(continued)

Example 3.2 (continued)

Solution: In this case we are looking at the heat equation with no sources and perfectly insulated boundaries.

So, we will start off by again assuming that our product solution will have the form:

$$u(x,t) = \varphi(x)T(t)$$

And because the differential itself has not changed here, we will get the same result from plugging this in as we did in the previous example, so the two ordinary differential equations that we will need to solve are:

$$\frac{dT}{dt} = -k\lambda T \frac{d^2\varphi}{dx^2} = -\lambda\varphi$$

Now, the point of this example was really to deal with the boundary conditions, so let us plug the product solution into them to get:

$$\frac{\partial(T(t)\varphi(x))(0,t)}{\partial x} = 0 \frac{\partial(T(t)\varphi(x))(L,t)}{\partial x} = 0$$

$$T(t)\frac{d\varphi(0)}{dx} = 0 T(t)\frac{d\varphi(L)}{dx} = 0$$

Now, just as with the first example, if we want to avoid the trivial solution, we cannot have $T(t) = 0$ for every t, and so we must have:

$$\frac{d\varphi(0)}{dx} = 0 \frac{d\varphi(L)}{dx} = 0$$

Here is a summary of what we get by applying separation of variables to this problem.

$$\frac{dT}{dt} = -k\lambda T \begin{array}{c} \dfrac{d^2\varphi}{dx^2} + \lambda\varphi = 0 \\ \dfrac{d\varphi(0)}{dx} = 0 \dfrac{d\varphi(L)}{dx} = 0 \end{array}$$

Next, let us see what we get if use periodic boundary conditions with the heat equation.

Example 3.3: Use separation of variable on the following partial differential equation.

$\frac{\partial u}{\partial t} = k\frac{\partial^2 u}{\partial t^2}$

$u(x,0) = f(x)u(-L,t) = u(L,t)\frac{\partial u(-L,t)}{\partial x} = \frac{\partial u(L,t)}{\partial x}$

Solution: First note that these boundary conditions really are homogeneous boundary conditions. If we rewrite them as:

$u(-L,t) - u(L,t) = 0\frac{\partial u(-L,t)}{\partial x} - \frac{\partial u(L,t)}{\partial x} = 0$

It is a little easier to see.

Now, again we have done this partial differential equation, so we will start off with:

$u(x,t) = \varphi(x)T(t)$

And the two ordinary differential equations that we will need to solve are:

$\frac{dT}{dt} = -k\lambda T\frac{d^2\varphi}{dx^2} = -\lambda\varphi$

Plugging the product solution into the rewritten boundary conditions gives:

$T(t)\varphi(-L) - T(t)\varphi(L) = T(t)[\varphi(-L) - \varphi(L)] = 0$

$T(t)\frac{\varphi(-L)}{dx} - T(t)\frac{\varphi(L)}{dx} = T(t)\left[\frac{d\varphi(-L)}{dx} - \frac{d\varphi(L)}{dx}\right] = 0$

And we can see that we will only get non-trivial solution if:

$\varphi(-L) - \varphi(L) = 0\frac{d\varphi(-L)}{dx} - \frac{d\varphi(L)}{dx} = 0$

$\varphi(-L) = \varphi(L)\frac{d\varphi(-L)}{dx} = \frac{d\varphi(L)}{dx}$

So, here is what we get by applying separation of variables to this problem:

$$\frac{dT}{dt} = -k\lambda T \quad \begin{array}{l} \dfrac{d^2\varphi}{dx^2} + \lambda\varphi = 0 \\ \varphi(-L) = \varphi(L) \quad \dfrac{\varphi(-L)}{dx} = \dfrac{\varphi(L)}{dx} \end{array}$$

http://tutorial.math.lamar.edu/Classes/DE/SolvingHeatEquation.aspx#PDE_HeatEqn_Ex1 [2]

3.6.2 Solving Heat Equation Using Fourier Transform Method

The heat equation is a partial differential equation. Prior to Fourier's work, there was no known solution to the heat equation in a general situation, although particular solutions were known if the heat source behaved in a simple way, in particular, if the heat source was a sine or cosine wave. These simple solutions are now sometimes called eigensolutions. Fourier's idea was to model a complicated heat source as a superposition (or linear combination) of simple sine and cosine waves and to write the solution as a superposition of the corresponding eigensolutions. This superposition or linear combination is called the Fourier series.

Although the original motivation was to solve the heat equation, it later became obvious that the same techniques could be applied to a wide array of mathematical and physical problems.

Example 3.4: Use Finite Fourier Transforms to Solve:

$$\frac{\partial^2 U(x,t)}{\partial x^2} = \frac{\partial U(x,t)}{\partial t}$$

$$\text{B.C.}\begin{cases} U(0,t) = 0 \\ U(4,t) = 0 \end{cases}$$

where $0 < x < 4$ and $t > 0$

$$\text{I.C.}\{U(x,0) = 2x$$

Solution: Take the finite Fourier sine transform (with $L = 4$) (see Appendix E and Sect. 1.15) of both sides of the partial differential equation to obtain:

$$\int_0^4 \frac{\partial^2 U(x,t)}{\partial x^2} \sin \frac{n\pi x}{4} dx = \int_0^4 \frac{\partial U(x,t)}{\partial t} \sin \frac{n\pi x}{4} dx$$

Writing $u = \mathbb{F}_{\text{sine}}\{T\}$ and using [see Appendix E and Sect. 1.21 Example-2 (a)] with the conditions $U(0,t) = 0$ and $U(4,t) = 0$, we find [4]:

$$\frac{du}{dt} = -\frac{n^2\pi^2}{16} u \tag{3.14}$$

where $u = u(n,t)$.

Taking the finite Fourier sine transform of the condition $T(x,0) = 2x$, we have as in Example-2 (a), Sect. 1.12 of Appendix E of reference [4]:

$$u(n,0) = f_{\text{sine}}(2x) = \int_0^4 2x \sin\left(\frac{n\pi x}{4}\right) dx$$

$$= \left\{(2x)\left(\frac{-\cos(n\pi x/4)}{n\pi/4}\right) - (2)\left(\frac{-\sin(n\pi x/4)}{n^2\pi^2/16}\right)\right\}\Bigg|_0^4 \tag{3.15}$$

$$= \frac{32(1 - \cos(n\pi))}{n\pi}$$

(continued)

Example 3.4 (continued)

Solving the differential Eq. (3.8), we find if c is an arbitrary constant, then we have:

$$u = u(n, t) = ce^{-n^2\pi^2 t/16} \tag{3.16}$$

But if we use the initial condition for $t = 0$, then $c = u(n, 0)$ which from (2) and (3) results:

$$u = \frac{32(1 - \cos(n\pi))}{n\pi} e^{-(n^2\pi^2)t/16}$$

Thus, from Example-1 (a) Section 1.12 of Appendix E, the inverse Fourier sine transform is:

$$U(x, t) = \frac{2}{4}\sum_{n=1}^{\infty} \frac{32(1 - \cos(n\pi))}{n\pi} e^{-(n^2\pi^2)t/16}$$

$$= \frac{2}{4}\sum_{n=1}^{\infty} \frac{(1 - \cos(n\pi))}{n\pi} e^{-(n^2\pi^2)t/16}$$

Physically $U(x, t)$ represents the temperature at any point x at any time t in a solid bounded by the point $x = 0$ and $x = 4$. The conditions $u(0, t) = 0$ and $u(4, t) = 0$ express the fact that the ends are kept at temperature zero, while $u(x, 0) = 2x$ expresses the initial temperature as a function of x. Equivalently, the solid can be replaced by a bar on the axis with endpoint at $x = 0$ and $x = 4$ whose surface is insulated.

Example 3.5: Solve $\frac{\partial U}{\partial t} = \frac{\partial^2 U}{\partial x^2}$ for $x > 0$, $t > 0$ subject to the condition:

$$U(0, t) = 0 U(x, 0) = \begin{cases} 1 & 0 < x < 1 \\ 0 & x \geq 1 \end{cases} U(x, t) \text{ is bounded}$$

Taking the Fourier sine transform of both sides of the given partial differential equation, we find:

$$\int_0^\infty \frac{\partial U(x, t)}{\partial t} \sin(\lambda x) dx = \int_0^\infty \frac{\partial^2 U}{\partial x^2} \sin(\lambda x) dx \tag{3.17}$$

Then if: $u = u(\lambda, t) = \int_0^\infty U(x, t) \sin(\lambda x) dx$

(continued)

Example 3.5 (continued)

This becomes:

$$\frac{du}{dt} = \left\{\frac{\partial U(x,t)}{\partial x}\sin(\lambda x) - \lambda U(x,t)\cos(\lambda x)\right\}\Bigg|_0^\infty - \lambda^2\int_0^\infty U(x,t)\sin(\lambda x)dx$$
$$= \lambda U(0,t) - \lambda^2 u \tag{3.18}$$

on integrating the right hand side of (1) by parts and assuming that $U(x, t)$ and $U(x, t)/\partial x$ approach zero as $x \to \infty$.

From the condition for $U(x, 0)$, we have on taking the Fourier sine transform:

$$u(\lambda, 0) = \int_0^\infty U(x,0)\sin(\lambda x)dx = \int_0^1 \sin(\lambda x)dx = \frac{1-\cos\lambda}{\lambda} \tag{3.19}$$

Solving (2) subject to the condition (3) and $U(0, t) = 0$, we find:

$$u(\lambda, t) = \frac{1-\cos\lambda}{\lambda}e^{-\lambda^2 t}$$

Then taking the inverse Fourier sine transform, we find the required solution:

$$U(x,t) = \frac{2}{\pi}\int_0^1 \frac{1-\cos\lambda}{\lambda}e^{-\lambda^2 t}\sin(\lambda x)dxd\lambda$$

Physically, this can represent the temperature in a solid $x > 0$. See Example 3.1.

3.6.3 *Solving Heat Equation Using the Laplace Transform Method*

Partial differential equations, like their one-variable counterpart, ordinary differential equations, are ubiquitous throughout the scientific spectrum. However, they are, in general, more difficult to

solve. Yet here again, we may apply the Laplace transform method to solve PDEs by reducing the initial problem to a simpler ODE.

Partial differential equations come in three types. For a function of two variables $u = u(x, t)$ where we have to replace T with u, the general second-order linear PDE has the form Eq. C-75 of Sect. 6.1 of Appendix C, and in this section, we are much interested in the parabolic aspect of PDE which is in the form of one-dimensional *heat equation* as follows [3]:

$$\frac{\partial u(x,t)}{\partial t} = c\frac{\partial^2 u(x,t)}{\partial x^2} \quad \text{(parabolic form)} \tag{3.20}$$

In its simple form, we have assumed the above partial differential equation is a function of one variable x and it is a linear class.

The Laplace transform method is thoroughly explained in Sect. 2.0, Appendix E, and we are reminded of it here in very brief steps. We consider the function $u = u(x, t)$, where $t \geq 0$ is a time variable. Denote by $U(x, s)$ the *Laplace transform of u with respect to t*, that is to say:

$$U(x,s) = \pounds\{u(x,t)\} = \int_0^\infty e^{-st} u(x,t)dt$$

Here x is the "gun transformed variable." There are a few examples that are given in Appendix E under Sect. 2 that readers should refer to for more familiarity. Here we provide a few examples given by J. Schiff [3].

The heat flow in a finite or semi-infinite thin rod is governed by the PDE:

$$\frac{\partial u(x,t)}{\partial t} = c\frac{\partial^2 u(x,t)}{\partial x^2}$$

where c is a constant (called the *diffusivity*) and $u(x, t)$ is the temperature at position x and time t over a cross section of this rod. The temperature over this cross section is taken to be uniform (see Fig. 3.2).

Example 3.6: Solve:

$$\frac{\partial u(x,t)}{\partial t} = \frac{\partial^2 u(x,t)}{\partial x^2} x > 0 \text{ and } t > 0 \text{Heat PDE} \tag{3.21}$$

For boundary and initial conditions:

$$\{\text{(i)} \quad u(x,0) = 1 \text{for } x > 0 \text{Initial Condition(I.C.)}$$

$$\begin{cases} \text{(ii)} \quad \dfrac{\partial u(0,t)}{\partial x} = 0 \text{ for } t > 0 \\ \text{(iii)} \quad \lim\limits_{x\to\infty} u(x,t) = 1 \end{cases} \text{Boundary Condition(B.C.)}$$

Solution: Take Laplace Transform of above heat PDE (1) yields

(continued)

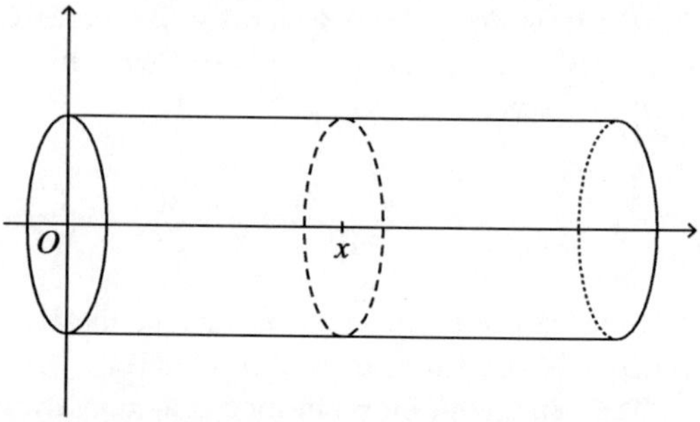

Fig. 3.2 Finite or semi-infinite thin rod

Example 3.6 (continued)

$$\frac{d^2U(x,s)}{dx^2} = sU(x,s) - u(x,0) = sU(x,s) - 1 \tag{3.22}$$

Laplace transform of B.C. (ii) and (iii) provides:

$$U(0,s) = \pounds\{u(0,t)\} = 0$$

$$\lim_{x\to\infty} U(x,s) = \lim_{x\to\infty} \pounds\{u(x,t)\} = \pounds\left\{\lim_{x\to\infty} u(x,t)\right\} = \frac{1}{s}$$

Now Eq. (3.9) is just an ordinary differential equation (ODE) whose solution is given by:

$$U(x,s) = c_1 e^{\sqrt{s}x} + c_2 e^{-\sqrt{s}x} + \frac{1}{s}$$

The boundary condition $\lim_{x\to\infty} U(x,s) = 1/s$ implies $c_1 = 0$, and $U(0,s) = 0$ implies:

$$U(x,s) = \frac{1}{s} - \frac{e^{\sqrt{s}x}}{s}$$

By Example-1 Section 2.10 of Appendix E, we have:

$$u(x,t) = erf\left(\frac{x}{2\sqrt{t}}\right) = \frac{2}{\sqrt{\pi}} \int_0^{x/2\sqrt{t}} e^{-u^2} du$$

Direct calculation shows that indeed satisfies Eq. (3.8) above and that the initial and boundary conditions are satisfied.

Example 3.7: Solve:

$$\frac{\partial u(x,t)}{\partial t} = \frac{\partial^2 u(x,t)}{\partial x^2} \text{ in domain} x > 0 \text{ and } t > 0$$

For:

$$\begin{cases} \text{(i)} u(x,0) = 0 \text{I.C.} \\ \text{(ii)} \quad u(0,t) = f(t) \text{ for } t > 0 \\ \text{(iii)} \lim_{x\to\infty} u(x,t) = 0 \end{cases}$$

Solution: The transformed equation is:

$$\frac{d^2U(x,s)}{dx^2} - sU(x,s) = 0$$

Whose solution is given by:

$$U(x,s) = c_2 e^{-\sqrt{s}x}$$

in view of condition (iii). By (ii):

$$U(0,s) = £\{ f(t)\} = F(s)$$

so that $c_2 = F(s)$ and:

$$U(x,s) = F(s)e^{-\sqrt{s}x}$$

Invoking the fact that $£^{-1}\{e^{-a\sqrt{s}}\} = \frac{a}{2\sqrt{\pi t^3}}e^{-a/4t}$ for $a > 0$ and the convolution theorem, we have the following results:

$$u(x,t) = \int_0^t \frac{x}{2\sqrt{\pi\tau^3}} e^{-x^2/4\tau} f(t-\tau)d\tau$$

Making the substitution $\sigma^2 = x^2/4\tau$, then we have:

$$u(x,t) = \frac{2}{\sqrt{\pi}} \int_{x/2\sqrt{t}}^{\infty} \frac{x}{2\sqrt{\pi\tau^3}} e^{-\sigma^2} f\left(t - \frac{x^2}{4\sigma^2}\right) d\sigma$$

which is the desired solution.

Example 3.8: Solve:

$$\frac{\partial u(x,t)}{\partial t}=\frac{\partial^2 u(x,t)}{\partial x^2}\text{in domain}0<x<1\text{and}t>0$$

For:

$$\{\text{(i)}\ u(x,0)=f(x)\mathbf{I.C.}$$

$$\begin{cases}\text{(ii)} & u(0,t)=0\ \text{for}\ t>0\\ \text{(iii)}\ u(1,t)=0\ \text{for}\ t>0\end{cases}\mathbf{B.C.}$$

Solution: Therefore:

$$\frac{d^2U(x,s)}{dx^2}-sU(x,s)=-f(x)$$

Here we solve this ODE by the Laplace transform method as well. To this end, let $Y(x)=U(x,s)$. Then $Y(0)=U(0,s)=0$, $Y(1)=U(1,s)=0$. Setting $a^2=s$, we obtain:

$$\sigma^2£(Y)-\sigma Y(0)-Y'(0)-a^2£(Y)=-£(f)=-F(\sigma)$$

that is,

$$£(Y)=\frac{Y'(0)}{\sigma^2-a^2}-\frac{F(\sigma)}{\sigma^2-a^2}$$

Inverting gives:

$$\begin{aligned}Y(x)=U(x,s)&=\frac{Y'(0)\sinh ax}{a}-\frac{1}{a}\int_0^x f(u)\sin ga(x-u)du\\&=\frac{Y'(0)\sinh\sqrt{s}x}{\sqrt{s}}-\frac{1}{\sqrt{s}}\int_0^x f(u)\sinh\sqrt{s}(x-u)du\end{aligned}$$

Thus:

$$\begin{aligned}U(x,s)=&\int_0^1 f(u)\frac{\sinh\sqrt{s}x\sinh\sqrt{s}(1-u)}{\sqrt{s}\sinh\sqrt{s}}du\\&-\int_0^x f(u)\frac{\sinh\sqrt{s}(x-u)}{\sqrt{s}}du\end{aligned}$$

We can write $\int_0^1=\int_0^x+\int_x^1$ and use the fact that sinh $(z\pm w)$ sinh z cosh $w\pm$ cosh z sinh w.

(continued)

Example 3.8 (continued)
Then:

$$U(x,s)=\int_0^x f(u)\left[\frac{\sinh\sqrt{s}x\sinh\sqrt{s}(1-u)}{\sqrt{s}\sinh\sqrt{s}}-\frac{\sinh\sqrt{s}(1-u)}{\sqrt{s}}\right]du$$

$$+\int_0^1 f(u)\frac{\sinh\sqrt{s}x\sinh\sqrt{s}(1-u)}{\sqrt{s}\sinh\sqrt{s}}du$$

$$=\int_0^x f(u)\frac{\sinh\sqrt{s}(1-x)\sinh\sqrt{s}u}{\sqrt{s}\sinh\sqrt{s}}du$$

$$+\int_0^1 f(u)\frac{\sinh\sqrt{s}x\sinh\sqrt{s}(1-u)}{\sqrt{s}\sinh\sqrt{s}}du$$

To determine the inverse, we use the complex inversion formula (see Appendix D Section 30.3). When it is applied to the first integral, we have:

$$\frac{1}{2\pi i}\int_{x_0-i\infty}^{x_0+i\infty} e^{ts}\left\{\int_0^x f(u)\frac{\sinh\sqrt{s}(1-x)\sinh\sqrt{s}u}{\sqrt{s}\sinh\sqrt{s}}du\right\}ds=\sum \operatorname{Re} s$$

There are simple poles in this case at $s_0=0$ and $s_n=-n^2\pi^2\ n-1,2,3,\cdots$

$$\operatorname{Res}(0)=\lim_{s\to 0} s\int_0^x f(u)\frac{\sinh\sqrt{s}(1-x)\sinh\sqrt{s}u}{\sqrt{s}\sinh\sqrt{s}}du=0$$

$$\operatorname{Res}\left(-n^2\pi^2\right)=\lim_{s\to -n^2\pi^2}\left(s+n^2\pi^2\right)e^{ts}\int_0^x f(u)\frac{\sinh\sqrt{s}(1-x)\sinh\sqrt{s}u}{\sqrt{s}\sinh\sqrt{s}}du$$

$$=\lim_{s\to -n^2\pi^2}\frac{s+n^2\pi^2}{\sinh\sqrt{s}}\lim_{s\to -n^2\pi^2} e^{ts}\int_0^x f(u)\frac{\sinh\sqrt{s}(1-x)\sinh\sqrt{s}u}{\sqrt{s}\sinh\sqrt{s}}du$$

$$=2e^{-n^2\pi^2 t}\int_0^x f(u)\frac{\sinh\left[(n\pi i)(1-x)\right]\sinh(n\pi i)u}{\cosh(n\pi i)}du$$

$$=2e^{-n^2\pi^2 t}\int_0^x f(u)\frac{\sin\left[n\pi(1-x)\right]\sin n\pi u}{-\cos n\pi}du$$

Where we have used the properties from Appendix D with $z=x+iy$

(continued)

Example 3.8 (continued)

$$\begin{aligned} \sinh z &= \cos y \sinh y + i \sin y \cosh x \\ \cosh z &= \cos y \cosh x + i \sin y \sinh x \end{aligned}$$

To obtain the last equality. There for have

$$\sum \text{Res} = 2\sum_{n=1}^{\infty} e^{-n^2\pi^2 t}\left(\int_0^x f(u) \sin n\pi u du\right) \sin n\pi x$$

Similarly, the inverse of the second integral is given by:

$$2\sum_{n=1}^{\infty} e^{-n^2\pi^2 t}\left(\int_0^1 f(u) \sin n\pi u du\right) \sin n\pi x$$

Finally:

$$u(x, t) = 2\sum_{n=1}^{\infty} e^{-n^2\pi^2 t}\left(\int_0^1 f(u) \sin n\pi u du\right) \sin n\pi x$$

The same result is obtained when we solve this example if using separation of variables method.

References

1. http://tutorial.math.lamar.edu/Classes/DE/DE.aspx
2. http://tutorial.math.lamar.edu/Classes/DE/SolvingHeatEquation.aspx
3. Schiff, J.L.: The Laplace Transform, theory and applications. Springer (1991)
4. Zohuri, B.: Directed energy weapons, physics of High Ener.gy Lasers (HEL), 1st edn. Springer (2016)

Chapter 4
Introduction to Electromagnetism

In order to understand laser interaction with matter, one needs to have some basic knowledge of electromagnetic field in addition to conduction of heat in solids that was described in the previous chapter. As part of the interaction of high-power laser energy with solid, or slab geometry, one needs to have some basic understanding of electromagnetic wave equation and consequently its behavior, when it comes to laser irradiation. Thus, in this chapter, we try to introduce Maxwell's equations and electric field of the electromagnetic wave, at a very holistic wave, and leave the granule aspect of it to the readers so they can refer themselves to more detailed related books.

4.1 Introduction

Electromagnetism is one of the four fundamental interactions of nature, along with strong interaction, weak interaction, and gravitation. It is the force that causes the interaction between electrically charged particles; the areas in which this happens are called electromagnetic fields as demonstrated in Fig. 4.1 here.

Electromagnetism is the force responsible for practically all the phenomena encountered in daily life (with the exception of gravity). Ordinary matter takes its form as a result of intermolecular forces between individual molecules in matter. Electromagnetism is also the force which holds electrons and protons together inside atoms, which are the building blocks of molecules. This governs the processes involved in chemistry, which arise from interactions between the electrons orbiting atoms.

Electromagnetic field theory has been and will continue to be one of the most important fundamental courses of electrical engineering curriculum. It is one of the best-established general theories that provides explanations and solutions to intricate electrical engineering problems when other theories are no longer applicable.

B. Zohuri, *Thermal Effects of High Power Laser Energy on Materials*,
https://doi.org/10.1007/978-3-030-63064-5_4

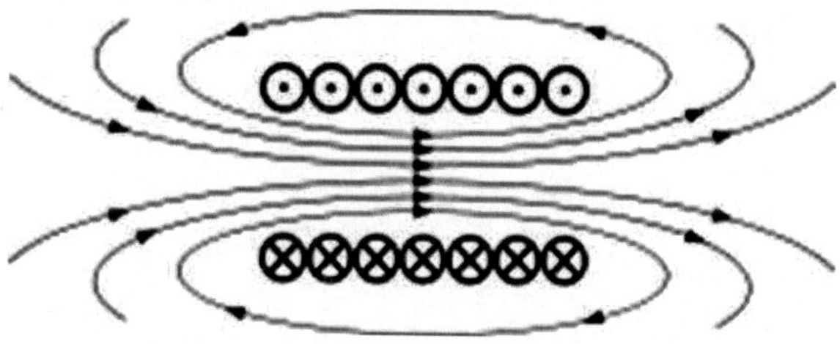

Fig. 4.1 Electricity and magnetism

4.2 Maxwell's Equations and Electric Field of the Electromagnetic Wave

Although Maxwell's equations were formulated over 100 years ago, the subject of electromagnetism has not remained static. Much of modern physics (and engineering) involves time-dependent electromagnetic fields in which Maxwell's displacement current plays a crucial role. Maxwell's equations contain all the information necessary to characterize the electromagnetic fields at any point in a medium.

In order to understand the behavior of material response to a high-power laser beam, one needs to consider the coupling of the laser energy to a material. Therefore, we need first to know the optical reflectivity R and the transmissivity T for light incident on a surface which divides two semi-infinite media. To understand reflectivity, we must use some general results from the theory of electromagnetic waves.

For the electromagnetic (EM) fields to exist, they must satisfy the four Maxwell equations at the source where they are generated, at any point in a medium through which they propagate, and at the load where they received or absorbed As the fields must satisfy the four coupled Maxwell equations involving four unknown variables, we first obtain an equation in terms of one unknown variables. Similar equations can then be obtained for the other variables. We refer to these equations as the general *wave equations*. It can be shown that the fields generated by time-varying sources propagate as *spherical waves*. However, in a small region far away from the radiating source, the spherical wave may be approximated as a *plane wave*, that is, one in which all the field quantities are in a plane normal to the direction of its propagation (the transverse plane). Consequently, a plane wave does not have any field component in its direction of propagation (the longitudinal direction).

We first seek the solution of a plane wave in an unbounded dielectric medium and show that the wave travels with the speed of light in free space. We then consider the general case of a finitely conducting medium. We show that the wave attenuates as a result of loss in energy as it travels in the conducting medium. Finally, we introduce the concept of reflection and transmission of a plane wave when it leaves one medium and enters another.

4.3 The Wave Equations for Electric and Magnetic Field

In regions of space where there is no charge or current, Maxwell's equations read as sets of Eq. 4.1:

$$\begin{aligned}&(\mathrm{i})\nabla\cdot\vec{E}=0(\mathrm{iii})\nabla\mathrm{x}\,\vec{E}=-\frac{\partial\vec{B}}{\partial t}\\&(\mathrm{ii})\nabla\cdot\vec{B}=0(\mathrm{iv})\nabla\mathrm{x}\,\vec{B}=\mu_0\varepsilon_0\frac{\partial\vec{E}}{\partial t}\end{aligned}\tag{4.1}$$

where:

E is the electric field
B is the magnetic field
μ_0 is the constant Biot–Savart's law known as permittivity of free space 4 π × 10^7 N/A^2
ε_0 is the constant Coulomb's law known as permittivity of free space 8.85 × 10^{-12} $C^2/(N\,.\,m^2)$

The above equations are constituting a set of coupled, first-order, partial differential equations for $\vec{E}$ and $\vec{B}$. They can be decoupled by applying the curl to (iii) and (iv):

$$\nabla\mathrm{x}\left(\nabla\mathrm{x}\,\vec{E}\right)=\nabla\left(\nabla\cdot\vec{E}\right)-\nabla\mathrm{x}\left(-\frac{\partial\vec{B}}{\partial t}\right)$$

$$=-\frac{\partial}{\partial t}\left(\nabla\mathrm{x}\,\vec{B}\right)=-\mu_0\varepsilon_0\frac{\partial^2\vec{E}}{\partial t^2}$$

$$\nabla\mathrm{x}\left(\nabla\mathrm{x}\,\vec{B}\right)=\nabla\left(\nabla\cdot\vec{B}\right)-\nabla^2\vec{B}=\nabla\mathrm{x}\left(\mu_0\varepsilon_0\frac{\partial\vec{E}}{\partial t}\right)$$

$$=\mu_0\varepsilon_0\frac{\partial}{\partial t}\left(\nabla\mathrm{x}\,\vec{E}\right)=-\mu_0\varepsilon_0\frac{\partial^2\vec{B}}{\partial t^2}$$

Or, since $\nabla\cdot\vec{E}=0$ and $\nabla\cdot\vec{B}=0$.

$$\nabla^2\vec{E}=\mu_0\varepsilon_0\frac{\partial^2\vec{E}}{\partial t^2}\,\nabla^2\vec{B}=\mu_0\varepsilon_0\frac{\partial^2\vec{B}}{\partial t^2}\tag{4.2}$$

Equation 4.2 is a demonstration of separations between $\vec{E}$ and $\vec{B}$, but they are second order [1]. In vacuum, then, each Cartesian component of $\vec{E}$ and $\vec{B}$ satisfies the *three-dimensional wave* Eq. 4.3

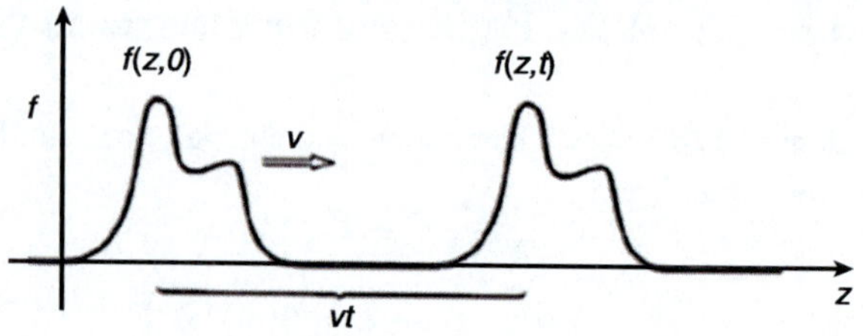

Fig. 4.2 Wave propagation in z-direction [1]

$$\nabla^2 f = \frac{1}{v^2}\frac{\partial^2 f}{\partial t^2} \tag{4.3}$$

This supports standard wave equation in general within Cartesian form, which is similar to a classical wave equation of small disturbance on the string, where v represents the speed of propagation and is given by [1]:

$$v = \sqrt{\frac{T}{\mu}} \tag{4.4}$$

where μ is the mass per unit length and this equation is admits as solutions all functions of the form as Eq. 4.5 for example in z direction of propagation [1].

$$f(z,t) = g(z - vt) \tag{4.5}$$

This mathematical derivation can be sort of depicted as Fig. 4.2 at two different two times, once at $t = 0$, and at later time t. Each point on the wave form simply shifts to the right by an amount vt, where v is the velocity.

Then we can see that Maxwell's equations imply that empty space supports the propagation of electromagnetic waves, traveling at a speed:

$$v = \frac{1}{\sqrt{\mu_0 \varepsilon_0}} = 3.00 \text{ x } 10^8 meter/\sec ond = 3.00 \text{ x } 10^8 m/s = c$$

which is precisely the velocity of light c [1], which Maxwell himself was astonished and he, himself put it: "wave can scarcely avoid the interference that light consists in the transverse undulations of the same medium which is the cause of electric and magnetic phenomena."

4.4 The Sinusoidal Waves

Of all possible wave forms, the sinusoidal one form like Eq. 4.6 below is for a good reason the most familiar. Figure 4.3 shows this function at time $t = 0$ with A as *amplitude* of the wave.

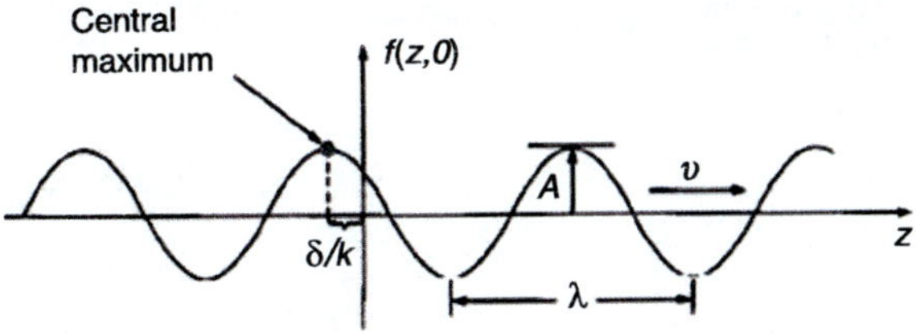

Fig. 4.3 A sinusoidal wave propagation [1]

$$f(z,t) = A\cos\left[k(z - vt) + \delta\right] \tag{4.6}$$

(it is positive and represents the maximum displacement from equilibrium). The argument of the cosine is called the *phase*, and δ is the *phase constant,* and one can add any integer multiple 2π to δ without changing $f(z, t)$. Ordinarily, one uses a value in the range $0 \le \delta \le 2\pi$. A point known as central maximum on the curve takes place at $z = vt - \delta/k$ when the phase is 0. If $\delta = 0$, the central maximum passes the origin at time $t = 0$; more generally, δ/k is the distance by which the central maximum (and therefore the entire wave) is "delayed". Finally, k is the *wave number*; it is related to the *wavelength* λ by the Eq. 4.7 for when z advances by $2\pi/k$, and the cosine executes one complete cycle.

$$\lambda = \frac{2\pi}{k} \tag{4.7}$$

As time passes, the entire wave train proceeds to the right, at speed v. At any fixed point z, the string vibrates up and down, undergoing one full cycle in a *period.*

$$T = \frac{2\pi}{kv} \tag{4.8}$$

Now introduce frequency ν as the number of oscillations per unit time, and show it in the form of Eq. 4.9, then we have:

$$\nu = \frac{1}{T} = \frac{kv}{2\pi} = \frac{v}{\lambda} \tag{4.9}$$

For our purpose it is nicer to write Eq. 4.6 in a nicer form and present that in terms of *angular frequency* ω given below in the analogous case of uniform circular motion that represents the number of radians swept out per unit time.

$$\omega = 2\pi\nu = kv \tag{4.10}$$

Then the new form of sinusoidal wave in terms of ω rather than v is given below:

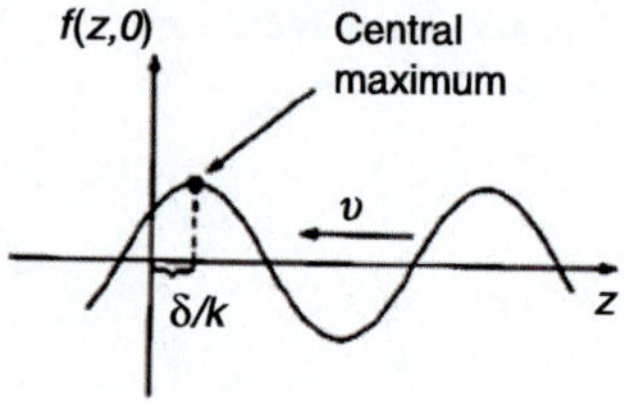

Fig. 4.4 Sinusoidal wave traveling to the left [1]

$$f(z,t) = A\cos(kz - \omega t + \delta) \tag{4.11}$$

If both k and ω are traveling to the left, then Eq. 4.11 can be written in the form of [1]:

$$f(z,t) = A\cos(kz + \omega t - \delta) \tag{4.12}$$

This is in correspondence to and consistent with our previous convention that δ/k shall represent the distance by which the wave is "delayed" since the wave is now moving to the *left*, and delay means a shift to the *right*. At $t = 0$, the wave looks like Fig. 4.4, and since the cosine is an *even function*, we can just as well write Eq. 4.12 as form of Eq. 4.13 as below.

$$f(z,t) = A\cos(-kz - \omega t - \delta) \tag{4.13}$$

Comparing both Eqs. 4.12 and 4.13, we conclude that *we could simply switch the sign of k* to produce a wave with the same amplitude, phase constant, frequency, and wavelength traveling in opposite direction.

What we have learned in Appendix D about complex in view of Euler's formula, we have:

$$e^{i\theta} = \cos\theta + i\sin\theta \tag{4.14}$$

We can now write Eq. 4.12 like below:

$$f(z,t) = \text{Re}\left[Ae^{i(kz-\omega t+\delta)}\right] \tag{4.15}$$

where Re[ξ] denotes the real part of any complex number such as ξ. This allows us to introduce the *complex wave function* of Eq. 4.16 as below:

$$\tilde{f}(z,t) = \tilde{A}e^{i(kz-\omega t)} \tag{4.16}$$

with introduction to complex amplitude $\tilde{A} = Ae^{i\delta}$ absorbing the phase constant. The actual wave function is the real part of $\tilde{f}$:

$$f(z,t) = \text{Re}\left[\tilde{f}(z,t)\right] \tag{4.17}$$

Knowing $\tilde{f}$, it is simple matter to find f.

Example 4.1 Combine two sinusoidal waves of f_1 and f_2

Solution Let us write the following function f_3 as:

$$f_3 = f_1 + f_2 = \text{Re}\left(\tilde{f}_1\right) + \text{Re}\left(\tilde{f}_2\right) = \text{Re}\left(\tilde{f}_1 + \tilde{f}_2\right) = \text{Re}\left(\tilde{f}_3\right)$$

with $\tilde{f}_3 = \tilde{f}_1 + \tilde{f}_2$. You may simply add the corresponding complex wave functions, and then take the real part. In particular, if they have the same frequency and wave number:

$$\tilde{f}_3 = \tilde{A}_1 e^{i(kz-\omega t)} + \tilde{A}_2 e^{i(kz-\omega t)} = \tilde{A}_3 e^{i(kz-\omega t)}$$

where:

$$\tilde{A}_3 = \tilde{A}_1 + \tilde{A}_2 \text{or} \tilde{A}_3 e^{i\delta_3} + \tilde{A}_1 e^{i\delta_1} + \tilde{A}_2 e^{i\delta_2}$$

Now we are going to figure out what A_3 and δ_3 are:

$$(A_3)^2 = \left(A_3 e^{i\delta_3}\right)\left(A_3 e^{-i\delta_3}\right) = \left(A_1 e^{i\delta_1} + A_2 e^{i\delta_2}\right)\left(A_1 e^{-i\delta_1} + A_2 e^{-i\delta_2}\right)$$
$$= (A_1)^2 + (A_2)^2 + A_1 A_2\left(e^{i\delta_1} e^{-i\delta_2} + e^{-i\delta_1} e^{i\delta_2}\right) = (A_1)^2 + (A_2)^2 + A_1 A_2 \cos(\delta_1 - \delta_2)$$

$$A_3 = \sqrt{(A_1)^2 + (A_2)^2 + A_1 A_2 \cos(\delta_1 \delta_2)}$$

$$A_3 e^{i\delta_3} = A_3(\cos\delta_3 + i\sin\delta_3) = A_1(\cos\delta_1 + i\sin\delta_1) = A_2(\cos\delta_2 + i\sin\delta_2)$$
$$= (A_1\cos\delta_1 + A_2\cos\delta_2) + i(A_1\sin\delta_1 + A_2\sin\delta_2)$$

$$\tan\delta_3 = \frac{A_3\sin\delta_3}{A_3\cos\delta_3} = \frac{A_1\sin\delta_1 + A_2\sin\delta_2}{A_1\cos\delta_1 + A_2\cos\delta_2}$$

$$\delta_3 = \tan^{-1}\left(\frac{A_1\sin\delta_1 + A_2\sin\delta_2}{A_1\cos\delta_1 + A_2\cos\delta_2}\right)$$

As we can see the combined wave still has the same frequency and wavelength is given by

$$f_3(z,t) = A_3\cos(kz - \omega t + \delta_3)$$

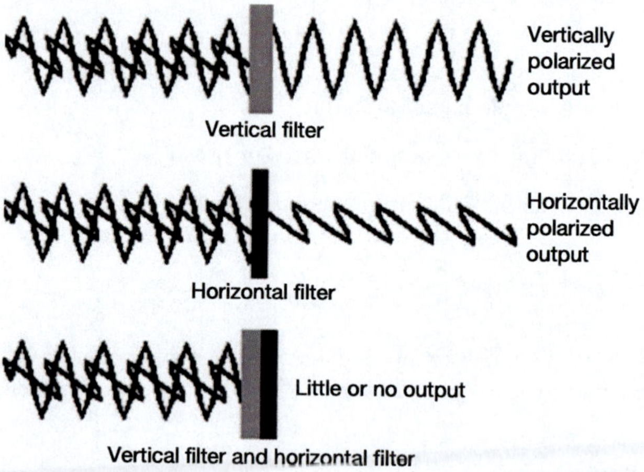

Fig. 4.5a Polarization wave

As we can see, the combined wave still has the same frequency, and wavelength is given by:

$$f_3(z,t) = A_3 \cos(kz - \omega t + \delta_3)$$

The Polarization of the Waves

Polarization, also called wave polarization, is an expression of the orientation of the lines of electric flux in an electromagnetic field (EM field). Polarization can be constant—that is, existing in a particular orientation at all times—or it can rotate with each wave cycle.

Polarization affects the propagation of EM fields at infrared (IR), visible, ultraviolet (UV), and even X-ray wavelengths. In ordinary visible light, there are numerous wave components at random polarization angles. When such light is passed through a special filter, the filter blocks all the light except that having a certain polarization. When two polarizing filters are placed so a ray of light passes through them both, the amount of light transmitted depends on the angle of the polarizing filters with respect to each other. The most lights are transmitted when the two filters are oriented, so they polarize light in the same direction. The least light is transmitted when the filters are oriented at right angles to each other (Fig. 4.5a) and in vector presentation Fig. 4.5b.

Electromagnetic waves are considered transverse waves in which the vibrations of electric and magnetic fields are perpendicular to each other and to the direction of propagation (Fig. 4.6). These two fields change with time and space in a sinusoid fashion. Generally, only the electric field is represented, related to the propagation direction, because it is with the electric field that detectors (eye, photographic film, CCD, etc.) interact.

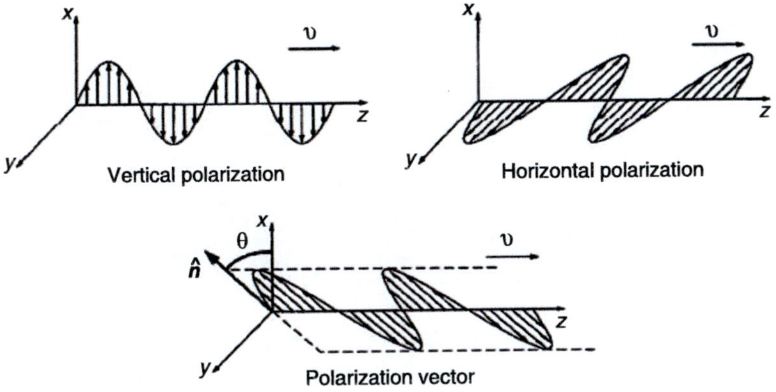

Fig. 4.5b Vector form of polarization wave [1]

Fig. 4.6 Electromagnetic wave propagation

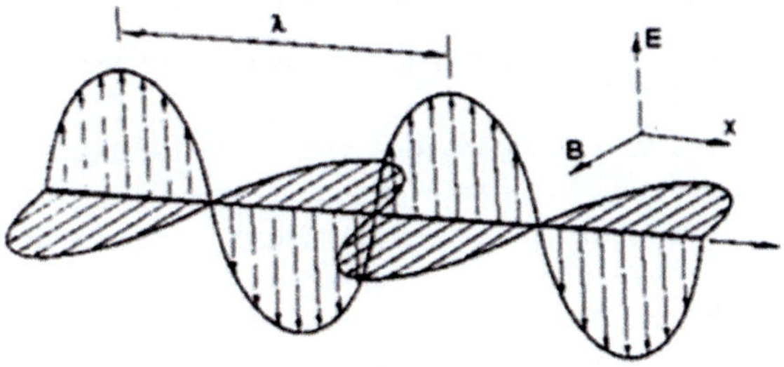

Table 4.1 Type of radiation vs. wavelength range

Type of radiation	Wavelength range
Gamma-rays	<1 pm
X-rays	1 nm–1 pm
Ultraviolet	400 nm–1 nm
Visible	750 nm–400 nm
Infrared	2.5 mm–750 nm
Microwaves	1 mm–25 mm
Radio waves	>1 mm

Visible light makes up just a small part of the full electromagnetic spectrum. Electromagnetic waves with shorter wavelengths and higher energies include ultraviolet light, X-rays, and gamma rays. Electromagnetic waves with longer wavelengths and lower energies include infrared light, microwaves, and radio and television waves. Table 4.1 is a presentation of the types of radiation along with their wavelength range.

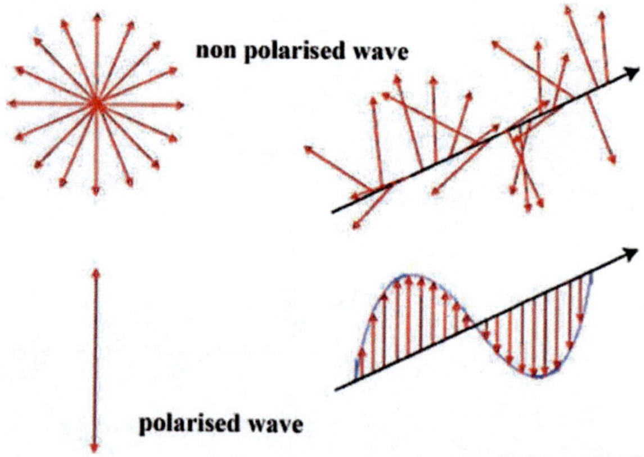

Fig. 4.7 Electric field orientation for polarized and non-polarized electromagnetic wave

The polarization of an electromagnetic wave refers to the orientation of its electric field $\vec{E}$. When the direction of electric field $\left(\vec{E}\right)$ is randomly varying with time on a very fast scale, related to the direction of propagation, the wave is considered non-polarized.

In case of a linearly polarized wave, the electric vector has a fixed orientation related to the propagation direction like in Fig. 4.7.

The polarization of electromagnetic wave can be produced by absorption, scattering, reflection, and birefringence.

A linear polarizer is a device that only allows electric field components parallel to a certain direction (called the polarization axis) to pass through. Any electromagnetic wave that comes through such a polarizer is polarized in the direction of the polarization axis. After leaving the polarizer, the wave's polarization ($\vec{E}$ -field direction) does not change, and the wave is considered to be linearly polarized.

If the linearly polarized light passes through a second polarizer, the transmitted intensity $I(\theta)$ of the waves as it leaves the second polarizer is given by the Malus law.

The specific manner in which a beam of electromagnetic wave at a specific polarization is reflected (and refracted) at interface between two different media can be used to determine the refractive index of the solid.

Specifically, for a particular interface, there is a particular angle of incidence (relative to the normal vector of the surface), called the Brewster angle, which is related to the refractive index of a material. At this angle, the reflection coefficient of light polarized parallel to the plane of incidence is zero. Thus, if the incident light is non-polarized and impinges on the material at the Brewster angle, the light reflected from the solid will be polarized in the plane perpendicular to the plane of incidence. If the incident light is polarized parallel to the plane of incidence, the intensity of the reflected light will be theoretically zero at the Brewster angle.

The proposed experiment used the polarization by reflection due to its simplicity, but other polarization methods can be used too.

Fig. 4.8 Polarization by reflection

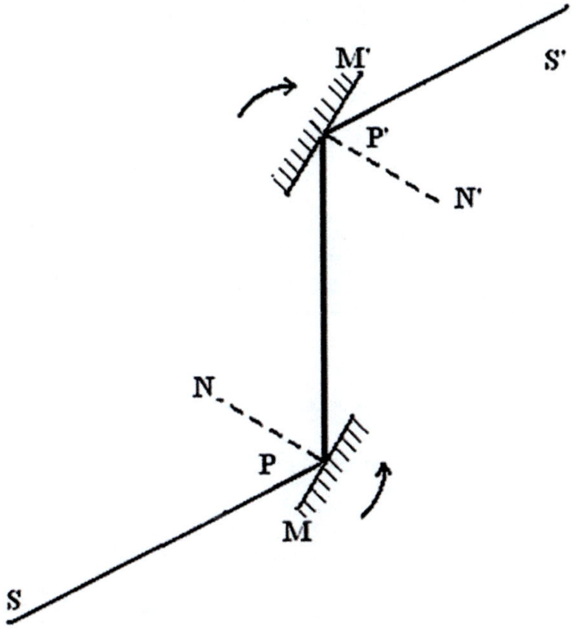

When a light fascicle falls on the M mirror at Brewster angle, the reflected fascicle is linearly polarized. Using a second rotating mirror M', the Malus law can be checked. If mirror M' is rotated around the PP' axe, the reflected P'S' fascicle has a variable intensity with two minim and two maxim values. When the second fascicle falls as the first fascicle at the Brewster angle, the S'P′ fascicle has a minimum value (see Fig. 4.8).

The experiments with visible light and mirrors were made more than one century ago, starting with Brewster, and repeated in a lot of laboratories.

To date some experiments were made in order to check the phenomena for IR, and the effect can be accepted as valid for UV and gamma rays. Related to this kind of experiment, it is worth to be reminded a paper from 1969: Rotatable polarizers for the near infra-red, R M Lambert et al., J. Phys. E: Sci. Instrum. 2799–801 doi: https://doi.org/10.1088/0022-3735/2/9/311.

Polarization is an important optical property inherent in all laser beams. Brewster windows, reflective phase retarders, and absorbing thin-film reflectors use the advantage of polarization. On the other hand, it can cause troublesome and sometimes unpredictable results when ignored. Since virtually all laser sources exhibit some degree of polarization, understanding this effect is necessary in order to specify components properly. The following text gives a basic polarization definition and presents the polarization types most commonly encountered.

Light is a transverse electromagnetic wave; this means that the electric and magnetic field vectors point perpendicular to the direction of wave travel

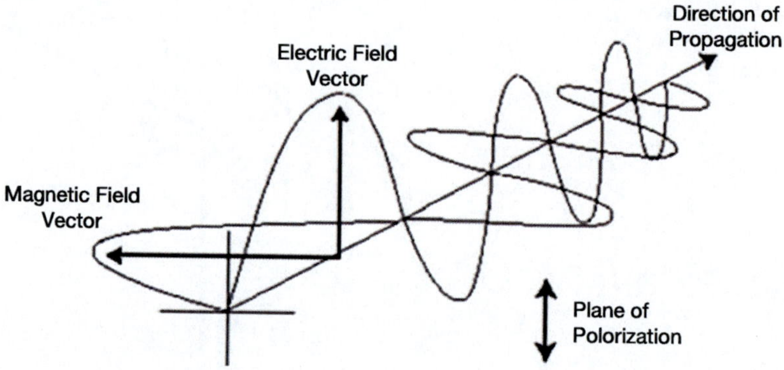

Fig. 4.9 Definition of a polarization vector

Fig. 4.10 Unpolarized light

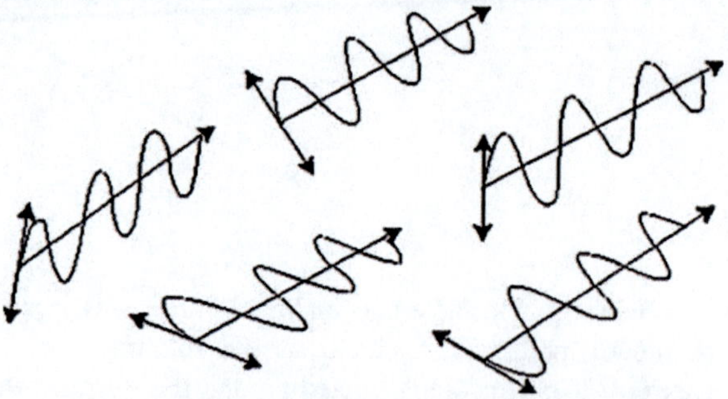

(Fig. 4.9). When all the electric field vectors for a given wave-train lie in a plane, the wave is said to be plane or linearly polarized. The orientation of this plane is the direction of polarization.

Unpolarized light refers to a wave collection which has an equal distribution of electric field orientations for all directions (Fig. 4.10). While each individual wave-train may be linearly polarized, there's no preferred direction of polarization when all the waves are averaged together.

Randomly polarized light is exactly what it says; the light is plane polarized, but the direction is unknown, and may vary with time. Random polarization causes problems in optical systems since some components are polarization sensitive. If the polarization state changes with time, then the components' transmission, reflection, and/or absorption characteristics will also vary with time.

Polarization is a vector that has both direction and amplitude. Like any vector, it's defined in an arbitrary coordinate system as the sum of orthogonal components. In Fig. 4.11, we see a plane polarized wave which points at 45° to the axes of our coordinate system. Thus, when described in this coordinate system, it has equal x and y components. If we then introduce a phase difference of 90° (or one-quarter wavelength) between these components, the result is a wave in which the electric field vector has a fixed amplitude but whose direction varies as we move down the

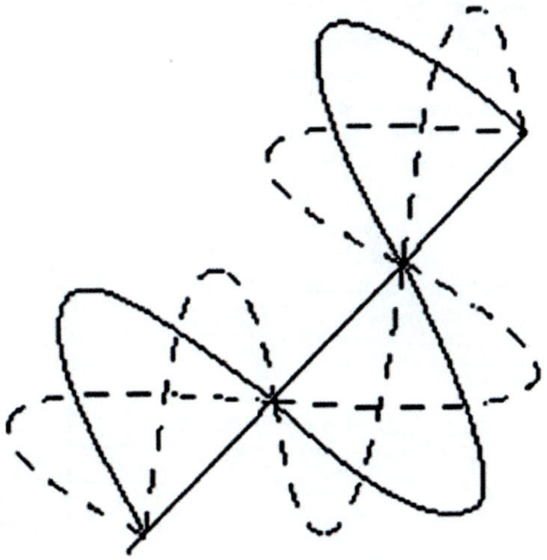

Fig. 4.11 Plane polarized wave which points at 45° to the axes

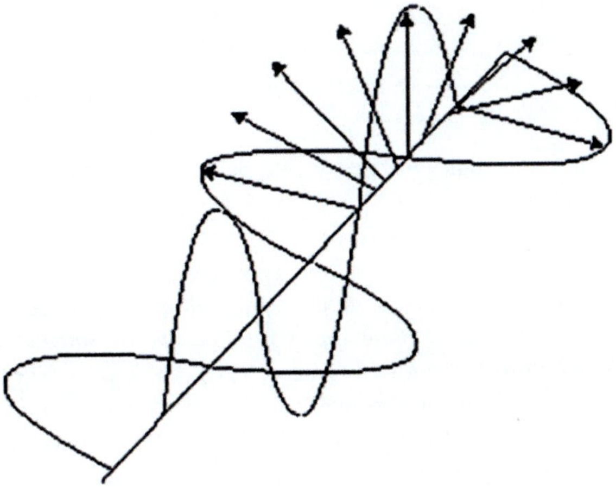

Fig. 4.12 Circularly polarized wave

wave-train (see Fig. 4.12). Such a wave is said to be circularly polarized since the tip of the polarization vector traces out a circle as it passes a fixed point.

If we have two wave-trains with unequal amplitude and with a quarter-wave phase difference, then the result is elliptical polarization. The tip of the polarization vector will trace out an ellipse as the wave passes a fixed point. The ratio of the major to the minor axis is called the ellipticity ratio of the polarization.

Always state the polarization orientation when ordering optical coatings for use at non-normal incidence. If you are unsure about how to determine the polarization state of your source, please contact our applications engineers for assistance.

When light strikes an optical surface, such as a beam splitter, at a non-perpendicular angle, the reflection and transmission characteristics depend upon polarization. In this case, the coordinate system we use is defined by the

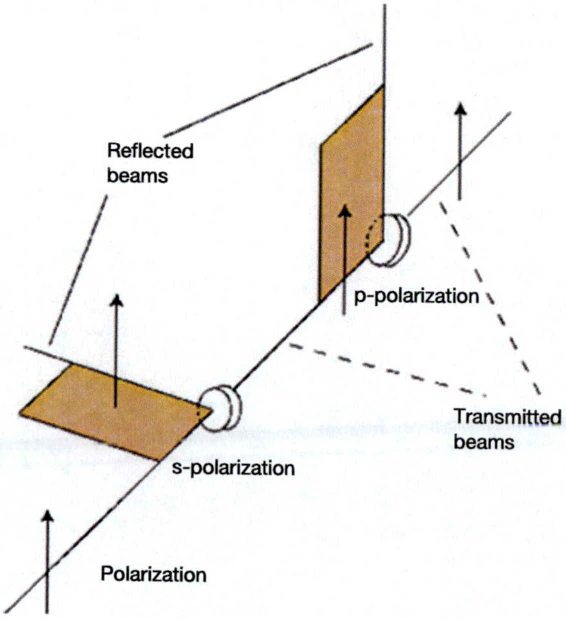

Fig. 4.13 For s-polarization, the input polarization is perpendicular to the plane (shown in color) containing the input and output beams. For p-polarization, the input polarization is parallel to the plane (shown in color) containing the input and output beams

plane containing the input and reflected beams. Light with a polarization vector lying in this plane is called p-polarized, and light, which is polarized perpendicular to this plane, is called s-polarized. Any arbitrary state of input polarization can be expressed as a vector sum of these s and p components. To understand the significance of s and p polarizations, examine the graph which shows the single surface reflectance as a function of angle of incidence for the s and p components of light at a wavelength of 10.6 μm striking a ZnSe surface (Fig. 4.13). Note that while the reflectance of the s component steadily increases with angle, the p component at first decreases to zero at 67° and then increases after that. The angle at which the p reflectance drops to zero is called Brewster's angle. This effect is exploited in several ways to produce polarizing components or uncoated windows which have no transmission loss such as the Brewster windows.

The angle at which p reflectance drops to zero, termed Brewster's angle and can be calculated from$\theta_B = \tan^{-1}(n)$, where θ_B is Brewster's angle and n is the material's index of refraction polarization state, is particularly important in laser cutting application as well as our consideration for purposes of this book and laser interaction with metal. Figure 4.14 shows polarization summary.

Also, by now we know that if the waves travel down the path of let say z-direction, it is called *transverse* because the displacement (in case of rope or string) is perpendicular to the direction of propagation, and it will be called *longitudinal* if displacement from equilibrium is along the direction of propagation. Sound waves, which are nothing but compression waves in air, are longitudinal, while electromagnetic waves as we shall see are transverse (Sect. 4.1 of this chapter).

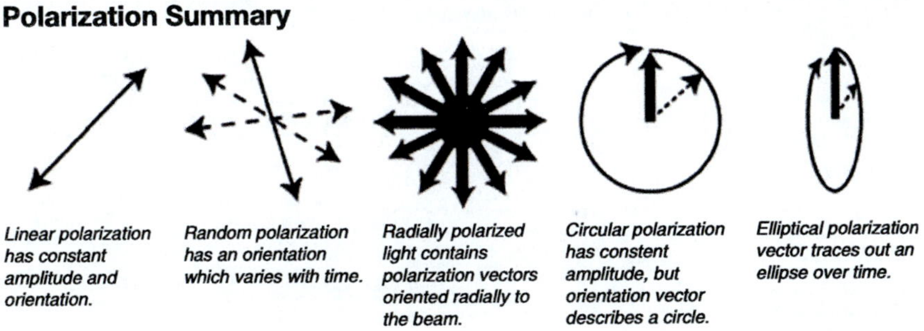

Fig. 4.14 Polarization summary

4.5 The Monochromatic Plane Waves

As we discussed in Sect. 4 of this appendix, we now pay our attention to sinusoidal waves of frequency ω where in visible range, each different frequency corresponds to different *colors*. Such waves are called *monochromatic waves* (see Table 4.2). Suppose furthermore that the waves are traveling in the z-direction and have no x or y dependencies; these are called *plane waves* [3], because the fields are uniform over every plane perpendicular to the direction of propagation (see Fig. 4.15). We are interested, and then in complex form of both magnetic and electric filed as follows:

$$\tilde{\vec{E}}(z,t) = \tilde{\vec{E}}_0 e^{i(kx-\omega t)} \text{and} \tilde{\vec{B}}(z,t) = \tilde{\vec{B}}_0 e^{i(kx-\omega t)} \tag{4.18}$$

where $\tilde{\vec{E}}_0$ and $\tilde{\vec{B}}_0$ are the complex amplitudes [1]. The physical fields, of course are the real parts of $\tilde{\vec{E}}$ and $\tilde{\vec{B}}$. Now, the wave equations for $\vec{E}$ and $\vec{B}$ Equation 4.2 were derived from Maxwell's equations. However, whereas every solution to Maxwell's equations (in empty space) must obey the wave equation, the converse is *not* true; Maxwell's equations impose extra constraints on $\tilde{\vec{E}}_0$ and $\tilde{\vec{B}}_0$. In particular, since $\nabla \cdot \vec{E} = 0$ and $\nabla \cdot \vec{B} = 0$, it follows that;

$$\left(\tilde{E}_0\right)_z = \left(\tilde{B}_0\right)_z = 0 \tag{4.19}$$

This is because the real part of $\tilde{\vec{E}}$ differs from the imaginary part only in the replacement of sine by cosine; if the former obeys Maxwell's equations, so does the latter, and hence $\tilde{\vec{E}}$ as well [1]. Equation 4.19 also indicates that *electromagnetic waves are transverse*: the electric and magnetic fields are perpendicular to the

Table 4.2 The electromagnetic spectrum

The electromagnetic spectrum		
Frequency (Hz)	Type	Wavelength (m)
10^{22}		10^{-13}
10^{21}	Gamma rays	10^{-12}
10^{20}		10^{-11}
10^{19}		10^{-10}
10^{18}	X-rays	10^{-9}
10^{17}		10^{-8}
10^{16}	Ultraviolet	10^{-7}
10^{15}	Visible	10^{-6}
10^{14}	Infrared	10^{-5}
10^{13}		10^{-4}
10^{12}		10^{-3}
10^{11}		10^{-2}
10^{10}	Microwave	10^{-1}
10^{9}		1
10^{8}	TV, FM	10
10^{7}		10^{2}
10^{6}	AM	10^{3}
10^{5}		10^{4}
10^{4}	RF	10^{5}
10^{3}		10^{6}
The visible range		
Frequency (Hz)	Color	Wavelength (m)
1.0×10^{15}	Near ultraviolet	3.0×10^{-7}
7.5×10^{14}	Shortest visible blue	4.0×10^{-7}
6.5×10^{14}	Blue	5.6×10^{-7}
5.6×10^{14}	Green	5.4×10^{-7}
5.1×10^{14}	Yellow	5.9×10^{-7}
4.9×10^{14}	Orange	6.1×10^{-7}
3.9×10^{14}	Longest visible red	7.6×10^{-7}
3.0×10^{14}	Near infrared	1.0×10^{-6}

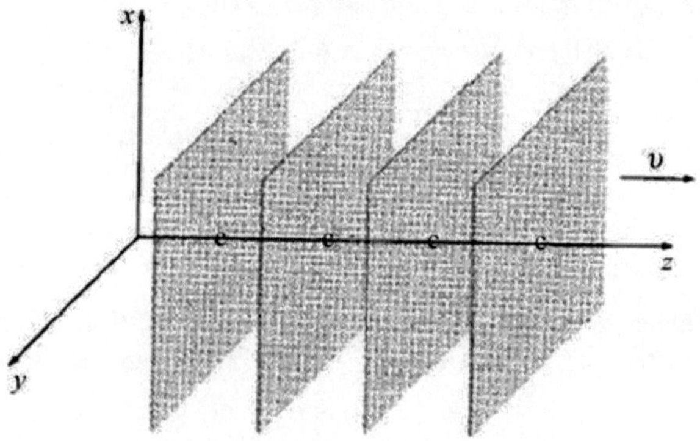

Fig. 4.15 Plane wave propagation [1]

direction of propagation. Moreover, Faraday's law, $\nabla \text{x}\, \vec{E} = -\partial\vec{B}/\partial t$, implies a relationship between the electric and magnetic amplitudes in the following equation:

$$-k\left(\widetilde{E}_0\right)_y = \omega\left(\widetilde{B}_0\right)_x \text{and} - k\left(\widetilde{E}_0\right)_x = \omega\left(\widetilde{B}_0\right)_y \tag{4.20}$$

or in compact complex variable presentation, we have:

$$\vec{\widetilde{B}}_0 = \frac{k}{\omega}\left(\widehat{z}\,\text{x}\,\vec{\widetilde{E}}_0\right) \tag{4.21}$$

From this equation, it is evident that $\vec{E}$ and $\vec{B}$ are *in phase and mutually perpendicular*; their (real) amplitudes are related by:

$$B_0 = \frac{k}{\omega}E_0 = \frac{1}{c}E_0 \tag{4.22}$$

The fourth of Maxwell's equation, $\nabla \text{x}\, \vec{B} = \mu_0\varepsilon_0\left(\partial\vec{E}/\partial t\right)$, does not yield an independent condition; it simply reduces to Eq. 4.20.

We can easily generalize to monochromatic plane waves traveling in an arbitrary direction. The notation is facilitated by the introduction of *propagation* (or *wave*) *vector* $\vec{k}$ pointing in the direction of propagation, whose magnitude is the wave number k. The scalar product $\vec{k}\cdot\vec{r}$ is the appropriate generalization of kzin Fig. 4.16, so we have:

$$\begin{aligned}\vec{\widetilde{E}}\left(\vec{r},t\right) &= \vec{\widetilde{E}}_0 e^{i\left(\vec{k}\cdot\vec{r}-\omega t\right)}\widehat{\vec{n}}\\ \vec{\widetilde{B}}\left(\vec{r},t\right) &= \frac{1}{c}\vec{\widetilde{E}}_0 e^{i\left(\vec{k}\cdot\vec{r}-\omega t\right)}\left(\widehat{\vec{k}}\,\text{x}\,\widehat{\vec{n}}\right) = \frac{1}{c}\left(\widehat{\vec{k}}\,\text{x}\,\vec{\widetilde{E}}\right)\end{aligned} \tag{4.23}$$

where $\widehat{\vec{n}}$ is the polarization vector. Because $\vec{E}$ is transverse:

$$\widehat{\vec{n}}\cdot\widehat{\vec{k}} = 0 \tag{4.24}$$

The transversality of $\vec{B}$ follows automatically from Eq. 4.23. The actual (real) electric and magnetic fields in a monochromatic plane wave with propagation vector $\vec{k}$ and polarization $\widehat{\vec{n}}$ are [1]:

$$\vec{E}\left(\vec{r},t\right) = E_0\cos\left(\vec{k}\cdot\vec{r} - \omega t + \delta\right)\widehat{\vec{n}} \tag{4.25}$$

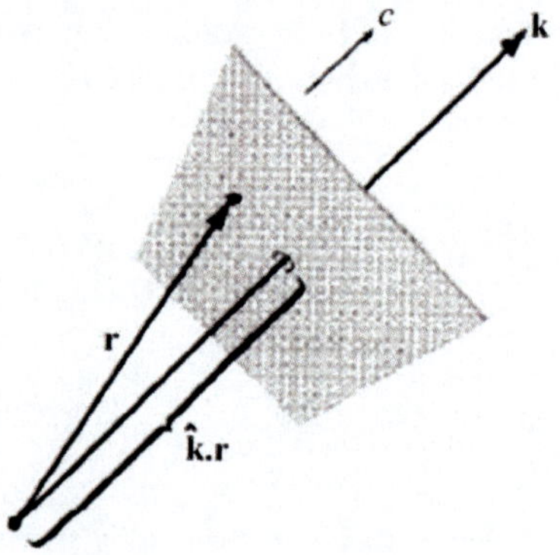

Fig. 4.16 Scalar product of $\vec{k}\cdot\vec{r}$

$$\vec{B}\left(\vec{r},t\right)=\frac{1}{c}E_0\cos\left(\vec{k}\cdot\vec{r}-\omega t+\delta\right)\left(\widehat{\vec{k}}\ \mathrm{x}\ \widehat{\vec{n}}\right) \tag{4.26}$$

Example 4.2 If $\vec{E}$ points in the x-direction, then $\vec{B}$ points in the y-direction (Eq. 4.21)

Solution Writing the following:

$$\tilde{\vec{E}}\ (z,t)=\tilde{\vec{E}}_0 e^{-\left(\vec{k}\cdot\vec{r}-\omega t\right)}\widehat{\vec{x}}\ \text{and}\tilde{\vec{B}}\ (z,t)=\frac{1}{c}\tilde{\vec{E}}_0 e^{i\left(\vec{k}\cdot\vec{r}-\omega t\right)}\widehat{\vec{y}}$$

or taking the real part we have

$$\begin{aligned}\vec{E}(z,t)&=E_0\cos\left(kz-\omega t+\delta\right)\widehat{\vec{x}}\\ \vec{B}(z,t)&=\frac{1}{c}E_0\cos\left(kz-\omega t+\delta\right)\widehat{\vec{y}}\end{aligned} \tag{4.27}$$

Example 4.3 Write down the (real) electric and magnetic fields for a monochromatic plane wave of amplitude E_0, frequency ω, and phase angle zero that is (a) traveling in the negative x-direction and polarized in the z-direction and (b) traveling in the direction from the origin to the point (I, I, I), with polarization parallel to the xz-plane. In each case, sketch the wave, and give the explicit Cartesian components of $\vec{k}$ and $\widehat{\vec{n}}$ (Fig. 4.17).

Solution We can start with the following for part (a):

$$\vec{k}=-\frac{\omega}{c}\widehat{\vec{x}}\ \vec{k}\cdot\vec{r}=\left(-\frac{\omega}{c}\widehat{\vec{x}}\right)\cdot\left(x\widehat{\vec{x}}+y\widehat{\vec{y}}+z\widehat{\vec{z}}\right)=-\frac{\omega}{c}x\widehat{\vec{k}}\ \mathrm{x}\ \widehat{\vec{n}}=-\widehat{\vec{x}}\ \mathrm{x}\ \widehat{\vec{z}}$$

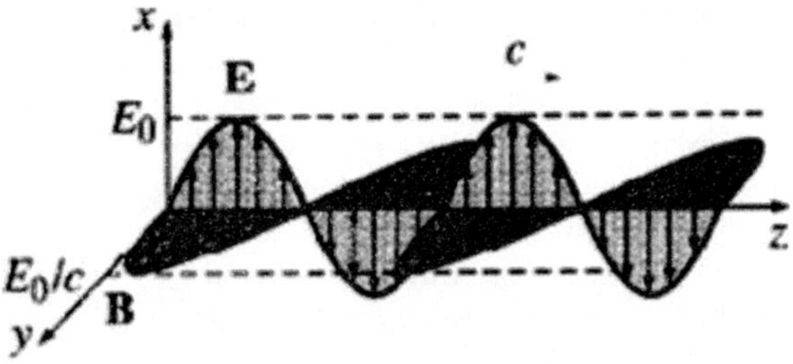

Fig. 4.17 Depiction of Example-1 solution

$$\vec{E}(x,t) = E_0 \cos\left(\frac{\omega}{c}x + \omega t\right)\hat{\vec{z}} \quad \vec{B}(x,t) = \frac{E_0}{c}\cos\left(\frac{\omega}{c}x + \omega t\right)\hat{\vec{y}}$$

and for part (b), we have the following:

$\vec{k} = \frac{\omega}{c}\left(\frac{\hat{\vec{x}}+\hat{\vec{y}}+\hat{\vec{z}}}{\sqrt{3}}\right)\hat{\vec{n}} = \frac{\hat{\vec{x}}-\hat{\vec{z}}}{\sqrt{2}}$ (since $\hat{\vec{n}}$ is parallel to the xz-plane, it must have the form $\alpha\hat{\vec{x}} + B\hat{\vec{z}}$; since $\hat{\vec{n}}\cdot\hat{\vec{k}} = 0, \beta = -\alpha$; and since it is a unit vector, $\alpha = 1/\sqrt{2}$)

$$\vec{k}\cdot\vec{r} = \frac{\omega}{\sqrt{3}c}\left(\hat{\vec{x}} + \hat{\vec{y}} + \hat{\vec{z}}\right)\cdot\left(x\hat{\vec{x}} + y\hat{\vec{y}} + z\hat{\vec{z}}\right) = \frac{\omega}{\sqrt{3}c}(x+y+z)$$

$\hat{\vec{k}} \times \hat{\vec{n}} = \frac{1}{\sqrt{6}}\begin{vmatrix} \hat{\vec{x}} & \hat{\vec{y}} & \hat{\vec{z}} \\ 1 & 1 & 1 \\ 1 & 0 & -1 \end{vmatrix} = \frac{1}{\sqrt{6}}\left(-\hat{\vec{x}} + 2\hat{\vec{y}} - \hat{\vec{z}}\right)$, therefore, we have the following as final solution for part (b)

$$\vec{E}(x,y,z,t) = E_0\cos\left[\frac{\omega}{\sqrt{3}c}(x+y+x) - \omega t\right]\left(\frac{\hat{\vec{x}} - \hat{\vec{z}}}{\sqrt{2}}\right)$$

$$\vec{B}(x,y,z,t) = E_0\cos\left[\frac{\omega}{\sqrt{3}c}(x+y+x) - \omega t\right]\left(\frac{\hat{\vec{x}} - 2\hat{\vec{y}} - \hat{\vec{z}}}{\sqrt{3}}\right)$$

4.6 Boundary Conditions: Reflection and Transmission (Refraction) Dielectric Interface

The important problem of the reflection and refraction (Transmission) of a wave at the interface of two media of different dielectric constants is a boundary value problem in which the physics is identical in principle to that involved in time harmonic problems involving conductors. If, for example, one medium is vacuum and a plane wave is incident on a second (dielectric) medium, the incident harmonic of plane wave generates oscillating time harmonic dipoles (or dipolar currents) that produce a field of their own [2]. The strength of the currents is, however, not known in advance and therein lays the essence of problem. Although this problem is a boundary value problem, it is especially simple because of the great symmetry and simple geometry. It turns out that it is solved by a adding to the incident plane waves only two other plane waves, a *reflected* and a *transmitted* (or refracted) one. The geometry is shown Fig. 4.19 with the plane $z = 0$ taken as the interface between the two media, labeled 1 and 2.

In terms of the *vector* propagation constant, $\vec{k}_i$, the incident fields are:

$$\vec{E}_r = \vec{R}e^{i\vec{k}_r\cdot\vec{r}}\,\text{and}\,\vec{B} = \left(\vec{k}_r/k\right)\,\text{x}\,\vec{E}_r\text{Reflected} \tag{4.28}$$

$$\vec{E}_t = \vec{T}e^{i\vec{k}_t\cdot\vec{r}}\,\text{and}\,\vec{B}_t = \left(\vec{k}_t/k\right)\,\text{x}\,\vec{E}_t\text{Transmitted} \tag{4.29}$$

The vector $\vec{k}_r$ and $\vec{k}_t$ must, for the moment, be considered arbitrary in direction, for although $\vec{k}_i$ is in the x, z-plane, we cannot assume the same a *priori* for $\vec{k}_r$ and $\vec{k}_t$. The magnitudes $k_i \equiv \left|\vec{k}_i\right|$ etc. of the wave vectors are, with n_1 and n_2 the indices of refraction of the two media and $k = \omega/c$,

$$\begin{aligned} k_i &= k_r = n_1 k \\ k_t &= n_2 k \end{aligned} \tag{4.30}$$

Considering the boundary conditions which states the tangential components of $\vec{E}$ and $\vec{B}$ are continuous we have to satisfy the following two steps;

First: If the tangential components of the three fields in Eqs. 4.28, 4.29, and 4.30 are to be matched at $z = 0$, it also should be clear that the spatial dependence given by the exponents must be identical. However, this is a necessary but not a sufficient condition.

Second: The vector coefficients $\vec{A}$, $\vec{R}$, and $\vec{T}$ must be determined.

The first condition, that the spatial variation of the three fields must be identical at $z = 0$, leads to the following equation:

$$\left(\vec{k}_i\cdot\vec{r}\right)_{z=0} = \left(\vec{k}_r\cdot\vec{r}\right)_{z=0} = \left(\vec{k}_t\cdot\vec{r}\right)_{z=0} \tag{4.31}$$

The first equality in Eq. 4.31 yields $k_{ix}x = k_{rx}x + k_{ry}y$. For this condition to hold for all x and y, we must have $k_{ry} = 0$, showing that $\vec{k}_r$ lies in the plane of incidence and also that $k_{ix}x = k_{rx}x$. Similarly, from the second equality in Eq. 4.31, $\vec{k}_t$ must also

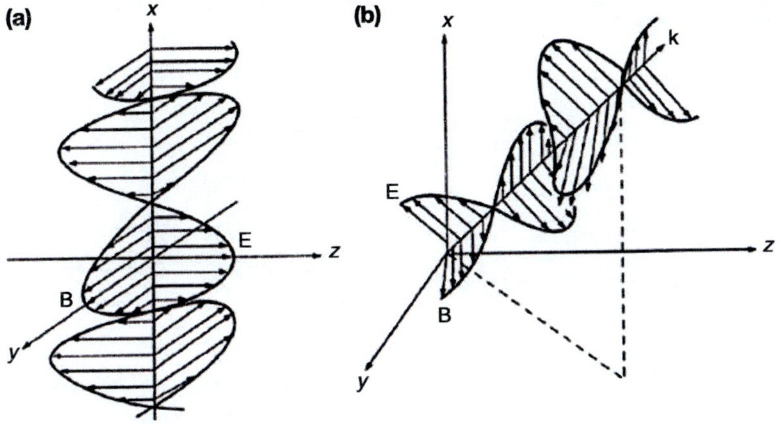

Fig. 4.18 Sketch of Example-2 solution

lie in this plane, so that $\vec{k}_i$, $\vec{k}_r$, $\vec{k}_t$ are all coplanar. Moreover, from the geometry of Fig. 4.18, we have from $k_{ix} = k_{rx}$ that $k_i \sin\theta_i = k_r \sin\theta_r$. Or since $k_i = k_r$, then we can write the following;

$$\sin\theta_i = \sin\theta_r \tag{4.32}$$

The *angle of incidence equals the angle of reflection*. Similarly, the equality of k_{ix} and k_{tx} yields:

$$k_i \sin\theta_i = k_t \sin\theta_t$$

Or using Eq. 4.30, we have:

$$\frac{\sin\theta_i}{\sin\theta_t} = \frac{n_2}{n_1} \tag{4.33}$$

Equation 4.33 is known as *Snell's law* of refraction. The conditions of Eqs. 4.32 and 4.33 are quite general ones that are independent of the detailed vectorial nature of the wave field. They hold for reflection and refraction of scalar waves. These conditions by themselves do not guarantee the continuity of tangential $\vec{E}$ and $\vec{B}$ across the boundary. To satisfy these conditions, more specification of polarization of the fields should be analyzed [2]. For convenience, we consider the general case of arbitrary incident polarization as a linear combination of a wave with polarization perpendicular to the plane of incidence and one with polarization parallel to this plane. The reflected and transmitted waves will then be similarly polarized. These two cases are sketched as parts (a) and (b) in Fig. 4.19.

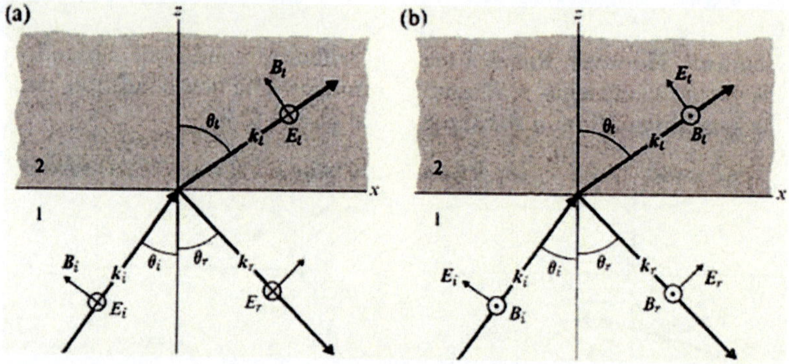

Fig. 4.19 Reflection and refraction at a dielectric interface. (**a**) Electric vector perpendicular to plane of incidence. (**b**) Electric vector parallel to plane of incidence

Consider first the case of $\vec{E}$ perpendicular to the plane of incidence, i.e., in the y-direction. The vector coefficients $\vec{A}$, $\vec{R}$, and $\vec{T}$ becomes scalar ones, with subscript $\perp$ to denote this case.

$$\begin{aligned} E_{i\perp} &= A_\perp e^{i\vec{k}_i \cdot \vec{r}} \\ E_{r\perp} &= A_\perp e^{i\vec{k}_r \cdot \vec{r}} \\ E_{t\perp} &= A_\perp e^{i\vec{k}_t \cdot \vec{r}} \end{aligned} \tag{4.34}$$

Then from the continuity of tangential $\vec{E}$ at the boundary, we have:

$$A_\perp + R_\perp = T_\perp \tag{4.35}$$

The condition on tangential $\vec{B}$becomes, with $B_{ix} = -A_\perp \cos\theta_i$, $B_{rx} = -R_\perp \cos\theta_r$, and$B_{tx} = -T_\perp \cos\theta_t$, also considering that $\cos\theta_i = -\cos\theta_r$ we then have:

$$n_1(A_\perp - R_\perp)\cos\theta_i = n_2 T_\perp \cos\theta_t \tag{4.36}$$

Solving Eqs. 4.35 and 4.36 for the ratios $R_\perp / A_\perp$ and $T_\perp / A_\perp$, we find, using Snell's law, that

$$\begin{aligned} \frac{R_\perp}{A_\perp} &= \frac{1 - \dfrac{tan\theta_i}{\tan\theta_r}}{1 + \dfrac{\tan\theta_i}{\tan\theta_r}} = -\frac{\sin(\theta_i - \theta_t)}{\sin(\theta_i + \theta_t)} \\ \frac{T_\perp}{A_\perp} &= \frac{2}{1 + \dfrac{tan\theta_i}{\tan\theta_r}} = \frac{2\cos\theta_i \sin\theta_t}{\sin(\theta_i + \theta_t)} \end{aligned} \tag{4.37}$$

For the second case, incident wave polarized parallel to the plane of incidence, we use $A_| = \left|\vec{A}\right|$ in Eq. 4.27 and, similarly $R_| = \left|\vec{R}\right|$ and $T_| = \left|\vec{T}\right|$ in Eqs. 4.28 and 4.29. The boundary condition yields:

$$\cos\theta_i\left(A_| - R_|\right) = \cos\theta_t T_|$$
$$n_1\left(A_| - R_|\right) = n_2 T_|$$

These equations lead to the results:

$$\begin{aligned} \frac{R_|}{A_|} &= \frac{\tan(\theta_i - \theta_t)}{\tan(\theta_i + \theta_t)} \\ \frac{T_|}{A_|} &= \frac{2\cos\theta_i \sin\theta_t}{\sin(\theta_i + \theta_t)\cos(\theta_i - \theta_t)} \end{aligned} \tag{4.38}$$

There are two phenomena worthy of note in connection with the above discussion. Consider the case of polarization in the plane of incidence. We see from Eq. 4.38 that $R_{\|}/A_{\|}$ will be 0 for $\theta_i + \theta_t = \pi/2$. Putting this condition into Snell's law, using $\sin\theta_t = \sin(\pi/2 - \theta_i) = \cos\theta_i$, we see that the angle of incidence θ_B (called the *Brewster's angle*) for which this happens is defined by:

$$\tan\theta_B = \frac{n_2}{n_1} \tag{4.39}$$

If wave with arbitrary polarization is incident on a dielectric interface, it can be considered to be a linear combination of a wave polarized parallel to, and a wave polarized perpendicular to, the plane of incident. At the Brewster's angle, the parallel component will not be reflected so that the reflected wave will be a plane polarized in a plane perpendicular to the plane of incidence. This effect can then be made the basis of a device for polarizing an unpolarized beam of radiation.

The second phenomenon is that of *total internal reflection*. In either (a) or (b) of Fig. 4.4, suppose that the index n_1 is greater than n_2. Then, from Snell's law, $\sin\theta_t = (n_1/n_2)\sin\theta_i$ is always greater than θ_i.There will then be some value of θ_i, call it θ_{int} for which $\theta_t = \pi/2$; this angle is defined by

$$\sin\theta_{\text{int}} = \frac{n_2}{n_1}$$

Since in general:

$$\vec{E}_t = \vec{T}e^{ikn_2(x\sin\theta_t + z\cos\theta_t)} \tag{4.40}$$

for $\theta_t = \pi/2$ there is no wave in the second medium; the z-dependence vanishes. Now if $\theta_i > \theta_{\text{int}}$, then $\sin\theta_t$, is larger than unity from Snell's law and, as a conse-quence, $\cos\theta_t$ is imaginary

$$\cos\theta_t = \sqrt{1-\left(\frac{n_1}{n_2}\right)^2 \sin^2\theta_i} = i\sqrt{\left(\frac{n_1}{n_2}\right)^2 \sin^2\theta_i - 1}$$

Equation 4.31 then becomes:

$$\vec{E}_t = \vec{T}e^{-kn_2 z\sqrt{(n_1/n_2)\sin^2\theta_i - 1}}e^{ikn_2 x\sin\theta_t}$$

This corresponds to a wave which is exponentially attenuated as a function of z, and which propagates as a function of x with propagation constant $kn_2 \sin\theta_t$. Such a wave is the prototype of a surface wave, and readers can see further discussion and more details in Reference 2 of this Chapter 2.

The interesting question is this: What happens when a wave passes from one transparent medium into another—air to water, say, or glass to plastic. As in the case of waves on a string, we expect to get a reflected wave and a transmitted wave. The details depend on the exact nature of the electrodynamics boundary conditions [1].

$$\left.\begin{array}{ll} \text{(i) } \varepsilon_1 E_1^\perp = \varepsilon_2 E_2^\perp & \text{(iii) } E_1^| = E_2^| \\ \text{(ii) } B_1^\perp = B_2^\perp & \text{(iv) } \frac{1}{\mu_1}B_1^| = \frac{1}{\mu_2}B_2^| \end{array}\right\} \tag{4.41}$$

These equations relate the electric and magnetic fields just to the left and just to the right of the interface between two linear media. In the following sections, we use them to deduce the laws governing reflection and refraction of electromagnetic waves.

4.7 Electromagnetic Waves in Matter

Some solutions to Maxwell's equations have been already discussed in previous sections. The present section extends the treatment of electromagnetic waves. Since most regions of interest are free of charge, it will be assumed that charge density $\rho = 0$. Moreover, linear isotropic (invariant with respect to direction) materials will be assumed, with the following relationships:

$$\begin{cases} \vec{D} = \varepsilon\vec{E} \\ \vec{B} = \mu\vec{H} \\ \vec{J} = \sigma\vec{E} \end{cases} \tag{4.42}$$

where:

$\vec{D}$ = flux density (C/m^2).

$\vec{E}$ = electric field (N/C).
ε = permittivity of the medium ($C^2/N.m^2$) or equivalent (F/m).
$\vec{B}$ = magnetic field (T).
$\vec{H}$ = magnetic field strength (A/m).
μ = mobility within materials ($m^2/V.s$).
$\vec{J}$ = current density (A/m^2).
σ = conductivity of materials (S/m).
ρ = charge density (C/m^3).

where:

A = ampere.
C = coulomb.
N = newton.
F = farad.
S = siemens.
T = tesla.

Inside matter, but in regions where there is no *free* charge or *free current*, Maxwell's equations become:

$$\left.\begin{array}{ll}(\mathrm{i})\nabla\cdot\vec{D}=0 & (\mathrm{iii})\nabla\mathrm{x}\,\vec{E}=-\dfrac{\partial\vec{B}}{\partial t}\\ (\mathrm{ii})\nabla\cdot\vec{B}=0 & (\mathrm{iv})\nabla\mathrm{x}\,\vec{H}=\dfrac{\partial\vec{D}}{\partial t}\end{array}\right\} \tag{4.43}$$

If the medium is *linear*:

$$\vec{D}=\varepsilon\vec{E}\mathrm{and}\vec{H}=\frac{1}{\mu}\vec{B} \tag{4.44}$$

and homogeneous (so ε and μ do not vary from point to point), Maxwell's equations reduce to:

$$\left.\begin{array}{ll}(\mathrm{i})\nabla\cdot\vec{E}=0 & (\mathrm{iii})\nabla\mathrm{x}\,\vec{E}=-\dfrac{\partial\vec{B}}{\partial t}\\ (\mathrm{ii})\nabla\cdot\vec{B}=0 & (\mathrm{iv})\nabla\mathrm{x}\,\vec{B}=\mu\varepsilon\dfrac{\partial\vec{E}}{\partial t}\end{array}\right\} \tag{4.45}$$

which remarkably differ from the vacuum analogs Eq. 4.1 only in the replacement of $\mu_0\varepsilon_0$ by $\mu\varepsilon$. This mathematically is obvious, yet the physical implications are astonishing [4]. As the wave passes through, the fields busily polarize and magnetize all the molecules, and the resulting (oscillating) dipoles create their own electric and magnetic fields. These combine with the original fields in such a way as to create a *single* wave with the same frequency but a different speed. This extraordinary conspiracy is responsible for the phenomenon of *transparency*. It is a distinctly *non-trivial* consequence of the *linearity* of the medium [4]. Looking at the set of Equation of 4.45, it is evident electromagnetic waves propagate through a linear homogeneous medium at a speed:

$$v = \frac{1}{\sqrt{\varepsilon\mu}} = \frac{c}{n} \tag{4.46}$$

where

$$n \equiv \sqrt{\frac{\varepsilon\mu}{\varepsilon_0\mu_0}} \tag{4.47}$$

is the index of refraction of the material. For most materials, μ is very close to μ_0, so

$$n \cong \sqrt{\varepsilon_r} \tag{4.48}$$

where ε_r is the *dielectric constant* or also known as *relative permittivity* [1] and is equal to $\varepsilon_r = \varepsilon/\varepsilon_0$. Since ε_r is almost always greater than 1, light travels *more slowly* through matter—a fact that is well known from optics [1].

4.8 Reflection and Transmission at Normal Incidence

Suppose the *xy*-plane forms the boundary between two linear media. A plane wave of frequency ω, traveling in the *z*-direction and polarized in the *x*-direction, approaches the interface from the left (Fig. 4.20):

$$\left.\begin{aligned} \vec{\widetilde{E}}_I(z,t) &= \widetilde{E}_{0_I} e^{i(k_1 z-\omega t)} \vec{\widehat{x}} \\ \vec{B}_I(z,t) &= \frac{1}{v_1}\widetilde{E}_{0_I} e^{i(k_1 z-\omega t)} \vec{\widehat{y}} \end{aligned}\right\} \tag{4.49}$$

It gives rise to a reflected wave:

$$\left.\begin{aligned} \vec{\widetilde{E}}_R(z,t) &= \widetilde{E}_{0_R} e^{i(k_1 z-\omega t)} \vec{\widehat{x}} \\ \vec{B}_R(z,t) &= -\frac{1}{v_1}\widetilde{E}_{0_R} e^{i(-k_1 z-\omega t)} \vec{\widehat{y}} \end{aligned}\right\} \tag{4.50}$$

which travels back to the left in medium (1) and transmitted waves:

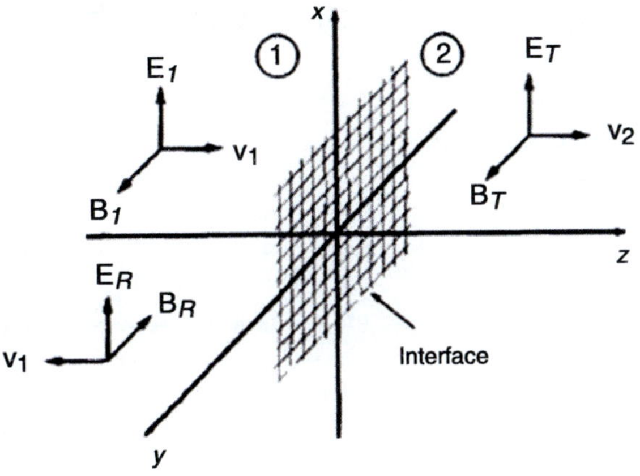

Fig. 4.20 Normal wave incidence [1]

$$\left.\begin{aligned}\vec{\tilde{E}}_T(z,t) &= \tilde{E}_{0_T} e^{i(k_2 z-\omega t)}\vec{\tilde{x}} \\ \vec{\tilde{B}}_T(z,t) &= \frac{1}{v_2}\tilde{E}_{0_T} e^{i(k_2 z-\omega t)}\vec{\tilde{y}}\end{aligned}\right\} \tag{4.51}$$

which continue on the right in medium (2). Note the minus sign in $\vec{B}_R$, as required by Eq. 4.23 or, if you prefer, by the fact that the Poynting vector aims in the direction of propagation.

At $z = 0$, the combined fields on the left, $\vec{\tilde{E}}_I + \vec{\tilde{E}}_R$ and $\vec{\tilde{B}}_I + \vec{\tilde{B}}_R$, must join the fields on the right, $\vec{\tilde{E}}_T$ and $\vec{\tilde{B}}_T$, in accordance with the boundary conditions Eq. 4.41. In this case, there are no components perpendicular to the surface, so (i) and (ii) are trivial. However, (iii) requires that:

$$\tilde{E}_{0_I} + \tilde{E}_{0_R} = \tilde{E}_{0_T} \tag{4.52}$$

while (iv) says:

$$\frac{1}{\mu_1}\left(\frac{1}{v_1}\tilde{E}_{0_I} - \frac{1}{v_1}\tilde{E}_{0_R}\right) = \frac{1}{\mu_2}\left(\frac{1}{v_2}\tilde{E}_{0_T}\right) \tag{4.53}$$

or

$$\tilde{E}_{0_I} - \tilde{E}_{0_R} = \beta\tilde{E}_{0_T} \tag{4.54}$$

where:

$$\beta \equiv \frac{\mu_1 v_1}{\mu_2 v_2} = \frac{\mu_1 n_2}{\mu_2 n_1} \tag{4.55}$$

Equations 4.52 and 4.54 are easily solved for the outgoing amplitudes, in terms of the incident amplitude:

$$\widetilde{E}_{0_R} = \left(\frac{1-\beta}{1+\beta}\right)\widetilde{E}_{0_I} \text{and} \widetilde{E}_{0_T} = \left(\frac{2}{1+\beta}\right)\widetilde{E}_{0_I} \tag{4.56}$$

These results are strikingly similar to the ones for waves on a string. Indeed, if the permittivities μ *are* close to their values in vacuum (as, remember, they are for most media), then $\beta = \frac{v_1}{v_2}$, and we have [1]:

$$\widetilde{E}_{0_R} = \left(\frac{v_2 - v_1}{v_2 + v_1}\right)\widetilde{E}_{0_I} \text{and} \widetilde{E}_{0_T} = \left(\frac{2v_2}{v_2 + v_1}\right)\widetilde{E}_{0_I} \tag{4.57}$$

In that case, as before, the reflected wave is in *phase* (right-side up) if $v_2 > v_1$ and *out of phase* (upside down) if $v_2 < v_1$; the real amplitudes are related by:

$$E_{0_R} = \left(\frac{v_2 - v_1}{v_2 + v_1}\right)E_{0_I} \text{and} E_{0_T} = \left(\frac{2v_2}{v_2 + v_1}\right)E_{0_I} \tag{4.58}$$

or, in terms of the indices of refraction [1]:

$$E_{0_R} = \left(\frac{n_1 - n_2}{n_1 + n_2}\right)E_{0_I} \text{and} E_{0_T} = \left(\frac{2n_1}{n_1 + n_2}\right)E_{0_I} \tag{4.59}$$

According to definition of intensity (average power per unit area) that is given by D. Griffiths [1], we should ask what fraction of the incident energy is reflected, and what fraction is transmitted?

$$I = \frac{1}{2}\varepsilon v E_0^2 \tag{4.60}$$

If again $\mu_1 = \mu_2 = \mu_0$, then the ratio of the reflected intensity to the incident intensity is:

$$R \equiv \frac{I_R}{I_I} = \left(\frac{E_{0_R}}{E_{0_I}}\right)^2 = \left(\frac{n_1 - n_2}{n_1 + n_2}\right)^2 \tag{4.61}$$

whereas the ratio of the transmitted intensity to the incident intensity is:

$$T \equiv \frac{I_T}{I_I} = \frac{\varepsilon_2 v_2}{\varepsilon_1 v_1}\left(\frac{E_{0_T}}{E_{0_I}}\right)^2 = \frac{4n_1 n_2}{(n_1+n_2)^2} \tag{4.62}$$

R is called the **reflection coefficient** and T the **transmission coefficient**; they measure the fraction of the incident energy that is reflected and transmitted, respectively. Notice that [1]:

$$R + T = 1 \tag{4.63}$$

as conservation of energy, of course, requires. For instance, when light passes from air ($n_1 = 1$) into glass ($n_2 = 1.5$), $R = 0.04$ and $T = 0.96$. Not surprisingly, most of the light is transmitted [1].

Example 4.4 Calculate the exact reflection and transmission coefficients, without assuming $\mu_1 = \mu_2 = \mu_0$. Confirm $R + T = 1$.

Solution From Eq. 4.61 we have $R = \left(\frac{E_{0_R}}{E_{0_I}}\right)^2$ substituting from Eq. 4.56 results where $\beta \equiv \frac{\mu_1 v_1}{\mu_2 v_2}$ and using Eq. 4.62 $T = \frac{\varepsilon_2 v_2}{\varepsilon_1 v_1}\left(\frac{E_{0_T}}{E_{0_I}}\right)^2$ results that $T = \beta\left(\frac{2}{1+\beta}\right)^2$ which is Eq. 4.56. Note that $\frac{\varepsilon_2 v_2}{\varepsilon_1 v_1} = \frac{\mu_1}{\mu_2}\frac{\varepsilon_2 v_2}{\varepsilon_1 v_1}\frac{v_2}{v_1} = \frac{\mu_1}{\mu_2}\left(\frac{v_1}{v_2}\right)^2 = \frac{\mu_1 v_1}{\mu_2 v_2} = \beta$

$$\begin{aligned} R &= \frac{2}{(1+\beta)^2}\left[4\beta + (1-\beta)^2\right] \\ &= \frac{1}{(1+\beta)^2}(4\beta + 1 - 2\beta + \beta^2) \\ &= \frac{1}{(1+\beta)^2}(1 + 2 + \beta + \beta^2) = 1 \end{aligned}$$

Example 4.5 In writing Eqs. 4.50 and 4.51, assumption was the reflected and transmitted waves have the same *polarization* as the incident wave—along the x-direction. Prove that this must be so. [Hint: Let the polarization vector of the transmitted and reflected wave be $\widehat{\vec{n}}_T = \cos\theta_T \widehat{\vec{x}} + \sin\theta_T \widehat{\vec{y}}$ and $\widehat{\vec{n}}_R = \cos\theta_R \widehat{\vec{x}} + \sin\theta_R \widehat{\vec{y}}$ and prove from the boundary condition that $\theta_T = \theta_R = 0$].

Solution Equation 4.42 is replaced by $\widetilde{E}_{0_I}\widehat{\vec{x}} + \widetilde{E}_{0_R}\widehat{\vec{n}}_R = \widetilde{E}_{0_T}\widehat{\vec{n}}_T$ and Eq. 4.54 $\widetilde{E}_{0_I}\widehat{\vec{y}} - \widetilde{E}_{0_R}\left(\widehat{\vec{z}} \times \widehat{\vec{n}}_R\right) = \beta\widetilde{E}_{0_T}\left(\widehat{\vec{z}} \times \widehat{\vec{n}}_T\right)$. The y component of the first equation is $\widetilde{E}_{0_R}\sin\theta_R = \widetilde{E}_{0_T}\sin\theta_T$; the x component of the second is $\widetilde{E}_{0_R}\sin\theta_R = -\beta\widetilde{E}_{0_T}\sin\theta_T$. Comparing these two, we conclude that $\sin\theta_R = \sin\theta_T = 0$, and hence $\theta_T = \theta_R = 0$.

4.9 Reflection and Transmission at Oblique Incidence

In the last section, reflection and transmission was treated at *normal* incidence—that is, when the incoming wave hits the interface head-on at a 90-degree incident angle [1]. The more general case of *oblique* incidence, in which the incoming wave meets the boundary at an arbitrary angle θ_I, also is treated by David Griffiths' reference [1] of this appendix. We just write the result of his conclusion here, and we encourage readers to refer to his book on section 9.3.3 page 386 for more details [1]. Although the angle of incidence $\theta_I = 0$ is a special case of the oblique incidence, we did treat that separately for some cases where we are dealing with high-power laser interaction with matter in this book. The treatment by Griffiths is done based on a monochromatic plane wave with such arbitrary angle of incidence. See Fig. 4.21.

Suppose that a monochromatic plane wave (incident wave) approaches from the left as follows:

$$\tilde{\vec{E}}_I\left(\vec{r},t\right) = \tilde{\vec{E}}_{0_I} e^{i\left(\vec{k}_I\cdot\vec{r}-\omega t\right)} \text{ and } \tilde{\vec{B}}_I\left(\vec{r},t\right) = \frac{1}{v_1}\left(\hat{\vec{k}}_I \times \tilde{\vec{E}}_I\right) \tag{4.64}$$

And giving rise to a reflected wave of the form in Eq. 4.65:

$$\tilde{\vec{E}}_R\left(\vec{r},t\right) = \tilde{\vec{E}}_{0_R} e^{i\left(\vec{k}_R\cdot\vec{r}-\omega t\right)} \text{ and } \tilde{\vec{B}}_R\left(\vec{r},t\right) = \frac{1}{v_1}\left(\hat{\vec{k}}_R \times \tilde{\vec{E}}_R\right) \tag{4.65}$$

and a transmitted set of wave equation of 4.66:

$$\tilde{\vec{E}}_T\left(\vec{r},t\right) = \tilde{\vec{E}}_{0_T} e^{i\left(\vec{k}_T\cdot\vec{r}-\omega t\right)} \text{ and } \tilde{\vec{B}}_T\left(\vec{r},t\right) = \frac{1}{v_2}\left(\hat{\vec{k}}_T \times \tilde{\vec{E}}_T\right) \tag{4.66}$$

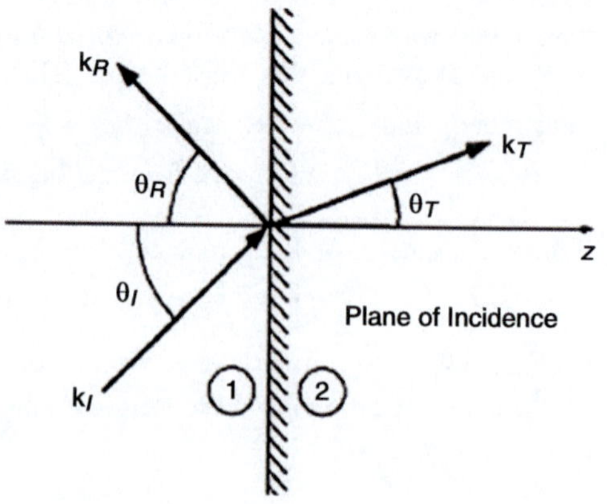

Fig. 4.21 Incidence at oblique angle [1]

All three waves have the same frequency ω that is determined once for all at the source (the laser beam, flashlight, or whatever, which produces the incident beam). The three wave numbers are related per Eq. 4.11:

$$k_I v_1 = k_R v_1 = k_T v_2 = \omega \text{or} k_I = k_R = \frac{v_2}{v_1} k_T = \frac{n_1}{n_2} k_T \tag{4.67}$$

The combined fields in medium (1), $\tilde{\vec{E}}_I + \tilde{\vec{E}}_R$ and $\tilde{\vec{B}}_I + \tilde{\vec{B}}_R$, must now be joined to the fields $\tilde{\vec{E}}_T$ and $\tilde{\vec{B}}_T$ in medium (2), using the boundary conditions Eq. 4.41. These all share the generic structure. *Boundary conditions should hold at all points on the plane and for all times for all above three plane wave equations when $z = 0$.* Otherwise, a slight change in x would destroy the equality (see Example 1) [1]. The time factors in this case are *already* equal; in fact we can regard this as an independent confirmation that the transmitted and reflected frequencies must match the incident frequency. So for spatial case, we have the following relation:

$$\vec{k}_I \cdot \vec{r} = \vec{k}_R \cdot \vec{r} = \vec{k}_T \cdot \vec{r} \text{whenz} = 0 \tag{4.68}$$

or more explicitly:

$$x(k_I)_x + y(k_I)_y = x(k_R)_x + y(k_R)_y = x(k_T)_x + y(k_T)_y \tag{4.69}$$

for all x and y. But Eq. 4.69 can *only* hold if the components are separately equal, for if $x = 0$, we get

$$(k_I)_y = (k_R)_y = (k_T)_y \tag{4.70}$$

While $y = 0$ gives

$$(k_I)_x = (k_R)_x = (k_T)_x \tag{4.71}$$

We may as well orient our axes so that $\vec{k}$ lies in the xy plane (i.e.$(k_I)_y = 0$); according to Eq. 4.70, so too will $\vec{k}_R$ and $\vec{k}_T$. *Conclusion*:

First Law: The incident, reflected, and transmitted wave vectors form a plane (called the *plane of incidence*), which also includes the normal to the surface (here, the z-axis).

Meanwhile, Eq. 4.71 implies that:

$$k_I \sin\theta_I = k_R \sin\theta_R = k_T \sin\theta_T \tag{4.72}$$

where θ_I is the *angle of incidence*, θ_R is the *angle of reflection*, and θ_T is the *angle of transmission*, more commonly known as the *angle of refraction*, all of them measured with respect to the normal (Fig. 4.21). In view of Eq. 4.67, then:

Second Law: The angle of incidence is equal to the angle of reflection:

$$\theta_I = \theta_T \tag{4.73}$$

This is the *law of reflection.*

As for the transmitted angle,

Third Law:

$$\frac{\sin\theta_T}{\sin\theta_I} = \frac{n_1}{n_2} \tag{4.74}$$

This is the *law of refraction*, or *Snell's law.*

These are the three fundamental laws of geometrical optics. It is remarkable how little actual electrodynamics went into them: we have yet to invoke any specific boundary conditions. Now that we have taken care of the exponential factors-they cancel, given Eq. 4.68 the boundary conditions Eq. 4.41 becomes [1]:

$$\left.\begin{array}{ll}
\text{(i)}\ \varepsilon_1\left(\vec{\tilde{E}}_{0_I} + \vec{\tilde{E}}_{0_R}\right)_z = \varepsilon_2\left(\vec{\tilde{E}}_{0_T}\right)_z & \\
\text{(ii)} & \left(\vec{\tilde{B}}_{0_I} + \vec{\tilde{B}}_{0_R}\right)_z = \left(\vec{\tilde{B}}_{0_T}\right)_z \\
\text{(iii)} & \left(\vec{\tilde{E}}_{0_I} + \vec{\tilde{E}}_{0_R}\right)_{x,y} = \left(\vec{\tilde{E}}_{0_T}\right)_{x,y} \\
\text{(iv)}\ \frac{1}{\mu_1}\left(\vec{\tilde{B}}_{0_I} + \vec{\tilde{B}}_{0_R}\right)_{x,y} = \frac{1}{\mu_2}\left(\vec{\tilde{B}}_{0_T}\right)_{x,y} &
\end{array}\right\} \tag{4.75}$$

where $\vec{\tilde{B}}_0 = (1/v)\vec{\tilde{k}}\ \text{x}\ \vec{\tilde{E}}_0$ in each case. The last two (iii) and (iv) represent pairs of equations, one for the *x*–direction and one for the *y*–direction.

Suppose that the polarization of the incident wave is *parallel* to the plane of incidence (the *xz* plane in Fig. 4.22); it follows (see Example-2) that the reflected and transmitted waves are also polarized in this plane. We will present that in an example to analyze the case of polarization perpendicular to the plane of incidence (see Example-2). Then (i) reads:

Fig. 4.22 Incident oblique wave

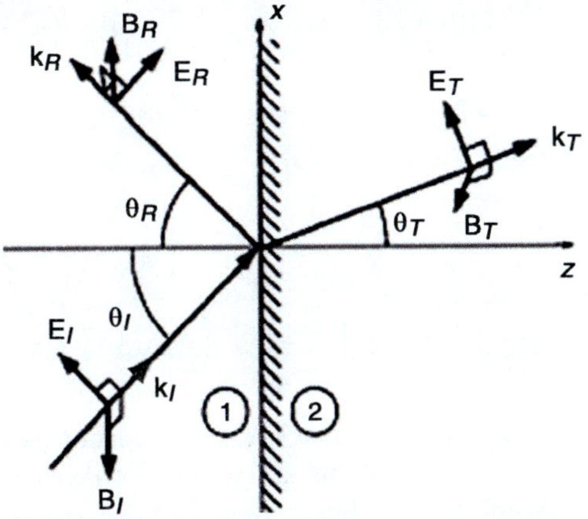

$$\varepsilon_1\left(-\tilde{\vec{E}}_{0_I}\sin\theta_I+\tilde{\vec{E}}_{0_R}\sin\theta_R\right)=\varepsilon_2\left(-\tilde{\vec{E}}_{0_T}\sin\theta_T\right) \tag{4.76}$$

(ii) adds nothing ($0 = 0$), since the magnetic fields have no z components; (iii) becomes [1];

$$\tilde{\vec{E}}_{0_I}\cos\theta_I+\tilde{\vec{E}}_{0_R}\cos\theta_R=\tilde{\vec{E}}_{0_T}\cos\theta_T \tag{4.77}$$

and (iv) says:

$$\frac{1}{\mu_1 v_1}\left(\tilde{\vec{E}}_{0_I}-\tilde{\vec{E}}_{0_R}\right)=\frac{1}{\mu_2 v_2}\left(\tilde{\vec{E}}_{0_T}\right) \tag{4.78}$$

Given the laws of reflection and refraction, Eqs. 4.75 and 4.78 both reduce to:

$$\tilde{\vec{E}}_{0_I}-\tilde{\vec{E}}_{0_R}=\beta\tilde{\vec{E}}_{0_T} \tag{4.79}$$

where again:

$$\beta\equiv\frac{\mu_1 v_1}{\mu_2 v_2}=\frac{\mu_1 n_2}{\mu_2 n_1} \tag{4.80}$$

Equation 4.77 says:

$$\tilde{\vec{E}}_{0_I}+\tilde{\vec{E}}_{0_R}=\alpha\tilde{\vec{E}}_{0_T} \tag{4.81}$$

where again:

$$\alpha = \frac{\cos\theta_T}{\cos\theta_I} \tag{4.82}$$

Solving Eqs. 4.79 and 4.81 for the reflected and transmitted amplitudes, we obtain [1]:

$$\begin{aligned} \tilde{\vec{E}}_{0_R} &= \left(\frac{\alpha-\beta}{\alpha+\beta}\right)\tilde{\vec{E}}_{0_I} \\ \tilde{\vec{E}}_{0_T} &= \left(\frac{2}{\alpha+\beta}\right)\tilde{\vec{E}}_{0_I} \end{aligned} \tag{4.83}$$

These are known as Fresnel's equations, for the case of polarization in the plane of incidence. (There are two other Fresnel equations, giving the reflected and transmitted amplitudes when the polarization is perpendicular to the plane of incidence (see Prob. 9.16.).) Notice that the transmitted wave is always in phase with the incident one; the reflected wave is either in phase ("right side up"), if $\alpha > \beta$, or 180° out of phase ("upside down"). if $\alpha < \beta$.

Note that there is an unavoidable ambiguity in the phase of the reflected wave. The convention that David Griffiths [1] has used in his book under Chapter 9, which is reflected here and adopted in Fig. 4.22 with $\tilde{\vec{E}}_R$ positive "upward," is consistent with some, but not all, of the standard optics text. Changing the sign of the polarization vector is equivalent to a 180° phase shift.

The amplitudes of the transmitted and reflected waves depend on the angle of incidence α because a is a function of θ_I:

$$\alpha = \frac{\sqrt{1-\sin^2\theta_T}}{\cos\theta_I} = \frac{\sqrt{1-[(n_1/n_2)\sin\theta_I]^2}}{\cos\theta_I} \tag{4.84}$$

In the case of normal incidence ($\theta_I = 0$), $\alpha = 1$, and we recover Eq. 4.56. At grazing incidence ($\theta_I = 90$), α diverges, and the wave is totally reflected (a fact that is painfully familiar to anyone who has driven at night on a wet road). Interestingly, there is an intermediate angle, θ_B (called *Brewster's angle*), at which the reflected wave is completely extinguished [1]. Because waves polarized perpendicular to the plane of incidence exhibit no corresponding quenching of the reflected component, an arbitrary beam incident at Brewster's angle yields a reflected beam that is totally polarized parallel to the interface. That's why Polaroid glasses, with the transmission axis vertical, help to reduce glare off a horizontal surface [1].

According to Eq. 4.83, this occurs when $\alpha = \beta$, or:

$$\sin^2\theta_B = \frac{1-\beta^2}{(n_1/n_2)^2-\beta^2} \tag{4.85}$$

For the typical case $\mu_1 \cong \mu_2$, so $\beta \cong n_2/n_1$, $\sin^2\theta_B \cong \beta^2/(1+\beta^2)$, and hence

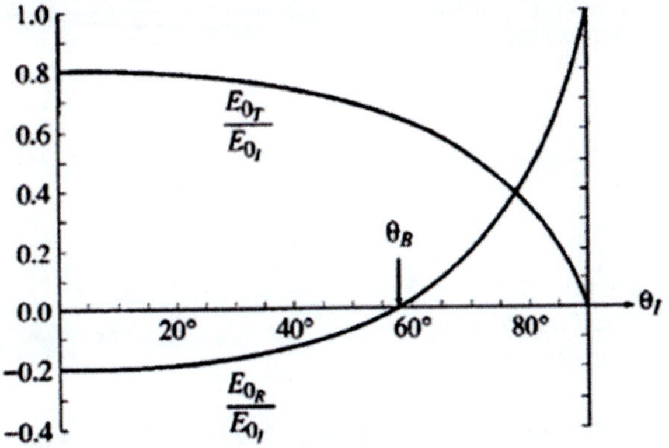

Fig. 4.23 Plot of the transmitted and reflected amplitude [1]

$$\tan \theta_B \cong \frac{n_2}{n_1} \tag{4.86}$$

Figure 4.23 shows a plot of the transmitted and reflected amplitudes as functions of eI, for light incident on glass ($n_2 = 1.5$) from air ($n_1 = 1.0$). (On the graph, a negative number indicates that the wave is 180^0 out of phase with the incident beam—the amplitude itself is the absolute value.)

The power per unit area striking the interface is $\vec{S} \cdot \hat{\vec{z}}$. Thus, the incident intensity is

$$I_I = \frac{1}{2}\varepsilon_1 v_1 E_{0_I}^2 \cos \theta_I \tag{4.87}$$

while the reflected and transmitted intensities are

$$I_I = \frac{1}{2}\varepsilon_1 v_1 E_{0_I}^2 \cos \theta_I \text{and} I_T = \frac{1}{2}\varepsilon_2 v_2 E_{0_I}^2 \cos \theta_T \tag{4.88}$$

(The cosines are there because we are talking about the average power per unit area of interface, and the *interface* is at an angle to the wave front.) [1]. The reflection and transmission coefficients for waves polarized parallel to the plane of incidence are:

$$R \equiv \frac{I_R}{I_I} = \left(\frac{E_{0_R}}{E_{0_I}}\right)^2 = \left(\frac{\alpha - \beta}{\alpha + \beta}\right)^2 \tag{4.89}$$

$$T \equiv \frac{I_T}{I_I} = \frac{\varepsilon_2 v_2}{\varepsilon_1 v_1}\left(\frac{E_{0_T}}{E_{0_I}}\right)^2 \frac{\cos \theta_T}{\cos \theta_I} = \alpha\beta\left(\frac{2}{\alpha + \beta}\right)^2 \tag{4.90}$$

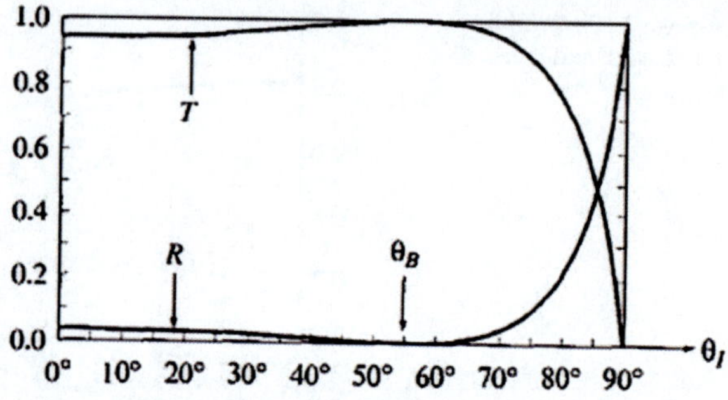

Fig. 4.24 Plot of reflection and transmission coefficient for polarized waves

They are plotted as functions of the angle of incidence in Fig. 4.24 (for the air/glass interface). R is the fraction of the incident energy that is reflected—naturally, it goes to 0 at Brewster's angle; T is the fraction transmitted—it goes to 1 at θ_B. Note that $R + T = 1$ as required by conservation of energy: the energy per unit time *reaching* a particular patch of area on the surface is equal to the energy per unit time *leaving* the patch [1].

Example 4.6 Suppose $Ae^{iax} + Be^{ibx} = Ce^{icx}$, for some nonzero constant A, B, C, a, b, c, and for all x. Prove that $a = b = c$ and $A + B = C$.

Solution $Ae^{iax} + Be^{ibx} = Ce^{icx}$, for all x, so (using $x = 0$), $A + B = C$.

Differentiate: $iaAe^{iax} + ibBe^{ibx} = icCe^{icx}$, so (using $x = 0$), $aA + bB = cC$.

Differentiating again: $-a^2Ae^{iax} - b^2Be^{ibx} = -c^2Ce^{icx}$, so (using $x = 0$), $a^2A + b^2B = c^2C$.

$a^2A + b^2B = c(cC) = c(aA + bB)$, or $(A + B)(a^2A + b^2B) = (A + B)c(aA + bB) = cC(aA + bB)$, or

$a^2A + b^2AB + a^2AB + b^2B^2 = (aA + bB)^2 = a^2A^2 + 2abAB + b^2B^2$, or

$(a^2 + b^2 - 2ab)AB = 0$, or $(a - b)^2AB = 0$. But A and B are nonzero, so $a = b$. Therefore $(A + B)e^{iax} = Ce^{icx}$. $a(A + B) = cC$, or $aC = cC$, so (since $C \neq 0$) $a = c$. Conclusion: $a = b = c$ is driven.

Example 4.7 Analyze the case of polarization *perpendicular* to the plane of incidence (i.e., electric fields in the y direction, in Fig. 4.25). Impose the boundary condition Eq. 4.75, and obtain the Fresnel equations for $\vec{\tilde{E}}_{0_R}$ and $\vec{\tilde{E}}_{0_T}$. Sketch $\left(\vec{\tilde{E}}_{0_R}/\vec{\tilde{E}}_{0_I}\right)$ and $\left(\vec{\tilde{E}}_{0_T}/\vec{\tilde{E}}_{0_I}\right)$ as functions of, for the case $\beta = n_2/n_1 = 1.5$. (Note that for this β, the reflected wave is always 180^0 out of phase). Show that there is no Brewster's angle for *any* n_1 and n_2: $\vec{\tilde{E}}_{0_R}$ is *never* 0 (unless, of course, $n_1 = n_2$ and $\mu_1 = \mu_2$, in which case, the two media are optically indistinguishable). Confirm that your Fresnel equations reduce to the proper forms at normal incidence. Compute the reflection and transmission coefficients, and check that they add up to 1.

Fig. 4.25 Example 4.7

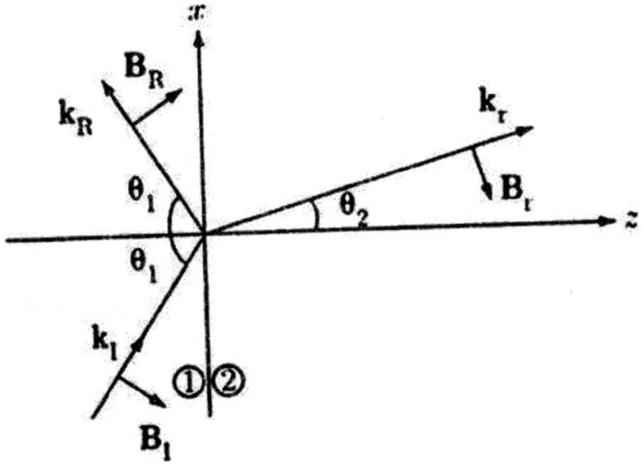

Solution We start with the following relationships:

$$\left\{\begin{array}{l}\widetilde{\vec{E}}_I=\widetilde{E}_{0_I}e^{i\left(\vec{k}_I\cdot\vec{r}-\omega t\right)}\widehat{\vec{y}}\\ \widetilde{\vec{B}}_I=\frac{1}{v_1}\widetilde{E}_{0_I}e^{i\left(\vec{k}_I\cdot\vec{r}-\omega t\right)}\left(-\cos\theta_1\widehat{\vec{x}}+\sin\theta_1\widehat{\vec{z}}\right)\end{array}\right\}$$

$$\left\{\begin{array}{l}\widetilde{\vec{E}}_R=\widetilde{E}_{0_R}e^{i\left(\vec{k}_R\cdot\vec{r}-\omega t\right)}\widehat{\vec{y}}\\ \widetilde{\vec{B}}_I=\frac{1}{v_1}\widetilde{E}_{0_R}e^{i\left(\vec{k}_R\cdot\vec{r}-\omega t\right)}\left(-\cos\theta_1\widehat{\vec{x}}+\sin\theta_1\widehat{\vec{z}}\right)\end{array}\right\}$$

$$\left\{\begin{array}{l}\widetilde{\vec{E}}_T=\widetilde{E}_{0_T}e^{i\left(\vec{k}_T\cdot\vec{r}-\omega t\right)}\widehat{\vec{y}}\\ \widetilde{\vec{B}}_T=\frac{1}{v_2}\widetilde{E}_{0_R}e^{i\left(\vec{k}_T\cdot\vec{r}-\omega t\right)}\left(-\cos\theta_2\widehat{\vec{x}}+\sin\theta_2\widehat{\vec{z}}\right)\end{array}\right\}$$

Boundary conditions:

$$\left\{\begin{array}{ll}\text{(i)}\quad \varepsilon_1E_1^{\perp}=\varepsilon_2E_2^{\perp} & \text{(iii)}\ E_1^{|}=E_2^{|}\\ \text{(ii)}\quad B_1^{\perp}=B_2^{\perp} & \text{(iv)}\ \frac{1}{\mu_1}B_1^{|}=\frac{1}{\mu_1}B_2^{|}\end{array}\right\}$$

Law of refractions:

$\frac{\sin\theta_2}{\sin\theta_1}=\frac{v_2}{v_1}$. [Note: $\vec{k}_I\cdot\vec{r}-\omega t=\vec{k}_R\cdot\vec{r}-\omega t=\vec{k}_T\cdot\vec{r}-\omega t$, at $z=0$, so we can drop all exponential factors in applying the boundary conditions].

Boundary condition (i): $0=0$ (trivial).

Boundary condition (iii): $\widetilde{E}_{0_I}+\widetilde{E}_{0_R}=\widetilde{E}_{0_T}$.

Boundary condition (ii): $\frac{1}{v_1}\widetilde{E}_{0_I}\sin\theta_1+\frac{1}{v_1}\widetilde{E}_{0_R}\sin\theta_1=\frac{1}{v_2}\widetilde{E}_{0_T}\sin\theta_2\Rightarrow$ $\widetilde{E}_{0_I}+\widetilde{E}_{0_R}=\left(\frac{v_1\sin\theta_2}{v_2\sin\theta_1}\right)\widetilde{E}_{0_T}$.

But the term in parentheses is 1, by the law of refraction, so this is the same as (ii).

Boundary condition (iv): $\frac{1}{\mu_1}\left[\frac{1}{v_1}\widetilde{E}_{0_I}(-\cos\theta_1)+\frac{1}{v_1}\widetilde{E}_{0_R}(-\cos\theta_1)\right]=\frac{1}{\mu_2 v_2}\widetilde{E}_{0_T}(-\cos\theta_2)$ or we can write$\widetilde{E}_{0_I}-\widetilde{E}_{0_R}=\left(\frac{\mu_1 v_1\sin\theta_2}{\mu_2 v_2\sin\theta_1}\right)\widetilde{E}_{0_T}$. Let $\alpha\equiv\frac{\cos\theta_2}{\cos\theta_1}$ and $\beta\equiv\frac{\mu_1 v_1}{\mu_2 v_2}$. Then we have $\widetilde{E}_{0_I}-\widetilde{E}_{0_R}=\alpha\beta\widetilde{E}_{0_T}$. Solving for $\widetilde{E}_{0_R}$ and $\widetilde{E}_{0_T}$: $2\widetilde{E}_{0_I}=(1+\alpha\beta)\widetilde{E}_{0_T}$ or $\widetilde{E}_{0_T}=\left(\frac{2}{1+\alpha\beta}\right)\widetilde{E}_{0_I}$ and $\widetilde{E}_{0_R}=\widetilde{E}_{0_T}-\widetilde{E}_{0_I}=\left(\frac{2}{1+\alpha\beta}-\frac{1+\alpha\beta}{1+\alpha\beta}\right)\widetilde{E}_{0_I}$ then we have $\widetilde{E}_{0_R}=\left(\frac{1-\alpha\beta}{1+\alpha\beta}\right)\widetilde{E}_{0_I}$. Since α and β are positive, it follows that $2/(1+\alpha\beta)$ is positive, and hence the *transmitted* wave is in *phase* with the incident wave, and the (real) amplitudes are related by $E_{0_T}=\left(\frac{2}{1+\alpha\beta}\right)E_{0_I}$. The *reflected* wave is in phase if $\alpha\beta<1$ and 180^0 out of phase if $\alpha\beta<1$; the (real) amplitudes are related by $E_{0_R}=\left(\frac{1-\alpha\beta}{1+\alpha\beta}\right)E_{0_I}$. These are the *Fresnel Equations* for polarization perpendicular to the plane of incidence. To construct the graphs, note that$\alpha\beta=\beta\frac{\sqrt{2.25-\sin^2\theta}}{\cos\theta}$, where θ is the angle of incidence, so, for $\beta=1.5$ and $\alpha\beta=\beta\frac{\sqrt{1-\sin^{-2}\theta}}{\cos\theta}=\frac{\sqrt{\beta^2-\sin^2\theta}}{\cos\theta}$. (Fig. 4.26)

Is there a Brewster's angle?. Well, $E_{0_R}=0$ would mean that $\alpha\beta=1$, and hence that $\alpha=\frac{\sqrt{1-(v_2/v_1)^2\sin^2\theta}}{\cos\theta}=\frac{1}{\beta}=\frac{\mu_2 v_2}{\mu_1 v_1}$ or , so $1-\left(\frac{v_2}{v_1}\right)^2\sin^2\theta=\left(\frac{\mu_2 v_2}{\nu_1 v_1}\right)^2\cos^2\theta$. Since $\mu_1\approx\mu_2$, this means $1\approx(v_2/v_1)^2$, which is

only true for optically indistinguishable media, in which case, there is of *course* no reflection—but that would be true at *any* angle, not just at a special "Brewster's angle." [If μ_2 were substantially different from μ_1, and the relative velocities were just right, it *would* be possible to get a Brewster's angle for this case at:

$$\left(\frac{v_2}{v_1}\right)^2=1-\cos^2\theta+\left(\frac{\mu_2}{\mu_1}\right)^2\cos^2\theta$$

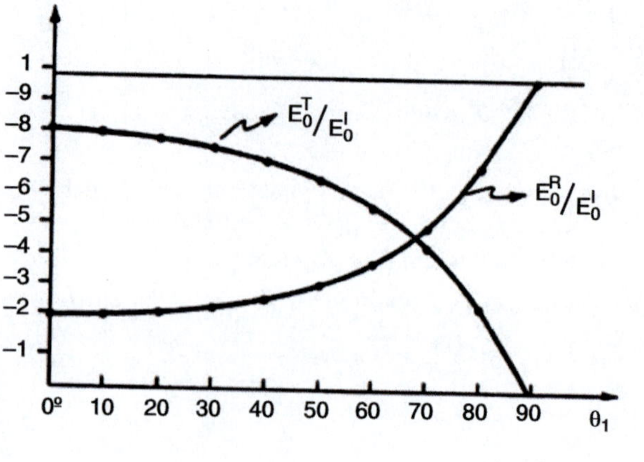

Fig. 4.26 Example of 4.7 solution image

$$\Rightarrow \cos^2\theta = \frac{(v_1/v_2)^2 - 1}{(\mu_2/\mu_1)^2 - 1} = \frac{(\mu_2\varepsilon_2/\mu_1\varepsilon_1) - 1}{(\mu_2/\mu_1)^2 - 1}$$
$$= \frac{(\varepsilon_2/\varepsilon_1) - (\mu_1/\mu_2)}{(\mu_2/\mu_1) - (\mu_1/\mu_2)}$$

But the media would be very peculiar. By the same token, δ_R is either always 0 or always π, for a given interface—it does not switch over as you change θ, the way it does for polarization in the plane of incidence. In particular, if $\beta = 3/2$, then $\alpha\beta > 1$, for $\alpha\beta = \frac{\sqrt{1-\sin^2\theta}}{\cos\theta} > 1$ if $2.25 - \sin^2\theta > \cos^2\theta$, or $2.25 > \sin^2\theta + \cos^2\theta = 1.0$. In general, for $\beta > 1$, $\alpha\beta > 1$, and hence $\delta_R = \pi$. For $\beta < 1$, $\alpha\beta < 1$, and $\delta_R = 0$. At *normal incidence*, $\alpha = 1$, so Fresnel's equations reduce to $E_{0_T} = \left(\frac{2}{1+\beta}\right)E_{0_I}$ and $E_{0_R} = \left(\frac{1-\beta}{1+\beta}\right)E_{0_I}$ consistent with Eq. 4.56. *Reflection* and *Transmission Coefficients*: $R = \left(\frac{E_{0_R}}{E_{0_I}}\right)^2 = \left(\frac{1-\alpha\beta}{1+\alpha\beta}\right)^2$. Referring to Eq. 4.90, $T = \frac{\varepsilon_2 v_2}{\varepsilon_1 v_1}\alpha\left(\frac{E_{0_T}}{E_{0_I}}\right)^2 = \alpha\beta\left(\frac{2}{\alpha+\beta}\right)^2$, therefore

$$R + T = \frac{(1-\alpha\beta)^2 + 4\alpha\beta}{(1+\alpha\beta)^2} = \frac{1 - 2\alpha\beta + \alpha^2\beta^2 + 4\alpha\beta}{(1+\alpha\beta)^2} = \frac{(1+\alpha\beta)^2}{(1+\alpha\beta)^2} = 1$$

Example 4.8 The index of refraction of diamond is 2.42. Construct the graph analogous to Fig. 4.27 for the air/diamond interface. (Assume $\mu_1 = \mu_2 = \mu_0$.) In particular, calculate (a) the amplitudes at normal incidence, (b) Brewster's angle, and (c) the "crossover" angle, at which the reflected and transmitted amplitudes are equal.

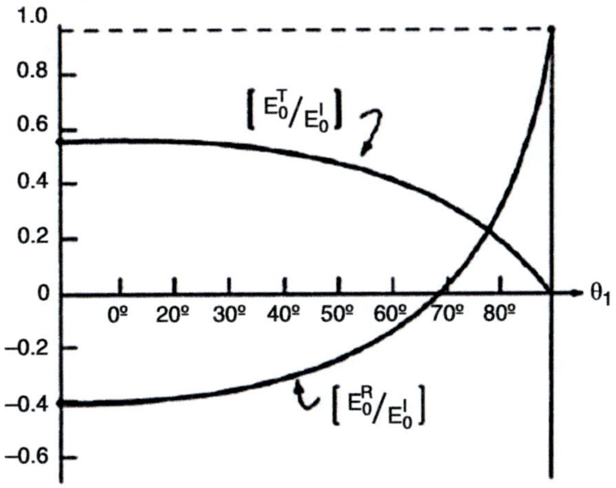

Fig. 4.27 Example of 4.8

Solution In Eq. 4.80 we see that $\beta = 2.42$, Eq. 4.84 $\alpha = \frac{\sqrt{1-(\sin\theta/2.42)^2}}{\cos\theta}$

(a) $\theta = 0 \Rightarrow \alpha = 1$. Equation 9.109 $\Rightarrow \left(\frac{E_{0_R}}{E_{0_I}}\right) = \frac{\alpha-\beta}{\alpha+\beta} = \frac{1-2.42}{1+2.42} = -\frac{1.42}{3.42} = -0.415$

$\left(\frac{E_{0_T}}{E_{0_I}}\right) = \frac{2}{\alpha+\beta} = \frac{2}{3.42} = 0.585$

(b) Eq. 4.86 $\Rightarrow \theta_B \tan^{-1}(2.42) = 67.5^0$

(c) $E_{0_R} = E_{0_T} \Rightarrow \alpha - \beta = 2; \alpha = \beta + 2 = 4.42$ and $(4.42)^2\cos^2\theta = 1 - \sin^2\theta/(2.42)^2$

$(4.42)^2(1 - \sin^2\theta) = (4.42)^2 - (4.42)^2\sin^2\theta = 1 - 0.171\sin^2\theta$

$19.5 - 1 = (19.5 - 017)\sin^2\theta$

$18.5 = 19.3\sin^2\theta; \sin^2\theta = 18.5/19.3 = 0.959 \Rightarrow \sin\theta = 0.979; \theta = 78.3^0$

4.10 Absorption and Dispersion

Expressions for the stored energy, energy flow, and power dissipated are derived for electromagnetic waves in terms of the complex permittivity $\tilde{\varepsilon}$ and permeability $\tilde{\mu}$ for a frequency-dispersive absorbing medium. This is shown to be possible when $\tilde{\varepsilon}$ and $\tilde{\mu}$ are known functions not only of frequency but also of all the loss factors—e.g., collision frequencies, etc. The derivation is not restricted to media with small losses.

In physics and electrical engineering, dispersion most often refers to frequency-dependent effects in wave propagation. Note, however, that there are several other uses of the word "dispersion" in the physical sciences. In the presence of dispersion, wave velocity is no longer uniquely defined, giving rise to the distinction of phase velocity and group velocity. A well-known effect of phase velocity dispersion is the color dependence of light refraction that can be observed in prisms and rainbows.

Dispersion relations describe the interrelations of wave properties like wavelength, frequency, velocities, refraction index, and attenuation coefficient. Besides geometry- and material-dependent dispersion relations, there are the overarching Kramers–Kronig relations that connect the frequency dependences of propagation and attenuation.

Dispersion may be caused either by geometric boundary conditions (waveguides, shallow water) or by interaction of the waves with the transmitting medium.

4.11 Electromagnetic Waves in Conductors

Previous sections were stipulating that the free charge density ρ_f and the free current density $\vec{J}_f$ are zero, and everything that followed was predicated on that assumption. Such a restriction is perfectly reasonable when you're talking about wave propagation through a vacuum or through insulating materials such as glass or (pure) water.

But in the case of conductors, we do not independently control the flow of charge, and in general $\vec{J}_f$ is certainly not 0. In fact, according to Ohm's law, the (free) current density in a conductor is proportional to the electric field:

$$\vec{J}_f = \sigma\vec{E} \tag{4.91}$$

With this, Maxwell's equations for linear media assume the form:

$$\left.\begin{array}{ll} \text{(i)}\nabla\cdot\vec{E} = \frac{1}{\varepsilon}\rho_f & \text{(iii)}\nabla\text{x}\,\vec{E} = -\frac{\partial\vec{B}}{\partial t} \\ \text{(ii)}\nabla\cdot\vec{B} = 0 & \text{(iv)}\nabla\text{x}\,\vec{B} = \mu\sigma\vec{E} + \mu\varepsilon\frac{\partial\vec{E}}{\partial t} \end{array}\right\} \tag{4.92}$$

Now the continuity equation for free charge:

$$\nabla\cdot\vec{J}_f = -\frac{\partial\rho_f}{\partial t} \tag{4.93}$$

together with Ohm's law and Gauss's law (i), gives:

$$\frac{\partial\rho_f}{\partial t} = -\sigma\left(\nabla\cdot\vec{E}\right) = -\frac{\sigma}{\varepsilon}\rho_f \tag{4.94}$$

for a homogeneous linear medium, from which it follows that:

$$\rho_f(t) = e^{-(\sigma/\varepsilon)t}\rho_f(0) \tag{4.95}$$

Thus, any initial free charge density $\rho_f(0)$ dissipates in a characteristic time $\tau \equiv \varepsilon/\sigma$. This reflects the familiar fact that if you put some free charge on a conductor, it will flow out to the edges. The time constant τ affords a measure of how "good" a conductor is: For a "perfect" conductor $\sigma = \infty$ and $\tau = 0$; for a "good" conductor, τ is much less than the other relevant times in the problem (in oscillatory systems, that means $\tau \ll 1/\omega$; for a "poor" conductor, τ is *greater* than the characteristic times in the problem ($\tau \gg 1/\omega$). D. Griffith[1] points out that for good conductor conductors, is absurdly short (10^{-19} s, for copper, whereas the time between collisions is $\tau_c = 10^{-14}$s). The problem is that Ohm's law itself breaks down on time scales shorter than τ_c; actually, the time it takes free charge to dissipate in a good conductor is of the order τ_c, not T. Moreover shows that it takes even longer for the fields and currents to equilibrate. But none of this is relevant to our present purpose; the free charge density in a conductor does eventually dissipate, and exactly how long the process takes is beside the point.

At present we're not interested in this transient behavior—we'll wait for any accumulated free charge to disappear. From then on $\rho_f = 0$, and we have:

$$\left.\begin{array}{ll} \text{(i)}\nabla \cdot \vec{E} = 0 & \text{(iii)}\nabla \text{x}\, \vec{E} = -\dfrac{\partial \vec{B}}{\partial t} \\ \text{(ii)}\nabla \cdot \vec{B} = 0 & \text{(iv)}\nabla \text{x}\, \vec{B} = \mu\sigma\vec{E} + \mu\varepsilon\dfrac{\partial \vec{E}}{\partial t} \end{array}\right\} \quad (4.96)$$

These differ from the corresponding equations for non-conducting media (Eq. 4.45) only in the addition of the last term in (iv). Applying the curl to (iii) and (iv), as before, we obtain modified wave equations for $\vec{E}$ and $\vec{B}$:

$$\nabla\vec{E} = \mu\varepsilon\frac{\partial^2\vec{E}}{\partial t^2} + \mu\sigma\frac{\partial\vec{E}}{\partial t}\text{and}\nabla\vec{B} = \mu\varepsilon\frac{\partial^2\vec{B}}{\partial t^2} + \mu\sigma\frac{\partial\vec{B}}{\partial t} \quad (4.97)$$

These equations still admit plane-wave solutions:

$$\tilde{\vec{E}}(z,t) = \tilde{\vec{E}}_0 e^{i\left(\tilde{k}z-\omega t\right)}\text{and}\tilde{\vec{B}}(z,t) = \tilde{\vec{B}}_0 e^{i\left(\tilde{k}z-\omega t\right)} \quad (4.98)$$

but this time the "wave number" $\tilde{k}$ is complex:

$$\tilde{k}^2 = \mu\varepsilon\omega^2 + i\,\mu\,\sigma\omega \quad (4.99)$$

as you can easily check by plugging Eq. 4.98 into Eq. 4.97. Taking the square root:

$$\tilde{k} = k + i\kappa \quad (4.100)$$

where:

$$k \equiv \omega\sqrt{\frac{\varepsilon\mu}{2}}\left[\sqrt{1+\left(\frac{\sigma}{\varepsilon\omega}\right)^2}+1\right]^{1/2}\text{and}\kappa \equiv \omega\sqrt{\frac{\varepsilon\mu}{2}}\left[\sqrt{1+\left(\frac{\sigma}{\varepsilon\omega}\right)^2}-1\right]^{1/2} \quad (4.101)$$

The imaginary part of k results in an attenuation of the wave (decreasing amplitude with increasing z):

$$\tilde{\vec{E}}(z,t) = \tilde{\vec{E}}_0 e^{-\kappa z} e^{i(kz-\omega t)}\text{and}\tilde{\vec{B}}(z,t) = \tilde{\vec{B}}_0 e^{-\kappa z} e^{i(kz-\omega t)} \quad (4.102)$$

The distance it takes to reduce the amplitude by a factor of $1/e$ (about a third) is called the *skin depth*:

$$d \equiv \frac{1}{\kappa} \tag{4.103}$$

It is a measure of how far the wave penetrates into the conductor [1]. Meanwhile, the real part of $\tilde{k}$ determines the wavelength, the propagation speed, and the index of refraction, in the usual way:

$$\lambda = \frac{2\pi}{k}, v = \frac{\omega}{k}, n = \frac{ck}{\omega} \tag{4.104}$$

The attenuated plane waves (Eq. 4.104) satisfy the modified wave (Eq. 4.97) for any $\tilde{\vec{E}}_0$ and $\tilde{\vec{B}}_0$. But Maxwell's equations (4.96) impose further constraints, which serve to determine the relative amplitudes, phases, and polarizations of $\vec{E}$ and $\vec{B}$. As before, (i) and (ii) rule out any z components: the fields are transverse. We may as well orient our axes so that $\vec{E}$ is polarized along the x-direction:

$$\tilde{\vec{E}}(z,t) = \tilde{E}_0 e^{-\kappa z} e^{i(kz-\omega t)} \hat{\vec{x}} \tag{4.105}$$

Then (iii) gives:

$$\tilde{\vec{B}}(z,t) = \frac{\tilde{k}}{\omega} \tilde{E}_0 e^{-\kappa z} e^{i(kz-\omega t)} \hat{\vec{y}} \tag{4.106}$$

Equation (iv) says the same thing. Once again, the electric and magnetic fields are mutually perpendicular. Like any complex number, $\tilde{k}$ can be expressed in terms of its modulus and phase [1]:

$$k = Ke^{i\phi} \tag{4.107}$$

where:

$$K = |\tilde{k}| = \sqrt{k^2 + \kappa^2} = \omega\sqrt{\varepsilon\mu\sqrt{1 + \left(\frac{\sigma}{\varepsilon\omega}\right)^2}} \tag{4.108}$$

and:

$$\phi \equiv \tan^{-1}(\kappa/k)$$

Fig. 4.28 Presentation of electric and magnetic fields

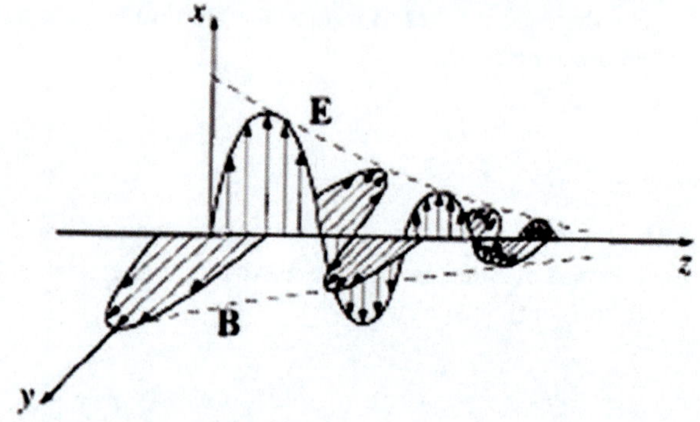

According to Eqs. 4.105 and 4.106, the complex amplitudes $\widetilde{E}_0 = E_0 e^{i\delta_E}$ and $\widetilde{B}_0 = B_0 e^{i\delta_B}$ are related by:

$$B_0 e^{i\delta_B} = \frac{K e^{i\phi}}{\omega} e^{i\delta_B} \tag{4.109}$$

Evidently the electric and magnetic fields are no longer in phase; in fact,

$$\delta_B - \delta_E = \phi \tag{4.110}$$

The magnetic field *lags behind* the electric filed [1]. Meanwhile, the (real) amplitudes of $\vec{E}$ and $\vec{B}$ are related by:

$$\frac{B_0}{E_0} = \frac{K}{\omega} = \sqrt{\varepsilon\mu\sqrt{1 + \left(\frac{\sigma}{\varepsilon\omega}\right)^2}} \tag{4.111}$$

The (real) electric and magnetic fields are, finally,

$$\left.\begin{aligned} \vec{E}(z,t) &= E_0 e^{-\kappa z} \cos(kz - \omega t + \delta_E)\widehat{\vec{x}} \\ \vec{B}(z,t) &= B_0 e^{-\kappa z} \cos(kz - \omega t + \delta_E + \phi)\widehat{\vec{y}} \end{aligned}\right\} \tag{4.112}$$

These fields are shown in Fig. 4.28 below.

Example 4.9 This example aims to do the following:

(a) Suppose you imbedded some free charge in a piece of glass. About how long would it take for the charge to flow to the surface?
(b) Silver is an excellent conductor, but it's expensive. Suppose you were designing a microwave experiment to operate at a frequency of 1010 Hz. How thick would you make the silver coatings?

Table 4.3 Resistivities in ohm-meters (all values are for 1 atm, 200 C) [7]

Material	Resistivity	Material	Resistivity
Conductors:		*Semiconductors:*	
Silver	1.59×10^{-8}	Salt water (saturated)	4.4×10^{-2}
Copper	1.68×10^{-8}	Germanium	4.6×10^{-1}
Gold	2.21×10^{-8}	Diamond	2.7
Aluminum	2.65×10^{-8}	Silican	2.5×10^{3}
Iron	9.61×10^{-8}	*Insulators:*	
Mercury	9.58×10^{-7}	Water (pure)	2.5×10^{5}
Nichrome	1.00×10^{-6}	Wood	10^{8}–10^{11}
Manganese	1.44×10^{-6}	Glass	10^{10}–10^{14}
Graphite	1.4×10^{-5}	Quartz (fused)	~10^{16}

(c) Find the wavelength and propagation speed in copper for radio waves at I MHz. Compare the corresponding values in air (or vacuum).

Solution

(a) Equation 4.95$\Rightarrow \tau = \varepsilon/\sigma$. But $\varepsilon = \varepsilon_0 \varepsilon_r$ and from Eq. 4.48, we have $\varepsilon_r \cong n^2$, and for glass the index of refraction is typically around 1.5, so $\varepsilon \simeq (1.5)^2$ x 8.85 x $10^{-12} C^2/Nm^2$, while $\sigma = 1/\rho \simeq 10^{-12} \Omega m$ (Table 4.3 below). Then $\tau = (2 \text{ x } 10^{-11})/10^{-12} = 20$ sec. (But the resistivity of glass varies enormously from one type to another, so this answer could be off by a factor of 100 in either direction).

(b) For silver, $\rho = 1.59$ x 10^{-8} (Table 4.3), and $\varepsilon \approx \varepsilon_0$, so $\omega\varepsilon = 2\pi \text{ x } 10^{10} \text{ x } 8.85 \text{ x } 10^{-12} = 0.56$. Since $\sigma = 1/\rho = 6.25 \text{ x } 10^7 \Omega m \gg \omega\varepsilon$, the skin depth (Eq. 4.105) is $d = \frac{1}{\kappa} \cong \sqrt{\frac{2}{\omega\sigma\mu}} = \sqrt{\frac{2}{2\pi \text{ x } 10^{10} \text{ x } 6.25 \text{ x } 10^7 \text{ x } 4\pi \text{ x } 10^{-7}}} =$ $6.4 \text{ x } 10^{-77} m = 6.4 \text{ x } 10^{-4} mm$ we place plate silver to a depth of about 0.001mm; there is no point in making it any thicker, since the fields do not penetrate much beyond this anyway.

(c) For copper, Table 4.3 above gives $\sigma = 1/(1.68 \text{ x } 10^{-8}) \approx 6 \text{ x } 10^7$, $\omega\varepsilon_0 = (2\pi \text{ x } 10^6) \text{ x } (8.85 \text{ x } 10^{-12}) = 8.85 \text{ x } 10^{-5}$. Since $\sigma \gg \omega\varepsilon$, Eq. 4.102-

$$\Rightarrow k \approx 2\pi\sqrt{\frac{2}{\omega\sigma\mu_0}} = 2\pi\sqrt{\frac{2}{2\pi \text{ x } 10^6 \text{ x } 6 \text{ x } 10^7 \text{ x } 4\pi \text{ x } 10^{-7}}} = 4 \text{ x } 10^{-4} m = 0.4 mm$$

From Eq. 4.103, the propagation speed is $\upsilon = \frac{\omega}{k} = \frac{\omega}{2\pi}\lambda = \lambda\nu = (4 \text{ x } 10^{-4}) \text{ x } 10^6 = 400 m/s$. In vacuum, $\lambda = \frac{c}{\nu} = \frac{4 \text{ x } 10^8}{10^6} = 300 m$; $\upsilon = c = 4 \text{ x } 10^8 m/s$. (But really, in a good conductor the skin depth is so small, compared to the wavelength, that the notions of "wavelength" and "propagation speed" lose their meaning.

4.12 Reflection at a Conducting Surface

The boundary conditions we used to analyze reflection and refraction at an interface between two dielectrics do not hold in the presence of free charges and currents. Instead, we have the more general relations (Eq. 4.41) with addition term of surface current $\vec{K}_f$ in (iv) [1]:

$$\left.\begin{array}{ll}\text{(i) } \varepsilon_1 E_1^{\perp} - \varepsilon_2 E_2^{\perp} = 0 & \text{(iii) } E_1^{|} - E_2^{|} = 0 \\ \text{(ii) } B_1^{\perp} - B_2^{\perp} = 0 & \text{(iv) } \frac{1}{\mu_1} B_1^{|} - \frac{1}{\mu_2} B_2^{|} = \vec{K}_f \times \hat{\vec{n}}\end{array}\right\} \tag{4.113}$$

In this case σ_f(not to be confused with conductivity) is the free surface charge, $\vec{K}_f$ the free surface current, and $\hat{\vec{n}}$ (not to be confused with the polarization of the wave) a unit vector perpendicular to the surface, pointing from medium (2) into medium (1). For ohmic conductors $\left(\vec{J}_f = \sigma\vec{E}\right)$, there can be no free surface current, since this would require an infinite electric field at the boundary [1].

Suppose now that the xy-plane forms the boundary between a non-conducting linear medium (1) and a conductor (2). A monochromatic plane wave, traveling in the z-direction and polarized in the x direction, approaches from the left, as in Fig. 4.19:

$$\tilde{\vec{E}}_I(z,t) = \tilde{E}_{0_I} e^{i(k_1 z - \omega t)} \hat{\vec{x}} \text{ and} \tilde{\vec{B}}_I(z,t) = \frac{1}{v_1} \tilde{B}_{0_I} e^{i(k_1 z - \omega t)} \tilde{\vec{y}} \tag{4.114}$$

This incident wave gives rise to a reflected wave [1]:

$$\tilde{\vec{E}}_R(z,t) = \tilde{E}_{0_R} e^{i(-k_1 z - \omega t)} \hat{\vec{x}} \text{ and} \tilde{\vec{B}}_R(z,t) = \frac{1}{v_1} \tilde{B}_{0_R} e^{i(-k_1 z - \omega t)} \tilde{\vec{y}} \tag{4.115}$$

Propagating back to the left in medium (1), and a transmitted wave:

$$\tilde{\vec{E}}_T(z,t) = \tilde{E}_{0_T} e^{i\left(\tilde{k}_2 z - \omega t\right)} \hat{\vec{x}} \text{ and} \tilde{\vec{B}}_T(z,t) = \frac{1}{v_1} \tilde{B}_{0_T} e^{i\left(\tilde{k}_2 z - \omega t\right)} \tilde{\vec{y}} \tag{4.116}$$

which is attenuated as it penetrates into the conductor.

At $z = 0$, the combined wave in medium (1) must join the wave in medium (2), pursuant to the boundary conditions in Eq. 4.113. Since $E^{\perp} = 0$ on both sides, boundary condition (i) yields $\sigma_f = 0$. Since $B^{\perp} = 0$, (ii) is automatically satisfied. Meanwhile, (iii) gives:

$$\widetilde{E}_{0_I} = \widetilde{E}_{0_R} = \widetilde{E}_{0_T} \tag{4.117}$$

and (iv) (with $K_f = 0$) says:

$$\frac{1}{\mu_1 v_1}\left(\widetilde{E}_{0_I} - \widetilde{E}_{0_R}\right) - \frac{\widetilde{k}_2}{\mu_2 \omega}\widetilde{E}_{0_T} = 0 \tag{4.118}$$

or:

$$\widetilde{E}_{0_I} - \widetilde{E}_{0_R} = \widetilde{\beta}\widetilde{E}_{0_T} \tag{4.119}$$

where:

$$\widetilde{\beta} \equiv \frac{\mu_1 v_1}{\mu_2 \omega}\widetilde{k}_2 \tag{4.120}$$

It follows that:

$$\widetilde{E}_{0_R} = \left(\frac{1-\widetilde{\beta}}{1+\widetilde{\beta}}\right)\widetilde{E}_{0_I} \text{and} \widetilde{E}_{0_T} = \left(\frac{2}{1+\widetilde{\beta}}\right)\widetilde{E}_{0_I} \tag{4.121}$$

These results are formally identical to the ones that apply at the boundary between *nonconductors* (Eq. 4.56), but the resemblance is deceptive since $\widetilde{\beta}$ is now a complex number.

For a *perfect* conductor ($\sigma = \infty$), $k_2 = \infty$ (Eq. 4.101), so $\widetilde{\beta}$ is infinite, and

$$\widetilde{E}_{0_R} = -\widetilde{E}_{0_I} \text{and} \widetilde{E}_{0_T} = 0 \tag{4.122}$$

In this case the wave is totally reflected, with a 180^0 phase shift. (That's why excellent conductors make good mirrors. In practice, you paint a thin coating of silver onto the back of a pane of glass—the glass has nothing to do with the reflection; it's just there to support the silver and to keep it from tarnishing. Since the skin depth in silver at optical frequencies is on the order of 100 A^0, you don't need a very thick layer).

Example 4.10 Calculate the reflection coefficient for light at an air-to-silver interface ($\mu_1 = \mu_2 = \mu_0$, $\varepsilon_1 = \varepsilon_0$, $\sigma = 6 \times 10^7 (\Omega \cdot m)^{-1}$), at optical frequencies ($\omega = 4 \times 10^{15} s^{-1}$)

Solution According to Eq. 4.121, $R = \left|\frac{\widetilde{E}_{0_R}}{\widetilde{E}_{0_I}}\right|^2 = \left|\frac{1-\widetilde{\beta}}{1+\widetilde{\beta}}\right|^2 = \left(\frac{1-\widetilde{\beta}}{1+\widetilde{\beta}}\right)\left(\frac{1-\widetilde{\beta}^*}{1+\widetilde{\beta}^*}\right)$, where $\widetilde{\beta}^*$ is complex conjugate of $\widetilde{\beta}$ and $\widetilde{\beta} = \frac{\mu_1 v_1}{\mu_2 \omega}\widetilde{k}_2 = \frac{\mu_1 v_1}{\mu_2 \omega}(k_2 + i\kappa_2)$ (Eqs. 4.100 and 4.118).

Since silver is a good conductor ($\sigma \gg \varepsilon\omega$), Eq. 4.101 reduces to $\kappa_2 \cong k_2 \cong \omega\sqrt{\frac{\varepsilon_2\mu_2}{2}}\sqrt{\frac{\sigma}{\varepsilon_2\omega}} = \sqrt{\frac{\sigma\omega\mu_2}{2}}$, so $\widetilde{\beta} = \frac{\mu_1 v_1}{\mu_2\omega}\sqrt{\frac{\sigma\omega\mu_2}{2}}(1+i) = \mu_1 v_1\sqrt{\frac{\sigma}{2\mu_2\omega}}(1+i)$. Now let $\gamma \equiv \mu_1 v_1\sqrt{\frac{\sigma}{2\mu_2\omega}} = \mu_0 c\sqrt{\frac{\sigma}{2\mu_0\omega}} = c\sqrt{\frac{\sigma\mu_0}{2\omega}} = (3 \text{ x } 10^8)\sqrt{\frac{(3 \text{ x } 10^7)(4\pi \text{ x } 10^{-7})}{(2)(4 \text{ x } 10^{15})}} = 29$

$R = \left(\frac{1-\gamma-i\gamma}{1+\gamma+i\gamma}\right)\left(\frac{1-\gamma+i\gamma}{1+\gamma-i\gamma}\right) = \frac{(1-\gamma)^2+\gamma^2}{(1+\lambda)^2+\gamma^2}$. Evidently 93% of the light is reflected.

4.13 The Frequency Dependence of Permittivity

In the preceding sections, we have seen that the propagation of electromagnetic waves through matter is governed by three properties of the material, which we took to be constants, and they are as follows:

1. The permittivity ε
2. The permeability μ
3. The conductivity σ

Actually, each of these parameters depends to some extent on the frequency of the waves you are considering. Indeed, if the permittivity were *truly* constant, then the index of refraction in a transparent medium, $n \cong \sqrt{\varepsilon_r}$, would also be constant. But it is well-known from optics that n is a function of wavelength (Fig. 4.29 shows the graph for a typical glass). A prism or a raindrop bends blue light more sharply than red and spreads white light out into a rainbow of colors. This phenomenon is called *dispersion*. By extension, whenever the speed of a wave depends on its frequency, the supporting medium is called *dispersive*. Conductors, incidentally, are dispersive; see Eqs. 4.101 and 4.103.

Fig. 4.29 Graph for typical class

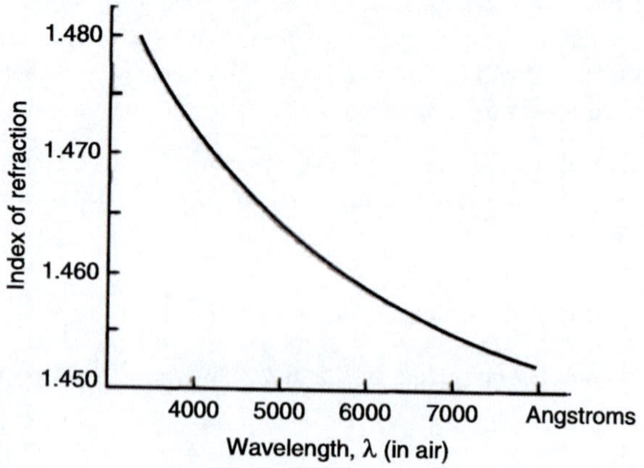

Because waves of different frequencies travel at different speeds in a dispersive medium, a wave form that incorporates a range of frequencies will change shape as it propagates. A sharply peaked wave typically flattens out, and whereas each sinusoidal component travels at the ordinary *wave* (or *phase*) velocity,

$$v = \frac{\omega}{k} \tag{4.123}$$

the packet as a whole (the "envelope") travels at the so-called *group velocity* [8]. Also, for more information, refer to Appendix G of this book.

$$v_g = \frac{d\omega}{dk} \tag{4.124}$$

Figure 4.30 is a typical depiction of these two wave velocities.

Our purpose in this section is to account for the frequency dependence of ε in nonconductors, using a simplified model for the behavior of electrons in dielectrics. The electrons in nonconductor are bound to specific molecules [1]. The actual binding forces can be quite complicated, but we shall picture each electron as attached to the end of an imaginary spring, with force constant k_{spring} (Fig. 4.31).

$$F_{binding} = -k_{spring}x = -m\omega_0^2 x \tag{4.125}$$

where x is displacement from equilibrium, m is the electron's mass, and ω_0 is the natural oscillation frequency, $\sqrt{k_{spring}/m}$. Utilizing Taylor series and expanding the potential energy about equilibrium point for sufficiently small displacement we have [1];

$$U(x) = U(0) + xU'(0) + \frac{1}{2}x^2U''(0) + \cdots$$

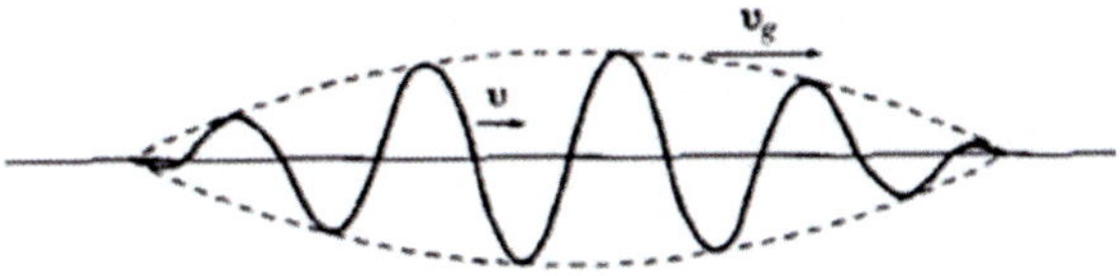

Fig. 4.30 Phase and group velocity[1]

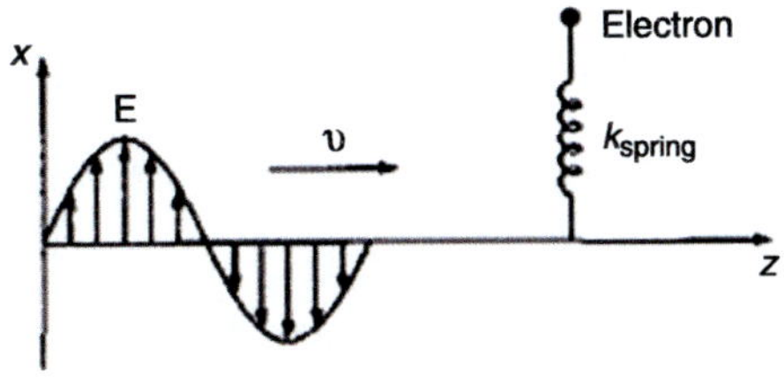

Fig. 4.31 Electron movement[1]

The first term is a constant, with no dynamical significance (you can always adjust the zero of potential energy so that $U(0) = 0$). The second term automatically vanishes, since $dU/dx = -F$, and by the nature of an equilibrium the force at that point is zero. The third term is precisely the potential energy of a spring with force constant $k_{spring} = d^2U/dx^2|_0$ (the second derivative is positive, for a point of stable equilibrium). As long as the displacements are small, the higher terms in the series can be neglected [10]. Meanwhile, there will presumably be some damping force on the electron:

$$F_{damping} = -m\gamma \frac{dx}{dt} \tag{4.126}$$

The damping must be opposite in direction to the velocity, and making it *proportional* to the velocity is the easiest way to accomplish this. The *cause* often damping does not concern us here—among other things, an oscillating charge radiates, and the radiation siphons off energy.

In the presence of an electromagnetic wave of frequency ω, polarized in the x (Fig. 4.28) direction, the electron is subject to a driving force:

$$F_{driving} = qE = qE_o \cos(\omega t) \tag{4.127}$$

where q is the charge of the electron and E_0 is the amplitude of the wave at the point z where the electron is situated. If we concentrate only at one point and one point only so that the maximum E occurs there at $t = 0$, then by Newton's second law, we have:

$$\begin{aligned} m\frac{d^2x}{dt^2} &= F_{total} = F_{binding} + F_{damping} + F_{driving} \\ m\frac{d^2x}{dt^2} + m\gamma\frac{dx}{dt} + n\omega_0^2 x &= qE_0 \cos(\omega t) \end{aligned} \tag{4.128}$$

Our model, then, describes the electron as a damped harmonic oscillator, driven at frequency ω. (Assumption is that the much more massive nuclei remain at rest.[1])

Equation 4.128 is easier to handle if we regard it as the real part of a *complex equation*:

$$\frac{d^2\tilde{x}}{dt^2} + \gamma\frac{d\tilde{x}}{dt} + \omega_0^2\tilde{x} = \frac{q}{m}E_0 e^{-i\omega t} \tag{4.129}$$

In steady state, the system oscillates at the driving frequency:

$$\tilde{x}(t) = \tilde{x}_0 e^{-i\omega t} \tag{4.130}$$

Inserting this into Eq. 4.129, we obtain:

$$\widetilde{x}_0 = \frac{q/m}{\omega_0^2 - \omega^2 - i\gamma\omega} E_0 \tag{4.131}$$

The dipole moment is the real part of:

$$\widetilde{p}(t) = q\widetilde{x}(t) = \frac{q/m}{\omega_0^2 - \omega^2 - i\gamma\omega} E_0 e^{-i\omega t} \tag{4.132}$$

The imaginary term in the denominator means that p is *out of phase* with E—lagging behind by angle $\tan^{-1}\left[\gamma\omega/\left(\omega_0^2 - \omega^2\right)\right]$ that is very small when $\omega \ll \omega_0$ and rises to π when $\omega \gg \omega_0$.

In general, differently situated electrons within a given molecule experience different natural frequencies and damping coefficients. Let's say there are f_j electrons with frequency ω_j and damping γ_j in each molecule. If there are N molecules per unit volume, the polarization **P** is given by the real part of

$$\widetilde{\vec{\mathbf{P}}} = \frac{Nq^2}{m}\left(\sum_j \frac{f_j}{\omega_j^2 - \omega^2 - i\gamma_j\omega}\right)\widetilde{\vec{\mathbf{E}}} \tag{4.133}$$

This applies directly to the case of a dilute gas; for denser materials, the theory is modified slightly, in accordance with the Clausius–Mossotti Equation. Note that we should not confuse the "polarization" of a medium, $\vec{\mathbf{P}}$, with the "polarization" of a *wave—same word*, but two completely unrelated meanings.

Now we define the electric susceptibility as the proportionality constant between $\vec{\mathbf{P}}$ and $\vec{\mathbf{E}}$ (specifically, $\vec{\mathbf{P}} = \varepsilon_0\chi_e\vec{\mathbf{E}}$) [1]. In the present case $\vec{\mathbf{P}}$ is *not* proportional to $\vec{\mathbf{E}}$ (this is not, strictly speaking, a linear medium) because of the difference in phase. However, the *complex* polarization $\widetilde{\vec{\mathbf{P}}}$ is proportional to the *complex* field $\widetilde{\vec{\mathbf{E}}}$, and this suggests that we introduce a *complex susceptibility*, $\widetilde{\chi}_e$:

$$\vec{\mathbf{P}} = \varepsilon_0\widetilde{\chi}_0\vec{\mathbf{E}} \tag{4.134}$$

From all these we conclude that the physical polarization is the real part of $\widetilde{\vec{\mathbf{P}}}$, just as the physical field is the real part of $\widetilde{\vec{\mathbf{E}}}$. In particular, the proportionality between $\widetilde{\vec{\mathbf{D}}}$ and $\widetilde{\vec{\mathbf{E}}}$ is the *complex permittivity* $\widetilde{\varepsilon} = \varepsilon_0(1 + \widetilde{\chi}_e)$, and the *complex dielectric constant* in this model is

$$\widetilde{\varepsilon}_r = 1 + \frac{Nq^2}{m\varepsilon_0}\sum_j \frac{f_j}{\omega_j^2 - \omega^2 - i\gamma_j\omega} \tag{4.135}$$

Ordinarily, the imaginary term is negligible; however, when ω is very close to one of the resonant frequencies (ω_j), it plays an important role, as we shall see. In a dispersive medium, the wave equation for a given frequency reads:

$$\nabla^2 \vec{\widetilde{\mathbf{E}}} = \widetilde{\varepsilon}\mu_0 \frac{\partial^2 \vec{\widetilde{\mathbf{E}}}}{\partial t^2} \tag{4.136}$$

This provides plane wave solutions, as before;

$$\vec{\widetilde{\mathbf{E}}}(z,t) = \vec{\widetilde{\mathbf{E}}}_0 e^{i\left(\widetilde{k}z-\omega t\right)} \tag{4.137}$$

with the complex wave number

$$\widetilde{k} = \sqrt{\widetilde{\varepsilon}\mu_0}\omega \tag{4.138}$$

Writing $\widetilde{k}$ in terms of its real and imaginary parts,

$$\widetilde{k} = k + i\kappa \tag{4.139}$$

Equation 4.137 becomes

$$\vec{\widetilde{\mathbf{E}}}(z,t) = \widetilde{E}_0 e^{-\kappa z} e^{i(kz-\omega t)} \tag{4.140}$$

Evidently the wave is *attenuated* which is not surprising, since the damping absorbs energy [1]. Because the intensity is proportional to E^2 and hence to $_0e^{-2\kappa z}$, the quantity

$$\alpha \equiv 2\kappa \tag{4.141}$$

is called the *absorption coefficient*. Meanwhile, the wave velocity is ω/k, and the index of refraction is

$$n = \frac{ck}{\omega} \tag{4.142}$$

However, in the present case k and κ have nothing to do with conductivity; rather, they are determined by the parameters of our damped harmonic oscillator. For gases, the second term in Eq. 4.135 is small, and we can approximate the square root (Eq. 4.138) by the first term in the binomial expansion, $\sqrt{1+\varepsilon} \cong 1 + \frac{1}{2}\varepsilon$. Then

$$\widetilde{k} = \frac{\omega}{c}\sqrt{\widetilde{\varepsilon}_r} \cong \frac{\omega}{c}\left[1 + \frac{Nq^2}{2m\varepsilon_0}\sum_j \frac{f_j}{\omega_j^2 - \omega^2 - i\gamma_j\omega}\right] \tag{4.143}$$

or

$$n = \frac{ck}{\omega} \cong 1 + \frac{Nq^2}{m\varepsilon_0}\sum_j \frac{f_j\left(\omega_j^2 - \omega^2\right)}{\left(\omega_j^2 - \omega^2\right)^2 - i\gamma_j^2\omega^2} \tag{4.144}$$

and

$$\alpha = 2\kappa \cong \frac{Nq^2\omega^2}{m\varepsilon_0 c}\sum_j \frac{f_j\gamma_j}{\left(\omega_j^2 - \omega^2\right)^2 - i\gamma_j^2\omega^2} \tag{4.145}$$

Griffiths [1] has plotted the index of refraction and the absorption coefficient in the vicinity of one of the resonances. *Most* of the time the index of refraction *rises* gradually with increasing frequency, consistent with our experience from optics (Fig. 4.26). However, in the immediate neighborhood of a resonance, the index of refraction drops sharply. Because this behavior is atypical, it is called **anomalous dispersion.** Notice that the region of anomalous dispersion ($\omega_1 < \omega < \omega_2$, in the figure) coincides with the region of maximum absorption; in fact, the material may be practically opaque in this frequency range. The reason is that we are now driving the electrons at their "favorite" frequency; the amplitude of their oscillation is relatively large, and a correspondingly large amount of energy is dissipated by the damping mechanism.

In Fig. 4.32, n runs below 1above the resonance, suggesting that the wave speed exceeds c. This is no cause for alarm, since energy does not travel at the wave velocity but rather at the *group velocity* (see Example-1). Moreover, the graph does not include the contributions of other terms in the sum, which add a relatively constant "background" that, in some cases, keeps $n > 1$ on both sides of the resonance.

If you agree to stay away from the resonances, the damping can be ignored, and the formula for the index of refraction simplifies:

$$n = 1 + \frac{Nq^2}{m\varepsilon_0}\sum_j \frac{f_j}{\omega_j^2 - \omega^2} \tag{4.146}$$

For most substances, the natural frequencies ω_j are scattered all over the spectrum in a rather chaotic fashion. But for transparent materials, the nearest significant resonances typically lie in the ultraviolet, so that $\omega < \omega_j$. In that case:

$$\frac{1}{\omega_j^2 - \omega^2} = \frac{1}{\omega_j^2}\left(1 - \frac{\omega^2}{\omega_j^2}\right)^{-1} \cong \frac{1}{\omega_j^2}\left(1 + \frac{\omega^2}{\omega_j^2}\right)$$

and Eq. 4.146 takes the form

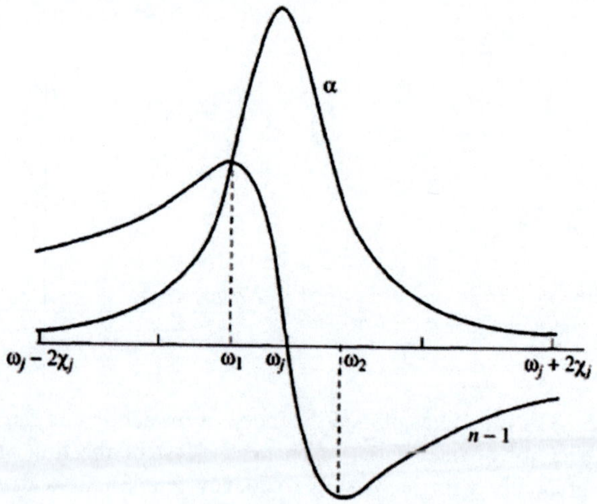

Fig. 4.32 Plot of the index of refraction and absorption coefficient in the vicinity of one of the resonances [1]

$$n = 1 + \left(\frac{Nq^2}{m\varepsilon_0}\sum_j \frac{f_j}{\omega_j^2}\right) + \omega^2\left(\frac{Nq^2}{2m\varepsilon_0}\sum_j \frac{f_j}{\omega_j^4}\right) \tag{4.147}$$

Or, in terms of the wavelength in vacuum ($\lambda = 2\pi c/\omega$):

$$n = 1 + A\left(1 + \frac{B}{\lambda^2}\right) \tag{4.148}$$

This is known as *Cauchy's formula*; the constant A is called the *coefficient of refraction,* and B is called the *coefficient of dispersion*. Cauchy's equation applies reasonably well to most gases, in the optical region.

What Griffiths [1] has described in here is certainly not the complete story of dispersion in nonconducting media. Nevertheless, it does indicate how the damped harmonic motion of electrons can account for the frequency dependence of the index of refraction, and it explains why n is ordinarily a slowly increasing function of ω, with occasional "anomalous" regions where it precipitously drops.

Example 4.11 Assuming negligible damping ($\gamma_j = 0$), calculate the group velocity ($v_g = \frac{d\omega}{dk}$) of the waves described by Eqs. 4.140 and 4.143. Show that $v_g < c$,even when $v < c$.

Solution $k = \frac{\omega}{c}\left[1 + \frac{Nq^2}{2m\varepsilon_0}\sum_j \frac{f_j}{\left(\omega_j^2 - \omega^2\right)^2}\right]$ and$v_g = \frac{d\omega}{dk} = \frac{1}{(dk/d\omega)}$

$$\frac{d\omega}{dk}=\frac{1}{c}\left[1+\frac{Nq^2}{2m\varepsilon_0}\sum_j\frac{f_j}{\left(\omega_j^2-\omega^2\right)^2}+\omega\sum f_j\frac{-(-2\omega)}{\left(\omega_j^2-\omega^2\right)^2}\right]$$

$$=\frac{1}{c}\left(1+\frac{Nq^2}{2m\varepsilon_0}\sum f_j\frac{\left(\omega_j^2+\omega^2\right)}{\left(\omega_j^2-\omega^2\right)^2}\right)$$

$v_g=c\left[1+\frac{Nq^2}{2m\varepsilon_0}\sum f_j\frac{\left(\omega_j^2+\omega^2\right)}{\left(\omega_j^2-\omega^2\right)^2}\right]$. Since the second term in square brackets is *positive*, it follows that $v_g<c$, whereas $v=\frac{\omega}{k}=c\left[1+\frac{Nq^2}{2m\varepsilon_0}\sum\frac{f_j}{\left(\omega_j^2-\omega^2\right)^2}\right]^{-1}$ is greater than c or less than c, depending on ω.

4.14 Electromagnetic (EM) Waves in Conductors

In our previous discussions of EM waves in matter, we have been assuming there are no free charges ($\rho_{free}=\rho_f$) or free currents $\left(\vec{J}_{free}=\vec{J}_f\right)$; the only charges present were the bound charges in the dielectric. When dealing with conductors, there are plenty of electrons that are free to move about when an external electric field is applied. There are still bound charges in a conductor (atoms in a conductor generally only contribute one or two conduction electrons each), but we will find that (in a good conductor) the interaction of the fields with the free charges and currents will dominate over everything else.

In ohmic materials which obey the Ohm's Law, we can write:

$$\vec{J}\left(\vec{r},t\right)=\sigma_c\vec{E}\left(\vec{r},t\right) \tag{4.149}$$

where σ_c is the conductivity of the conducting materials (not to be mistaken for the surface charge density!) and $\sigma_c=1/\rho_c$, which we will take to be uniform inside the conductor (ρ_c = resistivity of the metal conductor with dimension of Ohm-meter). For any free charge that might build up in the conductor due to these currents, we know from charge conservation:

$$\vec{\nabla}\cdot\vec{J}\left(\vec{r},t\right)=-\frac{\partial\rho_f}{\partial t} \tag{4.150}$$

Thus, inside such a conductor, we can assume that the linear/homogeneous/isotropic conducting medium has electric permittivity ε and magnetic permeability μ. Combine this with Gauss Law for electric fields in matter $\vec{\nabla}\cdot\left(\varepsilon\vec{E}\left(\vec{r},t\right)\right)=-\rho_f$, to find how quickly the free charge distribution would dissipate. Bear in your mind that Maxwell's equations are always true, in matter or not; splitting things up into free and bound charges (or for that matter current) can be convenient for solving for the fields, but we do not have to do things this way. Now substituting these relations into Eq. 4.150, we can write:

$$\frac{\partial \rho_f}{\partial t} = \vec{\nabla} \cdot \left(\varepsilon \vec{E}\left(\vec{r}, t\right)\right) = -\frac{\sigma_c}{\varepsilon}\left[\left(\varepsilon \vec{E}\left(\vec{r}, t\right)\right)\right] = -\frac{\sigma_c}{\varepsilon}\rho_f \tag{4.151}$$

or

$$\frac{\partial \rho_f}{\partial t} = -\frac{\sigma_c}{\varepsilon}\rho_f \tag{4.152}$$

The solution to this first-order differential equation is:

$$\rho_f(t) = \rho(0)\exp\left(-\sigma_c t/\varepsilon\right) = \rho\left(\vec{r}, t = 0\right)e^{-t/\tau_{relax}} \tag{4.153}$$

This is damped exponential type function. This also tells us that any free charge that might build up in a conductor will dissipate (due to the mutual repulsion of the charges) with a time constant $\tau_c = \varepsilon/\sigma_c$. The "time constant" represents the amount of time for the initial density to reduce to $1/e(\sim 37\%)$ of its original value. For a perfect conductor ($\sigma_c \to \infty$), the time constant goes to zero, meaning the charge density instantly dissipate. For a "good" conductor (like copper), where the conductivity is typically on the order of 10^8 $(\Omega.\text{m})^{-1}$, the time constant is somewhere around 10^{-19} seconds, which is actually much smaller than the typical time between collisions of the electrons with the atoms making up the conductor, which is $\tau_c \sim 10^{-14}$. For frequencies higher than $1/\tau_c$, Ohm's Law starts to break down, so τ_c is the important time constant here. Therefore, the assumption of dealing with the linear/homogeneous/isotropic conducting medium, is a valid assumption for the rest of this discussion and let us assume we are working with good conductors, and with frequencies that are at or below the optical range (10^{15} Hz, which is pushing the use of Ohm's law, but not so much that our results are no good). In other words, we should have $\omega \ll \sigma_c/\varepsilon$, or $\varepsilon\omega/\sigma_c \ll 1$. It turns out this assumption will make our calculation *much* easier as we go along.

Note that going forward with further analysis of electromagnetic wave through conduction, we write the symbol of σ_c for conductivity of conducting materials as σ_c; now that we know that is the case, it is not indeed the surface charge density. Also, we start using vector notation rather than bold character as well.

Now again assuming a conductor of linear, homogeneous, and isotropic conducting material that has electric permittivity ε and magnetic permeability μ, we substitute $\rho_f = 0$, which means free charges dissipate very rather quick and $\vec{J}\left(\vec{r}, t\right) = \sigma \vec{E}\left(\vec{r}, t\right)$, which also means that $\vec{J}\left(\vec{r}, t\right) \neq 0$, and then we can write Maxwell's equations inside such conductor as follows:

(1) $\vec{\nabla}\cdot\vec{E}\left(\vec{r},t\right)=\frac{\rho\left(\vec{r},t\right)}{\varepsilon}$

(2) $\vec{\nabla}\cdot\vec{B}\left(\vec{r},t\right)=0$

(3) $\vec{\nabla}\text{ x }\vec{E}\left(\vec{r},t\right)=\frac{\partial\vec{B}\left(\vec{r},t\right)}{\partial t}$ and using Ohm's law relation of $\vec{J}\left(\vec{r},t\right)=\sigma\vec{E}\left(\vec{r},t\right)$, we have the fourth equation of Maxwell:

(4) $\vec{\nabla}\text{ x }\vec{B}\left(\vec{r},t\right)=\mu\vec{J}\left(\vec{r},t\right)+\mu\varepsilon\frac{\partial\vec{E}\left(\vec{r},t\right)}{\partial t}=\mu\sigma\vec{E}\left(\vec{r},t\right)+\mu\varepsilon\frac{\partial\vec{E}\left(\vec{r},t\right)}{\partial t}$

Since electric charge is always conserved, thus the continuity equation inside the conductor ends up with a homogeneous differential equation of first order as Eq. 4.152 where the solution is provided as per Eq. 4.153 and depicted in Fig. 4.33, which is a damped exponential curve.

Example 4.12 Calculate the charge relaxation time for copper

Solution Assume the following data for cupper:

$$\rho_{Cu}={}^{1}/_{\sigma_{Cu}}=1.68\text{ x }10^{-8}\Omega-m\Rightarrow\sigma_{Cu}={}^{1}/_{\rho_{Cu}}=5.95\text{ x }10^{7}\text{Siemens/m}$$

If we assume $\varepsilon_{Cu}\approx3\varepsilon_0=3\text{ x }8.85\text{ x }10^{-8}F/m$ for cupper metal, then:

$$\tau_{Cu}^{relax}=\left(\varepsilon_{Cu}/\sigma_{Cu}\right)=\rho_{Cu}\varepsilon_{Cu}=4.5\text{ x }10^{-19}\text{ sec}$$

However, knowing that the characteristic/mean collision time of free electros in pure copper is $\tau_{Cu}^{coll}\simeq\lambda_{Cu}^{coll}/v_{thermal}^{Cu}$ where $\lambda_{Cu}^{coll}\simeq3.9\text{ x }10^{-8}m$ = mean free path (between successive collisions) in pure copper, and $v_{thermal}^{Cu}\simeq\sqrt{3k_BT/m_e}\simeq 12\text{ x }10^{5}m/\text{sec}$ and thus we obtain $\tau_{Cu}^{coll}\simeq3.2\text{ x }10^{-13}\text{ sec}$.

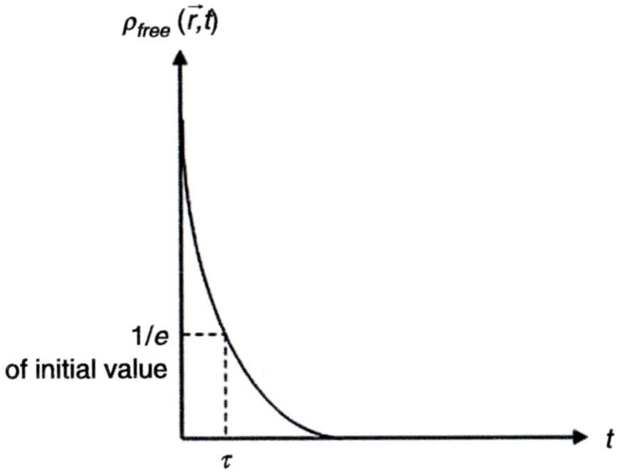

Fig. 4.33 Sketch of free density dissipated vs. characteristic time (solution of Eq. 4.153)

Hence we see that the calculated charge relaxation time in pure copper, $\tau_{Cu}^{relax} \simeq 4.5 \text{ x } 10^{-19}$ sec is much less than the calculated collision in pure copper, $\tau_{Cu}^{coll} \simeq 3.2 \text{ x } 10^{-13}$ sec. Furthermore, the experimentally measured charge relaxation time in pure copper is $\tau_{Cu}^{relax}(experimental) \simeq 4.0 \text{ x } 10^{-19}$ sec , which is roughly 5 fold magnitude larger than the calculated charge relaxation time $\tau_{Cu}^{relax} \simeq 4.5 \text{ x } 10^{-19}$ sec.

The problem here is that {the macroscopic} Ohm's Law is simply out of its range of validity on such short time scales! Two additional facts here are that both and C å ó are frequency-dependent quantities { i.e. $\varepsilon = \varepsilon(\omega)$ and $\sigma = \sigma(\omega)$}, which becomes increasingly important at the higher frequencies ($f = 2\pi/\omega \sim 1/\tau_{relax}$) associated with short time-scale, transient-type phenomena!

So, in reality, if we are willing to wait even a short time (e.g. $\Delta t \sim 1ps = 10^{-12}$sec) then any initial free charge density $\rho_{free}\left(\vec{r}, t=0\right)$ accumulated inside the conductor at $t = 0$ will have dissipated away/damped out, and from that time onwards, $\rho_{free}\left(\vec{r}, t=0\right)$ *can* be safely assumed.

Thus, after many charge relaxation time constants, e.g. $20\tau_{relax} \leq \Delta t \simeq 1ps = 10^{-12}$sec, then

Maxwell's equations for a conductor become {with $\rho_{free}\left(\vec{r}, t \geq \Delta t\right) = 0$ from then onwards}:

New sets of Maxwell's equation for a charged-equilibrated conductor

(1) $\vec{\nabla} \cdot \vec{E}\left(\vec{r}, t\right) = 0$

(2) $\vec{\nabla} \cdot \vec{B}\left(\vec{r}, t\right) = 0$

(3) $\vec{\nabla} \text{ x } \vec{E}\left(\vec{r}, t\right) = -\frac{\partial \vec{B}(\vec{r}, t)}{\partial t}$

(4) $\vec{\nabla} \text{ x } \vec{B}\left(\vec{r}, t\right) = \mu \vec{J}\left(\vec{r}, t\right) + \mu\varepsilon \frac{\partial \vec{E}(\vec{r}, t)}{\partial t} = \mu\left(\sigma \vec{E}\left(\vec{r}, t\right) + \varepsilon \frac{\partial \vec{E}(\vec{r}, t)}{\partial t}\right)$

Now because these equations are different from the previous derivation(s) of monochromatic plane electromagnetic (EM) waves propagating in free space/vacuum and/or in linear/homogeneous/isotropic nonconducting materials {only Eq. (4) has changed}, we re-derive the wave equations for $\vec{E}$ and $\vec{B}$ from scratch. As before, we apply curl operation to Eqs. (3) and (4), and using the following vector identity, we get:

$$\vec{\nabla} \text{ x } \left(\vec{\nabla} \text{ x } \vec{E}\right) = \left(\vec{\nabla} \cdot \vec{E}\right)\vec{\nabla} - \left(\vec{\nabla} \cdot \vec{\nabla}\right)\vec{E} + \left(\vec{E} \cdot \vec{\nabla}\right)\vec{\nabla} - \left(\vec{\nabla} \cdot \vec{\nabla}\right)\vec{E}$$

$$\vec{\nabla} \times \left(\vec{\nabla} \times \vec{E}\right) = -\frac{\partial}{\partial t}\left(\vec{\nabla} \times \vec{B}\right)$$
$$= \vec{\nabla}\left(\vec{\nabla} \cdot \vec{E}\right)^0 - \nabla^2\vec{E} = -\frac{\partial}{\partial t}\left(\mu\sigma\vec{E} + \mu\varepsilon\frac{\partial\vec{E}}{\partial t}\right)$$
$$= \vec{\nabla}^2\vec{E} = \mu\varepsilon\frac{\partial^2\vec{E}}{\partial t^2} + \mu\sigma\frac{\partial\vec{E}}{\partial t}$$

$$\vec{\nabla} \times \left(\vec{\nabla} \times \vec{B}\right) = \mu\left[\sigma\left(\vec{\nabla} \times \vec{E}\right)\right] + \varepsilon\frac{\partial}{\partial t}\left(\vec{\nabla} \times \vec{E}\right)$$
$$= \vec{\nabla}\left(\vec{\nabla} \cdot \vec{B}\right) - \nabla^2\vec{B} = \mu\sigma\frac{\partial\vec{B}}{\partial t} - \mu\varepsilon\frac{\partial^2\vec{B}}{\partial t^2}$$
$$= \nabla^2\vec{B} = \mu\varepsilon\frac{\partial^2\vec{B}}{\partial t^2} + \mu\sigma\frac{\partial\vec{B}}{\partial t}$$

Again, we can write:

$$\nabla^2\vec{E}\left(\vec{r},t\right) = \mu\varepsilon\frac{\partial^2\vec{E}\left(\vec{r},t\right)}{\partial t^2} + \mu\sigma\frac{\partial\vec{E}\left(\vec{r},t\right)}{\partial t} \quad \text{and} \quad \nabla^2\vec{B}\left(\vec{r},t\right)$$
$$= \mu\varepsilon\frac{\partial^2\vec{B}\left(\vec{r},t\right)}{\partial t^2} + \mu\sigma\frac{\partial\vec{B}\left(\vec{r},t\right)}{\partial t}$$

Note that these 3-D wave equations for $\vec{E}$ and $\vec{B}$ in a conductor have an additional term that has a single time derivative—which is analogous to a velocity-dependent damping term, e.g., for a mechanical harmonic oscillator.

The general solution(s) to the above wave equations are usually in the form of an oscillatory function * a damping term (i.e., a decaying exponential)—in the direction of the propagation of the EM wave, e.g., complex plane-wave type solutions for $\vec{E}$ and $\vec{B}$ associated with the above wave equation(s) are of the general form:

$$\tilde{\vec{E}}\,(z,t) = \tilde{\vec{E}}_0 e^{i\left(\tilde{k}x-\omega t\right)} \text{and} \tilde{\vec{B}}\,(z,t) = \tilde{\vec{B}}_0 e^{i\left(\tilde{k}z-\omega t\right)} = \left(\frac{\tilde{k}}{\omega}\right)\hat{k} \times \tilde{\vec{E}}\,(z,t)$$
$$= \frac{1}{\omega}\hat{\tilde{k}} \times \tilde{\vec{E}}\,(z,t) \tag{4.154}$$

With {frequency-dependent} complex wave number:
$\tilde{k}(\omega) = k(\omega) + i\kappa(w)$

Where $k(\omega) = \Re e\left(\tilde{k}(\omega)\right)$ and $\kappa(w) = \Im m\left(\tilde{k}(\omega)\right)$ and corresponding complex vector in the positive $+\hat{z}$ direction here is $\tilde{\vec{k}}\,(\omega) = \tilde{k}(\omega)\hat{k} = \tilde{k}(\omega)\hat{z}$, i.e. $\tilde{\vec{k}}\,(\omega) = [k(\omega) + i\kappa(\omega)]\hat{z}$.

We plug in $\tilde{\vec{E}}\,(z,t) = \tilde{\vec{E}}_0 e^{i\left(\tilde{k}x-\omega t\right)}$ and $\tilde{\vec{B}}\,(z,t) = \tilde{\vec{B}}_0 e^{i\left(\tilde{k}z-\omega t\right)}$ into their respective wave equations above, and obtain from each wave equation the same/identical *characteristic equation* {dispersion relation} between complex $\tilde{k}(\omega)$ and ω, and we get the following relation as:

$$\tilde{k}^2(\omega) = \mu\varepsilon\omega^2 + i\mu\sigma\omega \tag{4.155}$$

Thus, since $\widetilde{k}(\omega) = k(\omega) + i\kappa(\omega)$, then:

$$\boxed{\widetilde{k}^2(\omega) = [k(\omega) + i\kappa(\omega)]^2 = k^2(\omega) - \kappa^2(\omega) + 2ik(\omega)\kappa(\omega) = \mu\varepsilon\omega^2 + i\mu\sigma\omega} \tag{4.156}$$

If we temporarily suppress the ω-dependence of complex $\widetilde{k}(\omega)$, this relation becomes as:

$$\widetilde{k}^2(\omega) = (k + i\kappa)^2 = k^2 - \kappa^2 + 2ik\kappa = \mu\varepsilon\omega^2 + i\mu\sigma\omega \tag{4.157}$$

We can solve this relation to determine $(\omega) = \Re e\left(\widetilde{k}(\omega)\right)$ and $\kappa(\omega) = \Im m\left(\widetilde{k}(\omega)\right)$ as follows:

First, separate this relation into two relations by separating out its real and imaginary parts as:

$$k^2 - \kappa^1 = \mu\varepsilon\omega^2 \text{and} 2ik\kappa = i\mu\sigma\omega 2k\kappa = \mu\sigma\omega \tag{4.158}$$

Now we have two separate independent equations $k^2 - \kappa^2 = \mu\varepsilon\omega^2$ and $2k\kappa = \mu\sigma\omega$, and we have two unknowns such as k and κ. Hence, solving these two equations simultaneously we find the following results:

$$\begin{aligned} \kappa &= \frac{1}{2}\mu\sigma\omega/k \\ k^2 - \kappa^2 &= k^1 - \left(\frac{1}{2}\mu\sigma\omega/k\right)^2 k^2 - \frac{1}{k^2}\left(\frac{1}{2}\mu\sigma\omega\right)^2 = \mu\varepsilon\omega^2 \end{aligned} \tag{4.159}$$

Then multiply by k^2 and rearrange the terms to obtain the following relation:

$$k^4 - \left(\mu\varepsilon\omega^2\right)k^2 - \left(\frac{1}{2}\mu\sigma\omega\right)^2 = 0 \tag{4.160}$$

To solve this equation, we let $x \equiv k^2$, $a \equiv 1$, $b \equiv -\left(\mu\varepsilon\omega^2\right)$ and $c \equiv -\left(\frac{1}{2}\mu\sigma\omega\right)^2$, then Eq. 4.160 reduces to the form of a quadratic equation of $ax^2 + bx + c = 0$ with solution roots of

$$x = \frac{-b \pm \sqrt{b^1 - 4ac}}{2a} \quad \text{or} \quad k^2 = \frac{1}{2}\left[+\left(\mu\varepsilon\omega^2\right) \mp \sqrt{\left(\mu\varepsilon\omega^2\right)^2 + 4\left(\frac{1}{2}\mu\sigma\omega\right)^2}\right]$$

$$k^2 = \frac{1}{2}\left(\mu\sigma\omega^2\right)\left[1 \mp \sqrt{1 + 4\frac{(\mu^2\sigma^2\omega^2)}{4(\mu^2\varepsilon^2\omega^4)}}\right] = \frac{1}{2}\left(\mu\sigma\omega^2\right)\left[1 \mp \sqrt{1 + \frac{(\sigma^2)}{(\varepsilon^2\omega^2)}}\right]$$

$$= \frac{1}{2}\left(\mu\sigma\omega^2\right)\left[1 \mp \sqrt{1 + \left(\frac{\sigma}{\varepsilon\omega}\right)^2}\right]$$

Now we can see that on physical grounds ($k^2 > 0$), we ***must*** select the + sign, hence:

$$k^2 = \frac{1}{2}\left(\mu\sigma\omega^2\right)\left[1 + \sqrt{1 + \left(\frac{\sigma}{\varepsilon\omega}\right)^2}\right]$$

and thus:

$$k = \sqrt{k^2} = \omega\sqrt{\frac{\varepsilon\mu}{2}}\left[1 + \sqrt{1 + \left(\frac{\sigma}{\varepsilon\omega}\right)}\right]^{1/2} = \omega\sqrt{\frac{\varepsilon\mu}{2}}\left[\sqrt{1 + \left(\frac{\sigma}{\varepsilon\omega}\right)^2}\right]^{1/2} \tag{4.161}$$

Having thus solved for k (or equivalently k^2), then we can use either of our original two relations to solve for κ, e. g. $k^2 - \kappa^2 = \mu\varepsilon\omega^2$, then:

$$\kappa^2 = k^2 - \mu\varepsilon\omega^2 = \frac{1}{2}\left(\mu\varepsilon\omega^2\right)\left[\sqrt{1 + \left(\frac{\sigma}{\varepsilon\omega}\right)^2}\right] - \mu\varepsilon\omega^2$$

$$= \frac{1}{2}\left(\mu\varepsilon\omega^2\right)\left[\sqrt{1 + \left(\frac{\sigma}{\varepsilon\omega}\right)^2} - 1\right]$$

Thus, we obtain

$$\begin{aligned} k(\omega) &= \Re e\left(\tilde{k}(\omega)\right) = \omega\sqrt{\frac{\varepsilon\mu}{2}}\left[\sqrt{1 + \left(\frac{\sigma}{\varepsilon\omega}\right)^2} + 1\right]^{1/2} \\ \kappa(\omega) &= \Im m\left(\tilde{k}(\omega)\right) = \omega\sqrt{\frac{\varepsilon\mu}{2}}\left[\sqrt{1 + \left(\frac{\sigma}{\varepsilon\omega}\right)^2} - 1\right]^{1/2} \end{aligned} \tag{4.162}$$

Note that the imaginary part of $\tilde{k}$, $\kappa(\omega) = \Im m\left(\tilde{k}(\omega)\right)$ results in an *exponential* attenuation and damping of the monochromatic plan EM wave with increasing z as:

$$\begin{aligned}\vec{\tilde{E}}(z,t) &= \vec{\tilde{E}}_0 e^{-\kappa z} e^{i(kz-\omega t)} \text{ for Electric Field}\\ \vec{\tilde{B}}(z,t) &= \vec{\tilde{B}}_0 e^{-\kappa z} e^{i(kz-\omega t)} = \frac{1}{\omega}\vec{\tilde{k}} \text{ x } \vec{\tilde{E}}_0 e^{-\kappa z} e^{i(kz-\omega t)} \text{ for Magnetic Field}\end{aligned} \tag{4.163}$$

These solution (Eq. 4.163) satisfy the above wave equations for *any* choice of $\vec{\tilde{E}}_0$.

The characteristic distance over which $\vec{E}$ and $\vec{B}$ are attenuated and reduced to $1/e = e^{-1} = 0.3679$ of their initial value at $z = 0$ is known as the *skin depth* and shown as $\delta_{skin\ depth} = \delta_{sc} \equiv 1/\kappa(\omega)$ and SI unit has dimension of meter.

$$\delta_{sc}(\omega) = \frac{1}{\kappa(\omega)} = \frac{1}{\omega\sqrt{\frac{\varepsilon\mu}{2}\left[1+\left(\frac{\sigma}{c\omega}\right)^2-1\right]^{1/2}}} \Rightarrow \boxed{\begin{aligned}\vec{\tilde{E}}(z=\delta_{sc},t) &= \vec{\tilde{E}}_0 e^{-1} e^{i(kz-\omega t)}\\ \vec{\tilde{B}}(z=\delta_{sc},t) &= \vec{\tilde{B}}_0 e^{-1} e^{i(kz-\omega t)}\end{aligned}}$$

The real part of $\tilde{k}$, *i.e.* $k(\omega) = \Re e\left(\tilde{k}(\omega)\right)$ determines the spatial wavelength $\lambda(\omega)$, the propagation speed $\mathrm{v}(\omega)$ of the monochromatic ***EM*** plane wave (See Fig. 4.25) in the conductor, and also the index of refraction:

$$\lambda(\omega) = \frac{2\pi}{k(\omega)} = \frac{2\pi}{\Re e\left(\tilde{k}(\omega)\right)}$$

$$\mathrm{v}(\omega) = \frac{\omega}{k(\omega)} = \frac{\omega}{\Re e\left(\tilde{k}(\omega)\right)}$$

$$n(\omega) = \frac{\omega}{\mathrm{v}(\omega)} = \frac{ck(\omega)}{\omega} = \frac{c\Re e\left(\tilde{k}(\omega)\right)}{\omega}$$

The above plane wave solutions satisfy the above wave equations for any choice of $\vec{\tilde{E}}_0$. As we have also seen before, it can similarly be shown here that Maxwell's Eqs. (1) and (2) $\{\vec{\nabla}\cdot\vec{E} = 0$ and $\vec{\nabla}\cdot\vec{B} = 0\}$ rule out the presence of any (longitudinal) *z*-components for $\vec{E}$ and $\vec{B}$ for *EM* waves propagating in the +$\hat{z}$-direction $\Rightarrow \vec{E}$ and $\vec{B}$ are purely transverse waves.

If we consider, e.g., a linearly polarized monochromatic plane *EM* wave propagating in the +$\hat{z}$ -direction in a conducting medium, e.g., $\vec{\tilde{E}}(z,t) = \vec{\tilde{E}}_0 e^{-\kappa z} e^{i(kz-\omega t)}\hat{x}$, then:

$$\tilde{\vec{B}}(z,t) = \frac{1}{\omega}\tilde{\vec{k}} \times \tilde{\vec{E}}(z,t) = \left(\frac{\tilde{k}}{\omega}\right)\tilde{E}_0 e^{-\kappa z} e^{i(kz-\omega t)}\hat{y}$$

$$= \left(\frac{k+i\kappa}{\omega}\right)\tilde{E}_0 e^{-\kappa z} e^{i(kz-\omega t)}\hat{y} \tag{4.164}$$

$$\Rightarrow \tilde{\vec{E}}(z,t) \perp \tilde{\vec{B}}(z,t) \perp \hat{z}(+\hat{z} = \text{propagation direction})$$

The complex wave-number $\tilde{k} = k + ik = Ke^{i\phi_k}$ where $K \equiv |\tilde{k}| = \sqrt{k^2+\kappa^2}$ and $\phi_k \equiv \tan^1(\kappa/k)$. See Fig. 4.34.

Then we see that:

$$\tilde{\vec{E}}(z,t) = \tilde{\vec{E}}_0 e^{-\kappa z} e^{i(kz-\omega t)}\hat{x}\text{has}\tilde{E}_0 = E_0 e^{i\delta_E}$$

and that: $\tilde{\vec{B}}(z,t) = \tilde{B}_0 e^{-\kappa z} e^{i(kz-\omega t)}\hat{y} = \frac{\tilde{k}}{\omega}\tilde{E}_0 e^{-\kappa z} e^{i(kz-\omega t)}\hat{y}\text{has}\tilde{B}_0 = B_0 e^{-i\delta_B} = \frac{\tilde{k}}{\omega}\tilde{E}_0 = \frac{Ke^{i\phi_k}}{\omega}E_0 e^{i\delta_E}$

Thus, we see that: $B_0 e^{-i\delta_B} = \frac{Ke^{i\phi_k}}{\omega}E_0 e^{i\delta_E} = \frac{K}{\omega}E_0 e^{i(\delta_E+\phi_k)} = \frac{\sqrt{k^2+\kappa^2}}{\omega}E_0 e^{i(\delta_E+\phi_k)}$ i.e. inside a conductor, $\vec{E}$ and $\vec{B}$ are no longer in phase with each other.

Phases of $\vec{E}$ and $\vec{B}$: $\boxed{\delta_B = \delta_E + \phi_k}$

We also see that:

$\Delta\varphi_{B-E} \equiv \delta_B - \delta_E = \phi_k$ magnetic field ***lags*** behind the electric field.

We also see that: $\boxed{\frac{B_0}{E_0} = \frac{K}{\omega} = \left[\varepsilon\mu\sqrt{1+\left(\frac{\sigma}{\varepsilon\omega}\right)^2}\right]^{1/2} \neq \frac{1}{c}}$

The real physical $\vec{E}$ and $\vec{B}$ fields associated with linearly polarized monochromatic plane *EM* waves propagating in a conducting medium are *exponentially* damped:

$$\vec{E}(z,t) = \Re e\left(\tilde{\vec{E}}(z,t)\right) = E_0 e^{-\kappa z}\cos(kz - \omega t + \delta_E)\hat{x}$$

$$\vec{B}(z,t) = \Re e\left(\tilde{\vec{B}}(z,t)\right) = B_0 e^{-\kappa z}\cos(kz - \omega t + \delta_B)\hat{y} = B_0\cos(kz - \omega t + \{\delta_E + \phi_k\})\hat{y}$$

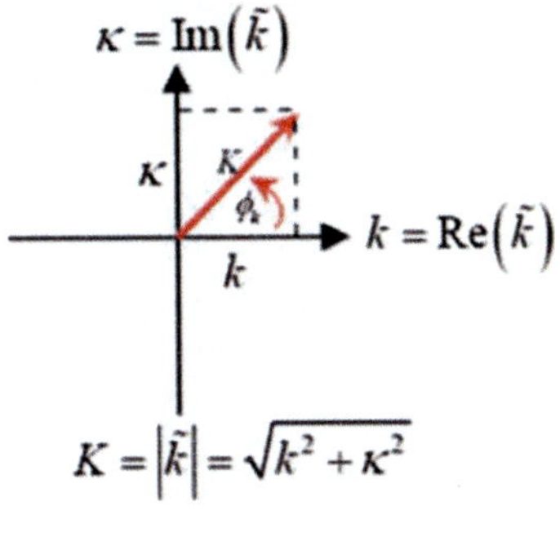

Fig. 4.34 In the complex $\tilde{k}$-plane

$$\frac{B_0}{E_0} = \frac{K(\omega)}{\omega} = \left[\varepsilon\mu\sqrt{1 + \left(\frac{\sigma}{\varepsilon\omega}\right)^2}\right]^{1/2}$$

$$K(\omega) \equiv \left|\widehat{k}(\omega)\right| = \sqrt{k^2(\omega) + \kappa^2(\omega)} = \omega\left[\varepsilon\mu\sqrt{1 + \left(\frac{\sigma}{\varepsilon\omega}\right)^2}\right]^{1/2} \tag{4.165}$$

$$\delta_B = \delta_E + \phi_k \qquad \vec{\tilde{k}}(\omega) = [k(\omega) + i\kappa(\omega)]\hat{z}$$

$$\phi_k(\omega) \equiv \tan^{-1}\left(\frac{\kappa(\omega)}{k(\omega)}\right) \text{ and } \quad \tilde{k}(\omega) = \left|\vec{\tilde{k}}(\omega)\right| = k(\omega) + i\kappa(\omega)$$

Definition of the skin depth in a conductor:

$$\boxed{\delta_{sc}(\omega) \equiv \frac{1}{\kappa(\omega)} = \frac{1}{\sigma\sqrt{\frac{\varepsilon\mu}{2}}\left[\sqrt{1 + \left(\frac{\sigma}{\varepsilon\omega}\right)^2} - 1\right]^{1/2}}} =$$

Distance over which the $\vec{E}$ and $\vec{B}$ fields fall to $1/e = e^{-1} = 0.3679$ of their initial values

Example 4.13 What is the skin depth of fine silver at microwave frequency of 1010 HZ, which is a common microwave region? Assume the silver has conductivity of $g = 10^{10}$HZ.

Solution The skin depth is presented by the following equation:

$$\delta = \sqrt{\frac{2}{\mu_0\omega g}} = \sqrt{\frac{2}{(2\pi \text{ x } 10^{10})(4\pi \text{ x } 10^{-7})(3 \text{ x } 10^3)}} = 9.2 \text{ x } 10^{-5} \text{ cm}$$

Thus, at microwave frequencies, the skin depth in silver is very small, and consequently, the difference in performance between a pure silver component and a silver-plated brass component would be expected to be negligible.

Example 4.14 For the case of seawater, we calculate the frequency at which the skin depth is 1 meter. For seawater, $\mu = \mu_0$ and $g \approx 4.3$ S/m.

Solution The expression for the frequency corresponding to a given skin depth δ is:

$$\omega = \frac{2}{g\mu_0\delta^2} = \frac{2}{4.3 \text{ x } 4\pi \text{ x } 10^{-1}\delta^2} = \frac{3.70 \text{ x } 10^5}{\delta^2}(\sec)^{-1}$$

which yields:

$$f = 58 \text{ x } 10^3 \text{HZ}$$

or frequency of 60 kHz for a skin depth of 1 meter. If a submarine is equipped with a very sensitive receiver and if a very powerful transmitter is used, it is possible to communicate with a submerged submarine. However, a low radiofrequency must be used, and even then an extremely sever attenuation of the signal occurs. At five skin depths (5 meters in the case calculated above), only 1 percent of the initial electric field remains and only 0.01% of the incident power.

Reference

1. David Griffiths "Introduction to electrodynamics", 3rd Edition, Prentice Hall, 1999

Chapter 5
Response of Materials to High-Power Laser Energy Radiation

The word laser is an acronym for light amplification by stimulated emission of radiation, although common usage today is to use the word as a noun—laser—rather than as an acronym, LASER. A laser is a device that creates and amplifies a narrow, intense beam of coherent light. Atoms emit radiation. We see it every day when the "excited" neon atoms in a neon sign emit light. Normally, they radiate their light in random directions at random times. The result is incoherent light—a technical term for what you would consider a jumble of photons going in all directions. The trick in generating coherent light—of a single or just a few frequencies going in one precise direction—is to find the right atoms with the right internal storage mechanisms and create an environment in which they can all cooperate, to give up their light at the right time and all in the same direction.

5.1 Introduction

In this chapter we will discuss the response of materials to a high-power radiation laser with a one-dimensional mathematical modeling and presentation of solution to different cases of heat conduction partial differential equation along with given boundary and initial conditions under different scenarios. The topics in this chapter also will include discussion of optical reflectivity of metals at infrared radiation (IR) wavelengths, laser-induced heat flow in materials, the effects of melting and vaporization, the impulse generated in materials by both pulsed and continuous radiation, and the influence of the absorption of laser radiation in the blow-off and melting region in front of the irradiated materials.

LASER stands for "light amplification by stimulated emission of radiation", which is almost totally a coherent beam. In order to produce a lasing beam, we look at an atom that will emit a photon of light when it decays from an excited energy state to a lower state. The energy difference between two states can be designated as ΔE which can be described in terms of frequency v of this light in the form of

B. Zohuri, *Thermal Effects of High Power Laser Energy on Materials*,
https://doi.org/10.1007/978-3-030-63064-5_5

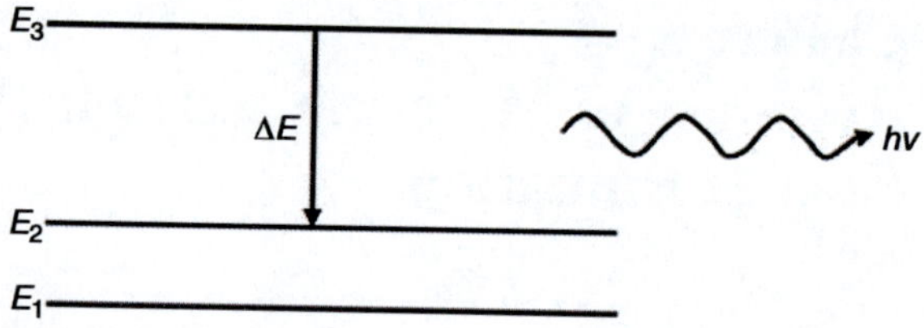

Fig. 5.1 Energy levels

Eq. 1.1, which is also known as Plank–Einstein relation in quantum mechanics, for the above photon emitting light due to its decay in energy level.

$$\nabla E = h\nu \tag{5.1}$$

where h is Planck's constant, and its value is given as $h = 6.62606957 \times 10^{-34}$ J. s = $4.135667516 \times 10^{-15}$ eV. Equation 5.1 in terms of energy level is depicted below (Fig. 5.1).

If we assume the wavelength of emitted photon light designated with symbol λ and speed of light with c, then relationship between wavelength and speed of light can be written as:

$$c = \lambda\nu \tag{5.2}$$

Substituting Eq. 5.2 into Eq. 5.1 for frequency ν, we get a new expression for the Plank–Einstein relationship as:

$$\Delta E = \frac{hc}{\lambda} \tag{5.3}$$

The above equation leads to another relationship involving Plank's constant h. Given p for the linear momentum of a particle (not only a photon, but other particles as well), the de Broglie wavelength λ of the particle is given by:

$$\lambda = \frac{h}{p} \tag{5.4}$$

In some applications where it makes sense to use the angular frequency, where the frequency is expressed in terms of radians per second instead of rotation per second or Hertz, it is customary to absorb a factor of 2π into the Planck constant. The resulting constant is called *reduced Planck constant* or *Dirac's constant.* It is equal to the Plank constant divided by 2π and is denoted as ħ and pronounced ***h-bar***.

$$\hbar = \frac{h}{2\pi} \tag{5.5}$$

Therefore, the energy of photon with angular frequency ω, where $\omega = 2\pi v$, is given as:

$$\Delta E = \hbar\omega \tag{5.6}$$

The reduced Plank's constant is the quantum of angular momentum in quantum mechanics.

The numerical value of reduced Plank constant is given as $\hbar = 1.054571726 \times 10^{-34}$ J.s $= 6.58211928 \times 10^{-16}$ eV.s.

The above conditions and circumstances are the case for any light source, whether laser, flame, incandescent body, etc.

Atoms emit photon for any conventional light source in a random, sporadic manner and spontaneously decay in lower-energy state when excited by heat or any other heat-generated source, such as electric current. On the other hand physics of laser indicates that the photons are emitted in phase and the electromagnetic radiation behavior types are encountered and more or less, we can describe it as a simply wave propagation of sinusoidal radiation filed takes place and at a microscopic level can be defined by the following mathematical solution of wave equation in conductor and taking the real part of the solution of the wave equation (i.e., we assumed the general solution is complex quantity type that includes both real and imaginary terms as part of solution) under consideration, then the relationship is presented as below:

$$\mathcal{E}(z,t) = \mathfrak{Re}\left[\mathcal{E}_0 e^{-2\pi k z/\lambda} e^{i\omega(t - nz/c)}\right] \tag{5.7}$$

where:

$\mathcal{E}$ = the electric field of the radiation.
$\mathfrak{Re}$ = the real part of the complex quantity in brackets.
$\mathcal{E}_0$ = the maximum amplitude.
k = the extinction coefficient and vacuum, $k = 0$.
z = the direction in which the wave is propagating.
λ = the wavelength.
t = time.
n = the index of refraction and in vacuum $n = 1$.
c = the speed of light in vacuum.

Equation 5.7 is just an electric field solution to set of Maxwell's equation inside a linear, homogeneous, and isotropic conducting medium that has electric permittivity ε and magnetic permeability μ. The solution in general form of vector presentation for the electric field using complex notation is:

$$\tilde{\vec{E}}(z,t) = \tilde{\vec{E}}_0 e^{(-kz)} e^{[i(kz - \omega t)]} \tag{5.8}$$

Readers can refer to Appendix F of reference [69] to find how the solution of wave equation in conductor will result in Eq. 5.7, which is nothing more than the standard representation of the electric field of traveling light wave. You can also look at the quick approach to derive Eq. 5.7 in the next page under wave equation as well.

Wave Equation

The macroscopic Maxwell's equations are as follows:

$$\vec{\nabla} \times \vec{E} = \frac{\partial \vec{B}}{\partial t} \tag{1}$$

$$\vec{\nabla} \times \vec{H} = \vec{j} + \frac{\partial \vec{D}}{\partial t} \tag{2}$$

$$\vec{\nabla} \cdot \vec{D} = \rho \tag{3}$$

$$\vec{\nabla} \cdot \vec{B} = 0 \tag{4}$$

The relationship between $\vec{E}$ (electric field), $\vec{D}$ (electric displacement), $\vec{B}$ (magnetic field), and $\vec{H}$ (magnetic field strength) generally speaking is nonlinear, but in our case of interest for high-power laser interaction with materials, we can approximate them by linear models, and the relationship in general will depend on the frequency of the radiation field. Note that these parameters and relationship between them describe the material behavior. In case of time harmonic fields, Fourier transformed field quantities are related to the following sets of equations:

$$\vec{D}\left(\vec{r}, \omega\right) = \varepsilon_0 \varepsilon(\omega) \vec{E}\left(\vec{r}, \omega\right) \tag{5}$$

$$\vec{B}\left(\vec{r}, \omega\right) = \mu_0 \mu(\omega) \vec{H}\left(\vec{r}, \omega\right) \tag{6}$$

With Eqs. (1), (2), (5), and (6), the wave equation can be established as:

$$\vec{\nabla}\left(\vec{\nabla} \cdot \vec{E}\right) - \Delta \vec{E} = -\mu_0 \varepsilon_0 \varepsilon \frac{\partial^2 \vec{E}}{\partial t^2} \tag{7}$$

In homogeneous media and with zero space charge $\vec{\nabla} \cdot \vec{E} = 0$, and with $\mu_0 \varepsilon_0 = 1/c^2$, Eq. (7) reduces to the following form:

(continued)

$$\Delta\vec{E} = \frac{\varepsilon}{c^2}\frac{\partial^2\vec{E}}{\partial t^2} \tag{8}$$

A solution of this partial differential equation (Eq. (8)) is the equation of plane wave:

$$\vec{E} = \vec{E}_0 e^{[i(kz-\omega t)]} \tag{9}$$

With:

k = complex wave number.
ω = real angular frequency.

The complex wave number is:

$$k = k_0\sqrt{\varepsilon} = k_0 n = k_{Real} + ik_{Imaginary} = k_r + ik_i \tag{10}$$

where n is the complex index of refraction. The plane wave solution can also be cast into the form of:

$$\vec{E} = \vec{E}_0 e^{(-k_i z)} e^{[i(k_r z-\omega t)]} \tag{11}$$

If the imaginary part of the complex wave number $k_i > 0$, the wave decays exponentially within the material.

Equation 5.7 is a valid relationship to measure the electric field $\mathcal{E}$ in a point in space for a laser emission, where the individual photons are in phase. This equation also allows not only to measure the electric filed $\mathcal{E}$ at a point in space for laser light; it also results in predicating the oscillation of $\mathcal{E}$ as well. This is not true for an light from a conventional source, and the Eq. 5.7 does not hold for the measuring such electric field at some point in space in order to express the sinusoidal variation, for the atoms emitting the light that are doing so in random, and the sinusoidal variation due to the emission from each atom needs to be averaged out to some, time-dependent value.

The fact that laser is a very coherent source of light where this coherency is created by taking advantage of simulated emission in materials in which metastable states can be induced, then by selected rules of quantum mechanics, we know that the lifetime of an atom in an excited energy state depends on these rules for transition to a lower state. Bear in mind that there are states from which transition to a lower level is extremely impossible; such states are called metastable states, and an atom that is not going to be distributed by outside influence will remain in a metastable state for a very long time.

If a metastable atom interacts with a photon of frequency such that Eq. 5.1 holds, whereas we said ΔE is the energy difference between the atom's normal and metastable states, stimulated emission will occur. The atom will decay to its normal state by emitting another photon of frequency ν, so that the net result is two photons, and the second photon will have the same phase temporally and spatially as the first.

5.2 How Laser Works

LASER stands for light **a**mplification by stimulated **e**mission of **r**adiation. Spontaneous emission is the process by which an excited atom spontaneously emits a photon. Electrons go from excited to a resting state when a photon of energy is released. Photon emission can be stimulated by an external source of energy that will increase the population of excited electrons, a process known as pumping. A laser contains a laser chamber, a lasing medium (solid, liquid, or gas), and an external source of energy. Stimulated emission occurs when the external source of energy causes electrons to be excited in the lasing medium. A cascade reaction is generated when these excited electrons release photons, which then collide with other excited electrons in the lasing medium and cause a release of many identical photons at the same time. Laser light continues to be generated as long as the above cascade perpetuates.

Laser light has the following properties:

1. **Coherence:** Laser beam are both temporally and spatially coherent. This phenomenon results from stimulated emission and allows laser beams to have a high power density
2. **Collimation:** Laser beams are parallel to each other (i.e., ignoring thermal blooming while traveling through atmospheric environment for a high-energy beam) and therefore exhibit collimation. A collimated beam is created in the laser chamber when light is reflected between two mirrors, and only the exit of parallel waves is allowed. Collimation allows laser light to travel a long distance without loss of intensity. In practice, a lens on a laser focuses the parallel light beam down to the smallest possible spot size, or the diffraction-limited spot, to allow the light to focus on the target.
3. **Radiometry:** The four main concepts in understanding laser light and target interactions are:

   (a) Energy
   (b) Power
   (c) Fluence
   (d) Irradiance

The amount of light emitted from a laser can be quantified by both energy and power. Energy represents work that is measured in joules, while power is measure in watts or joules per second which is rate at which energy is expended. The intensity of the laser beam on the target is a function of the area of the target over which it is spread known as *spot size*.

$$\text{Spot size} = \text{cross} - \text{sectional area of the laser beam}$$

Fluence which is measured in joules per square centimeter is the energy density of a laser beam.

$$\begin{aligned}\text{Fluence} &= \text{watts} \times \text{seconds/cm}^2 \\ &= \text{joules/cm}^2 + \text{laser output} \times \text{pulse duration/spot size}\end{aligned}$$

Irradiance measured in watts per square centimeter refers to the power density of a continuous wave laser beam, and it is inversely proportional to the square root of the radius of the spot size.

$$\text{Irradiance} = \text{watts/cm}^2 = \text{laser output/spot size}$$

Exposure time, fluence, and irradiance of a laser can be altered depending on the particular desired laser dueling target and conditions and circumstances that laser engages the target.

Laser interacts with the target in four possible ways:

1. **Reflection** R**:** takes place when light "bounces off" the target surface without entry into the target thickness secondary to difference in the refractive index at engagement point and the environment that incoming laser beam is traveling through. Increasing the angle incidence increases the amount of light reflected. Damages to the target surface or target itself occur with particular lasers if adequate reflection of laser beams occurs and there is proper protection employed.
2. **Absorption** A**:** absorptivity is the ratio of power deposited within the workpiece and the power of the incident radiation
3. **Transmission** T**:** occurs when the laser beam passes through the transparent target without altering either the target surface or the light itself
4. **Scattering** S**:** this refers to fragmentation of light after it has entered the target skin, and it results from the interaction of light with varied elements that make up target layers. When scattering occurs, light is dispersed over a larger area within the target, and the depth of penetration (skin depth) of the light beam is reduced at the same time.

As we said above, if a metastable atom interacts with a photon of frequency such that $\Delta E = h\nu$, where ΔE is the energy difference between the atom's normal and metastable states, stimulated emission will occur. The atom will decay to its normal state by emitting another photon of frequency ν, so that the net result is two photons, and the second photon will have the same phase temporally and spatially as the first.

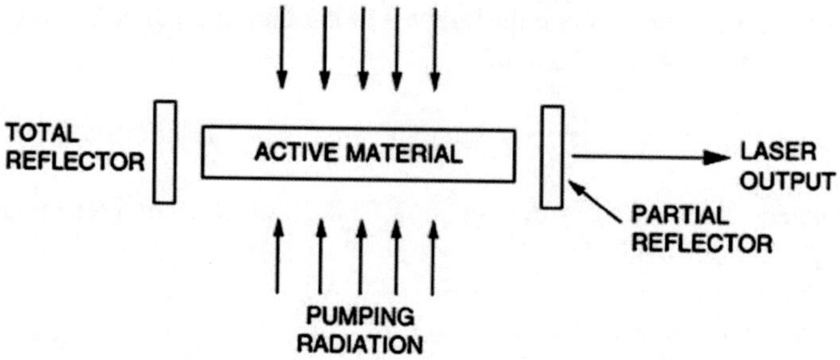

Fig. 5.2 Schematic representation of a laser

In laser, then, one establishes a large number of atoms in metastable states and arranges the optics to increase the likelihood of stimulated emission. Schematically, a typical laser oscillator looks like Fig. 5.2. The pumping radiation (e.g., light from a flash lamp) excites the atoms in the lasing medium (e.g., Cr^{+++} ions in ruby).

In the decay process (if we have a successful laser), a large number of ions are left in a metastable state; this is called a population inversion. As some atoms begin to decay, they stimulate others to decay. But this alone would not provide a laser, since the emission would occur in random directions. The role of reflection is very important; the photons moving perpendicular to the reflectors pass through the medium many times, and on each pass, more and more atoms are caused to emit. This results in the buildup of a very strong coherent light signal that travels in a single direction. Useful light output is obtained by making one of the mirrors a partial reflector [1].

It is interesting to look at a few examples of the intensity of laser light. In a typical ruby laser, the concentration [5] of Cr^{+++} ions is about 2×10^{19} cm^{-3}, and population inversions are of the order of 3×10^{16} cm^{-3}. Crudely speaking, we can think of creating 3×10^{16} quanta/cm^3 in the lasing medium. Since we have arranged the laser so that the output is in a single direction, and since photons move with the speed of light, we obtain $3 \times 10^{16} \times 3 \times 10^{10} = 9 \times 10^{26}$ quanta/cm^2.sec from the laser. For ruby, the lasing wavelength is 6943A, and since the energy of each quanta is $h\nu$, one can readily calculate that the output is about 2.5×10^8 W/cm^2.

Before we go forward to the next step, let us pause and briefly present Plank's energy density distribution using Rayleigh's energy distribution by focusing on the understanding the nature of the electromagnetic radiation inside the cavity, by considering the radiation to consist of standing waves having a temperature T with nodes at the metallic surface. By arguing that, these standing waves are equivalent to harmonic oscillators, for they result from the harmonic oscillations of a large number of electrical charges, electrons that are present in the walls of metallic surface of cavity. When cavity is in thermal equilibrium, the electromagnetic energy density inside the cavity is equal to the energy density of the charged particles in the walls of the cavity, and the average total energy of the radiation leaving cavity for the average energy of the oscillators along with the number of standing waves or mode of radiation in the frequency interval v to $v + dv$ is written as:

$$N(v) = \frac{8\pi v^2}{c^3} \tag{5.9}$$

where $c = 3 \times 10^8$ m/sec is the speed of light and the quantity $(8\pi v^2/c^3)dv$ gives the number of modes of oscillation per unit volume in the frequency range v to $v + dv$ is given by:

$$u(v, T) = N(v)\langle E \rangle = \frac{8\pi v^2}{c^3}\langle E \rangle \tag{5.10}$$

where $\langle E \rangle$ is the average energy of the oscillators present on the walls of the cavity or of the electromagnetic radiation in that frequency interval and the temperature dependence of $u(v, T)$ is buried in $\langle E \rangle$.

The question is how can we calculate the average energy $\langle E \rangle$? According to classical thermodynamics and equipartition theorem, all oscillators in the cavity have the same mean energy, irrespective of their frequencies [1].

$$\langle E \rangle = \frac{\int_0^\infty E e^{-E/(kT)} dE}{\int_0^\infty e^{-E/(kT)} dE} = kT \tag{5.11}$$

where $k = 1.3807 \times 10^{-23}$ J/K is the Boltzmann constant. An insertion of Eq. 5.11 into Eq. 5.10 leads to the Rayleigh–Jeans formula:

$$u(v, T) = \frac{8\pi v^2}{c^3} kT \tag{5.12}$$

Equation 5.12 except for low frequencies is in total disagreement with experimental data: $u(v, T)$ as given by Eq. 5.12 *diverges* for high values of v, whereas experimentally it must be finite per Fig. 5.3. Moreover, integrating Eq. 5.12 overall frequencies, the integral *diverges* which is indication of that the cavity contains an *infinite* amount of energy. Historically, this was called the *ultraviolet catastrophe*, for Eq. 5.12 diverges for *high* frequencies within the ultraviolet range.

Now, studying Plank's energy density divergence, an interpolation between Wien's rule and the Rayleigh–Jeans rule—Plank succeeded in avoiding the ultraviolet catastrophe and proposed an accurate description of blackbody radiation. He considered that the energy exchange between radiation and matter must be *discrete* rather than *continuum*. His postulation indicates that the energy radiation of frequency v emitted by oscillating charges from the walls of the cavity must come *only* in *integer multiples* of hv:

$$E = nhv \qquad n = 0,\ 1,\ 2,\ 3,\ \ldots \tag{5.13}$$

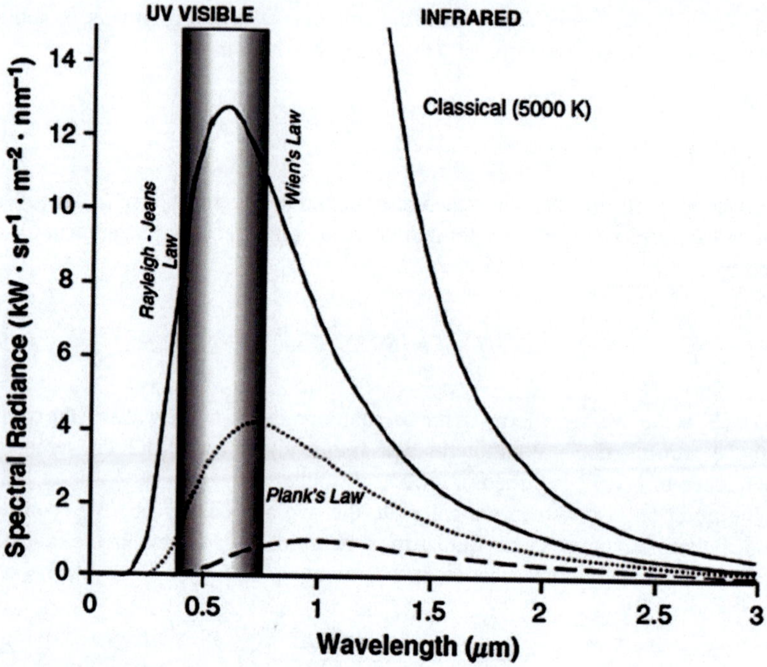

Fig. 5.3 Comparison of various spectral densities
Plank's law dotted curve
Rayleigh–Jeans and Wien's law solid curve

where h is a universal Plank's constant, $h\nu$ is the energy of a "quantum" of radiation, and ν represents the frequency of the oscillating charge particle in a cavity's wall as well as the frequency of the radiation emitted from the walls. This is because the frequency of the radiation emitted by an oscillating charged particle is equal to the frequency of oscillation of the particle itself [8]. Equation 5.13 is known as Plank's quantization rule for energy or Plank's postulate.

Therefore, assuming that the energy of an oscillator is quantized, Plank showed that the *correct* thermodynamic relation for the average energy can be obtained by merely replacing the responding to the discreteness of the oscillator's energies [2]:

$$\langle E\rangle = \frac{\sum_{n=0}^{\infty} nh\nu e^{-nh\nu/kT}}{\sum_{n=0}^{\infty} h\nu e^{-nh\nu/kT}} = \frac{h\nu}{e^{-h\nu/kT}-1} \tag{5.14}$$

and hence by inserting Eq. 5.14 into Eq. 5.10, the energy density per unit frequency of the radiation emitted from the hole of a cavity is given by:

$$u(v, T) = \frac{8\pi v^2}{c^3} \frac{hv}{e^{hv/kT} - 1} \tag{5.15}$$

Equation 5.15 is known as *Plank's distribution* and gives an exact fit to various experimental radiation distribution, as displayed in Fig. 5.3; the numerical value of h obtained by fitting Eq. 5.15 with the experimental data is 6.626×10^{-34} J sec.

We should note that, as shown in Fig. 5.3, we can rewrite Plank's energy density Eq. 5.15 to obtain the energy density per unit wavelength:

$$\widetilde{u}(\lambda, T) = \frac{8\pi hc}{\lambda^5} \frac{1}{e^{hc/\lambda kT} - 1} \tag{5.16}$$

In the above we claimed for ruby laser of wavelength 6943 °A, the energy of each quanta is hv, and one can calculate that the output is about 2.5×10^8 W/cm^2. If we compare this degree of to the power that a black body produces, let us assume sun in this case that emits at the same wavelength with a similar bandwidth, using the Planck radiation law, in term of angular frequency $\omega = 2\pi v = 2\pi c/\lambda$ we can calculate:

$$U_\omega = \frac{h\omega^3}{\pi^2 c^3} \frac{1}{e^{h\omega/kT} - 1} \tag{5.17}$$

U_ω is the energy, per unit volume and per unit bandwidth, radiated by a blackbody at temperature T and k is Boltzmann constant. The radiation leaves the blackbody source at rate c, so the power radiated per unit area of the source, per unit bandwidth, is:

$$I_\omega = \frac{cU_\omega}{4} = \frac{h\omega^3}{\pi^2 c^2} \frac{1/4}{e^{h\omega/kt} - 1} \tag{5.18}$$

If we use the sun's temperature of 6000 K, and $\lambda = 6943$ °A, then we can show that:

$$I_\omega \approx 2 \times 10^{-5} \mathrm{erg/cm^2}$$

For the ruby laser, a typical line width is 3 °A, so $\Delta\omega \approx 2 \times 10^{12}\ \mathrm{sec}^{-1}$. Thus, the power density at the source is:

$$I \approx 2 \times 10^7 \mathrm{erg/cm^2 \cdot sec} \approx 2.5\ \mathrm{W/cm^2}$$

Thus, the power density for comparable narrow-bandwidth, nearly single-frequency light is much greater at a laser source than at a conventional hotbody source, because laser light is coherent.

5.3 Laser Light Propagation

The preparation of laser light through the atmosphere possesses a complex problem, and it will not be discussed here. Suffice to say, anyone who has driven on a foggy night certainly realizes light is certainly scattered in the atmosphere. Lasers of high power density poses even more difficult propagation problems because the high intensity warms the air and creates a density change across the beam. This variation in density refracts the light and causes beam spreading, or "thermal blooming."

Consider briefly the propagation of laser light in free space or in vacuum. Under these ideal conditions, the only change in the power density is due to simple beam divergence. Since the typical laser emits light that is nearly unidirectional, the beam divergence is small. In fact, one feature of a laser is that the divergence is nearly at the diffraction limit, which is of the order of λ/a, where a is the diameter of the output aperture of the laser. For the ruby laser discussed above, this gives a divergence angle of:

$$\theta \approx \frac{6943 \times 10^{-8}}{1} \approx 7 \times 10^{-2} \text{ mrad}$$

for, say, a 1 cm aperture. In practice, one needs to go to much trouble to realize this limit of divergence, but it has been done. More commonly, an "off-the-shelf" ruby laser might have a beam divergence of a few mrad.

The newcomer to lasers has usually heard about diffraction-limited beams and the consequent extreme directionality of laser light lie is usually surprised to discover that at long distances front the source these beams have power densities that vary as the reciprocal of the square of the distance, like all radiating source. To see this, consider a source of power P watt, diameter a, and divergence angle θ, as shown in Fig. 5.4.

At distance r from the source, the power density is:

$$I = \frac{P}{\frac{\pi}{4}(a + 2r\tan\theta)^2}$$

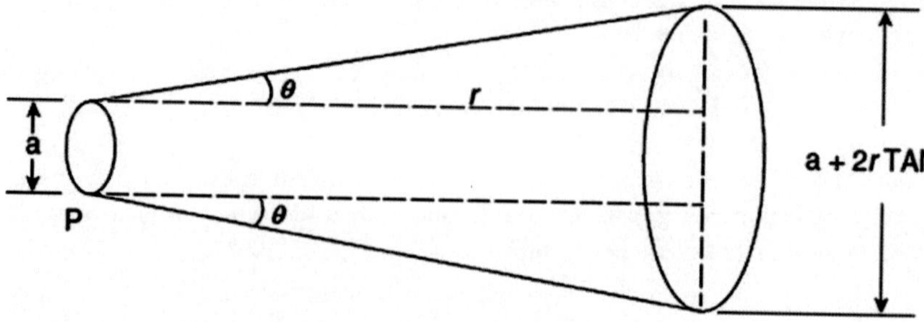

Fig. 5.4 Simplified sketch of laser beam divergence

or, since θ is very small and $\tan\theta \approx \theta$, then:

$$I = \frac{P}{\frac{\pi}{4}(a + 2r\theta)^2}$$

or:

$$I = \frac{P}{\frac{\pi a^2}{4}\left(1 + \frac{2r}{a}\theta\right)^2} \tag{5.19}$$

From this expression, it is apparent that for large distance, such that $2r/a \gg 1$:

$$I = \frac{P}{\frac{\pi a^2}{4}\frac{4r^2\theta^2}{a^2}}$$

Or:

$$I = \frac{P}{\pi r^2\theta^2}$$

or, since $\theta \approx \lambda/a$, then we can write:

$$I = \frac{P}{\pi r^2}\frac{a^2}{\lambda^2}$$

For example, consider a 10 kW beam of 10.6 μm (CO_2 laser) wavelength and 10 Cm aperture at 1 mi (i.e., a high power CO_2 laser):

$$I = \frac{10^4 \times 10^2}{\pi(5280 \times 12 \times 2.54)^2\left(10 \times 10^{-4}\right)^2}$$

Or:

$$I \approx 12\ \mathrm{W/cm^2}$$

From Eq. 5.19, if we substitute I_0, the power density at the source, for $P/(\pi a^2/4)$ and recall that $\theta \approx \lambda/a$, then we can write:

$$I = I_0\frac{1}{\left(1 + 2r\frac{\lambda}{a^2}\right)^2} \tag{5.20}$$

From this expression we can see that if r is small there is little change in power density emitted by the source. The distances at which this is true are referred to as "near field," and the fine details of the beam pattern, such as local variation in intensity, hot spots, etc., are preserved in the near field. It is apparent from Eq. 5.20 that this near field distance will be limited to r such that $I \approx I_0$, or:

$$\frac{2r\lambda}{a^2} \ll 1$$

Or:

$$r_{\text{near field}} \ll a^2/\lambda$$

For lasers with exceptionally good optics that have a Gaussian distribution of power density across the beam, the near field pattern will persist for distances on the order of (a^2/λ) [9–11].

As a final comment on power densities at distances from laser sources, let us use Eq. 5.20 to calculate the distance at which the power density is halved:

$$\frac{I}{I_0} = \frac{1}{2} = \frac{1}{\left(1 + \frac{2r\lambda}{a^2}\right)^2}$$

And:

$$r_{1/2} = \frac{a^2}{2\lambda}\left(\sqrt{2} - 1\right)$$

For our illustration of a CO_2 laser with a 10 cm aperture, $r \approx 680$ *ft*, or a little more than 0.1 mi.

5.4 Physics of Laser Absorption in Metals

Lasers provide the ability to accurately deliver large amounts of energy into confined regions of a material in order to achieve a desired response. For opaque materials, this energy is absorbed near the surface, modifying surface chemistry, crystal structure, and/or multi-scale morphology without altering the bulk. This chapter covers a brief introduction to the fundamental principles governing laser propagation and absorption as well as the resulting material responses.

The description of the interaction of electromagnetic radiation with matter can be done at different model levels. Matter for the purpose of this approach does consist of electrons and atomic kernel. Thus, the materials that laser radiation will interact with are assumed to be a linear, homogeneous, and isotropic conducting medium. For spatial dimension r in this matter, we assume $r \gg 10^{-15}$ m, and the atomic kernels can be considered to be point charges; for spatial dimensions greater than the classical electron radius $r \gg r_0 \sim 2.8 \times 10^{-15}$ m, and the electrons can be considered to be point charges also. These point charges interact with electromagnetic fields and excite spatially and temporally fast fluctuating fields on their part. The framework of quantum electrodynamics governs the interaction between electromagnetic fields and electrically charged particle rigorously. Therefore, the quantum theoretical treatment of electromagnetic fields leads to the concept of photons (i.e., there can only be an integer number of photons being emitted or absorbed).

During materials processing with laser radiation, the result of laser treatment is predominantly determined by the power that is absorbed within the workpiece. A measure of the power that is available for the material treatment process $P_{\text{absorption}} = P_{\text{abs}}$ is the absorption A or absorptivity. The absorptivity is the ratio of power that is deposited within the workpiece and the power of the incident radiation:

$$A = \frac{P_{\text{abs}}}{P} \tag{5.21}$$

The absorptivity A can have any value between 0 and 1.0. The absorbed radiation energy is in general transformed to heat energy through the heat conduction process. This energy conversion can pass through several stages that can possibly be utilized for materials processing. The absorption of laser radiation can, for example, lead to the dissociation of molecules. Before this non-equilibrium state relaxes to an equilibrium state, i.e., transformation of the absorbed energy to heat energy, the dissociated molecules can be removed. Considering these conditions then we can express the material is in the ablation stage. Definition of absorptivity as such is a global value that in general contains no information that where in the workpiece the deposition radiation energy is taking place. But we know in case of metallic interaction with laser, the absorption always occurs in a thin surface layer of the metal, which is called *localized absorption*. Bear in mind that this information is not included in the absorptivity A.

The absorptivity can be determined directly by measuring the power of the incident laser beam radiation and the temperature increases of the workpiece knowing the heat capacity or indirectly by measuring the power of the reflected light P_r and the power that is transmitted through the workpiece P_t as:

$$P_{\text{abs}} = P - P_r - P_t \tag{5.22}$$

Here what we mean for P_t as the transmitted power is the power that is transmitted through the whole workpiece not where P_t is the power that is transmitted through the surface of the workpiece. [6].

If the radiation that is transmitted through the workpiece surface is totally absorbed within the workpiece, then, the absorptivity is given by:

$$A = 1 - R \tag{5.23}$$

where R represents the surface reflectivity of the workpiece. In the case of non-conducting materials or very thin metal films part of the radiation that penetrates into the workpiece can leave the workpiece and we should count on transmittance part of laser beam radiation and is describe by variable T, thus new form of Eq. 5.23 is going to take place as:

$$A = 1 - R - T \tag{5.24}$$

In general, for heat conduction calculation and mathematical approach of high-power laser as directed energy weapon, we can ignore the transmittance part; therefore Eq. 5.23 holds for purpose of these kinds of analysis. But we need to take under serious consideration the following laser interaction material as parameters:

- Index of refraction n normally real part of a complex number for general index
- Index of absorption k normally imaginary part of a complex number for general index
- Electric conductivity of materials σ that laser is dueling with
- Heat conductivity of materials K
- Specific heat of materials at constant pressure c_p

All of the above are material specific parameters. They solely depend on the properties of the target material at hand, and the radiation wavelength can be calculated approximately for simple materials.

Furthermore, the absorptivity not only depends on the above material properties, but it is influenced by the following factors:

- Properties of the laser beam such as wavelength and polarization
- The ambient conditions (i.e., process gas, materials that surround the workpiece, etc.)
- The surface properties (i.e., roughness, morphology, etc.)
- The geometry of the workpiece (i.e., thickness, boundaries of the workpiece, etc.)
- The changes of the workpiece such as phase due to latent heat of vaporization and introduction of plasma at the ablation surface where the frequency of laser beam and plasma are equal to each other and the environment that is induced by the absorbed laser power (i.e., local heating or spot size, phase changes, laser-induced plasma)

5.4.1 *Description of the Phenomena*

The interaction between electromagnetism and matter can be described by assuming that the matter consists of electrons and atomic kernels. For spatial dimensions $r \gg 10^{-15}$ m, the atomic kernel can be considered as a point charge, while the electron can also be considered as a point charge because of spatial dimension greater than the classical electron radius of $r > > r_0 \sim 2.8 \times 10^{-15}$. These point charges interact with electromagnetic fields or simply a propagating sinusoidal radiation field governed by Eq. 5.8 and get excited spatially and temporally fast fluctuating fields of their own part. The interaction between electromagnetic field and electrically charged particles is treated rigorously within the framework of classical quantum electrodynamics [6] as well as what is shown in Appendix F of this volume. The quantum theoretical treatment of electromagnetic fields leads to the concept of photons, i.e., there can only be an integer number of photons being emitted or absorbed (i.e., what was described in Sect. 5.1). If the particles that interact with electromagnetic fields of Eq. 5.8 are bound, these too can only absorbed or emit certain energy quanta. From an atomistic point of view, this is described as the absorption of a photon and the creation of a phonon or energy quantum of lattice oscillations.

Classical description of the spatio-temporal evolution of electromagnetic field, which is defined by the microscopic aspect of Maxwell Equations in vacuum is interacting, with these point charges of positively charged atomic kernels and negatively charged electrons.

The atoms can be defined as dipoles that are excited by external radiation, which emits radiation on their part that interferes with the primary radiation. This process can be interpreted as coherent scattering. If radiation hits a surface of a solid matter or slab, the radiation that is emitted by the dipoles at this solid surface consists of three parts, and they are as follows:

1. The first one corresponds to the reflected wave.
2. The second one is emitted in the same direction as the incident wave and according to the Ewald-Oseen [7] extinction theorem modulus and phase of this wave are such that the incident wave and this wave extinct each other exactly within region 2. See Fig. 5.5.
3. The third part corresponds to the refracted wave.

Further evaluation of electromagnetic phenomena revels to average over macroscopic small but microscopic large spatial regions. Therefore, in this manner, one gets the macroscopic Maxwell's equations. The macroscopic Maxwell's equations treat the matter as a continuum whose electromagnetic properties are described by material parameters of interest target. These parameters can either be measured or calculated using microscopic approaches.

$$\begin{aligned} \vec{\nabla} \times \vec{E} &= \frac{\partial \vec{B}}{\partial T} & \vec{\nabla} \times \vec{H} &= \vec{j} + \frac{\partial \vec{D}}{\partial T} \\ \vec{\nabla} \cdot \vec{D} &= \rho & \vec{\nabla} \cdot \vec{B} &= 0 \end{aligned} \tag{5.25}$$

The macroscopic Maxwell's equations (5.25) are presented here, and solutions of these sets of equation are provided in Chap. 4 of this book.

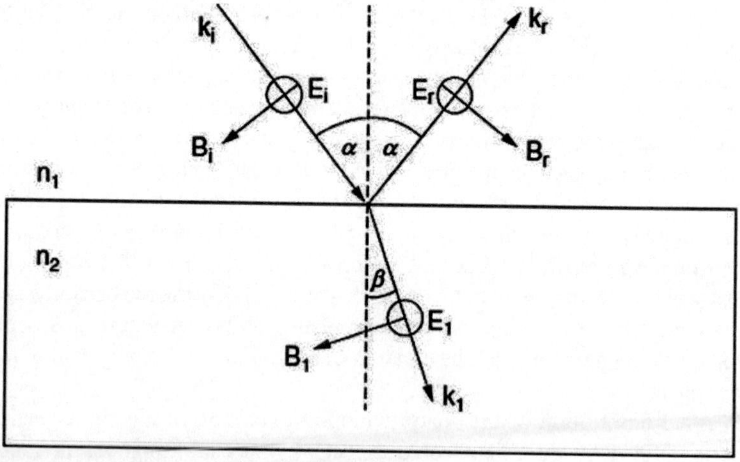

Fig. 5.5 Polarized perpendicular to the plane of incidence

5.5 The Behavior of Electromagnetic Radiation: An Interface

One of the major advantages of laser as a directed-energy weapon for engaging target materials processing is the ability to precisely control where in the material and at what rate energy is deposited. This control is exercised through the proper selection of laser processing parameters to achieve the desired material modification. In this section, we discuss the principles and equations that describe the propagation and absorption of laser energy and heat flow.

Most of the processes that occur during the interaction of laser radiation with matter start at the surfaces. For small intensities, the interaction of the electromagnetic field with the target surface is driven by Fresnel formula. The Fresnel formula describes the reflection and transmission of a plane harmonic wave incident on an infinitely extended ideal plane surface. Reflection r and transmission t of the field amplitudes are defined by the following relations:

$$r = \frac{E_r}{E_i} \tag{5.26}$$

$$t = \frac{E_t}{E_i} \tag{5.27}$$

with:

E_i = electric field of the incident wave.
E_r = electric field of the reflected wave.
E_t = electric field of the transmitted wave.

Although the Fresnel formula is establishing both Eqs. 5.26 and 5.27 and behavior at the surface of the metal target but is not capable describing anything about transmission or absorption within specimen.

The Fresnel formula is driven via Maxwell's equations (see Appendix F as well), considering the boundary conditions for the fields at the surface of interaction with target materials. This interaction can be divided into two groups of *perpendicular* ($\perp$) and *parallel* ($\|$) polarization that is also known as s-polarized (Senkrecht = perpendicular) and p-polarized (parallel), and we explain them further down in this chapter.

In case of perpendicular polarization, the electric field vector is perpendicular to the plane that is spanned by the incident and reflected wave vectors, whereas in case of parallel polarization, the field vectors are parallel to this plane. The wave vector $\vec{k}_i$, $\vec{k}_r$, and $\vec{k}_t$ all lie in one plane. The vector of the electric field can have components perpendicular ($\perp$) or parallel ($\|$) to this plane. Reflection and transmission depend on the direction of the electric field vector relative to the plane of incidence as illustrated in Fig. 5.6.

For practical applications, the polarization of the laser radiation is often chosen so that the field strength is either perpendicular or parallel to the plane of incidence, and in very special cases, circular polarization is used. In these cases, the decomposition of the field vectors in their perpendicular and parallel components can be omitted, and the mathematical treatment is simplified. Fig. 5.5 shows the situation in the case of normal incidence.

The angle α that the wave vectors of the incident and reflected wave, respectively, make with the surface normal and angle β of refracted wave are connected by Snell's law:

$$n_1 \sin \alpha = n_2 \sin \beta \tag{5.28}$$

n_1 and n_2 are the indices of refraction of the two media. The Fresnel formula for perpendicular polarization read [69]:

$$\frac{E_r}{E_i} = r_s = \frac{n_1 \cos \alpha - \frac{\mu_1}{\mu_2}\sqrt{n_2^2 - n_1^2 \sin^2 \alpha}}{n_1 \cos \alpha + \frac{\mu_1}{\mu_2}\sqrt{n_2^2 - n_1^2 \sin^2 \alpha}} \tag{5.29}$$

$$\frac{E_t}{E_i} = t_s = \frac{2n_1 \cos \alpha}{n_1 \cos \alpha + \frac{\mu_1}{\mu_2}\sqrt{n_2^2 - n_1^2 \sin^2 \alpha}} \tag{5.30}$$

with μ_1 and μ_2 for magnetic permeability of media 1 and 2, respectively.

Fig. 5.6 Incident, reflected, and transmitted beam lie in one plane. The electric field vector can be split into two components: one component is parallel, and the other component is perpendicular to this plane

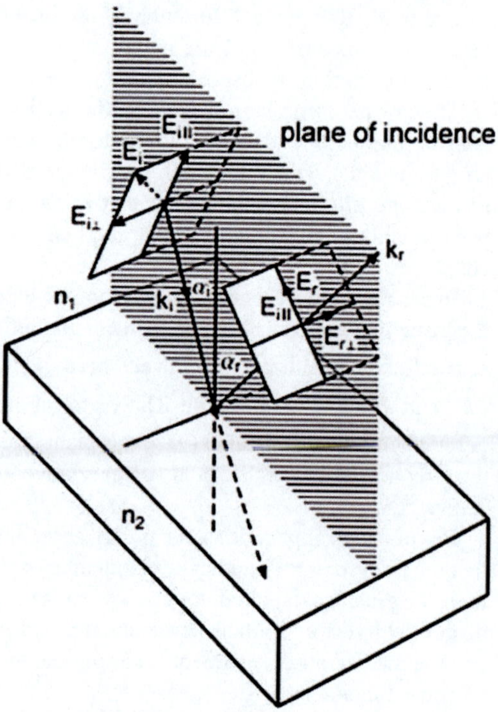

Figure 5.7 shows the situation for parallel polarization. The Fresnel formula for parallel polarization is read as (see Appendix G of Reference [69]):

$$\frac{E_r}{E_i} = r_p = \frac{\frac{\mu_1}{\mu_2} n_2^2 \cos\alpha - n_1\sqrt{n_2^2 - n_1^2 \sin^2\alpha}}{\frac{\mu_1}{\mu_2} n_2^2 \cos\alpha + n_1\sqrt{n_2^2 - n_1^2 \sin^2\alpha}} \tag{5.31}$$

$$\frac{E_t}{E_i} = t_p = \frac{2n_1 n_2 \cos\alpha}{\frac{\mu_1}{\mu_2} n_2^2 \cos\alpha + n_1\sqrt{n_2^2 - n_1^2 \sin^2\alpha}} \tag{5.32}$$

In the case of normal incidence, i.e., $\alpha = 0$, the plane of incidence cannot be defined uniquely anymore, and the difference between perpendicular and parallel polarization vanishes. In that case, Eqs. 5.29 and 5.31 and Eqs. 5.30 and 5.32 give the same results. The amplitudes of the reflected and transmitted waves in case of normal incidence are given by:

$$\frac{E_r}{E_i} = \frac{n_1 - n_2}{n_2 + n_1} \tag{5.33}$$

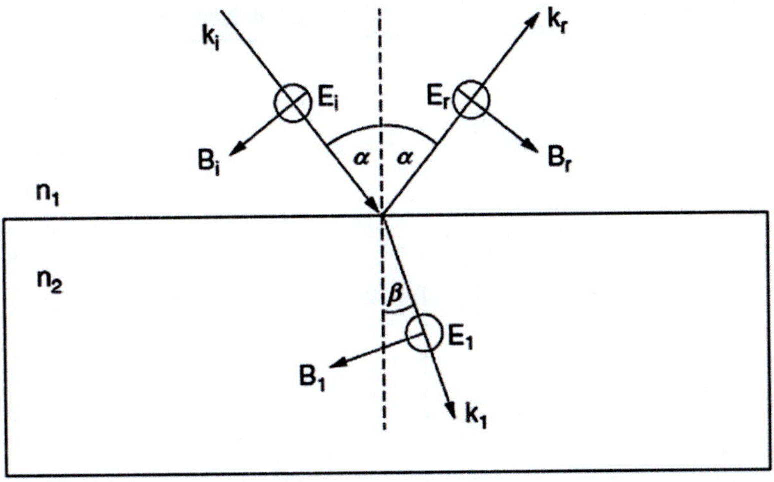

Fig. 5.7 The field vector of the reflected wave is often given in the opposite direction. In that case, Eq. 5.32 must be multiplied by -1

$$\frac{E_t}{E_i} = \frac{2n_1}{n_2 + n_1} \tag{5.34}$$

When $n_2 > n_1$ the reflected wave undergoes a phase change of 180°, and this holds for real indices of refraction.

5.5.1 *Light Propagation in Materials*

The primary interaction between laser radiation and a solid is photochemical excitation of electrons from their equilibrium states to some excited states by the absorption of photons.

Some of these transitions are schematically shown in Fig. 5.8. Interband transitions take place when photon energy is larger than band gap of the material. In this process, electron–hole pairs are generated. The free electrons may jump back from conduction band to valence band through thermal (dashed lines) or photochemical processes. If the photon energy is less than band gap of the material, the energy can be absorbed by defect levels in the band gap or produce intra-band transitions. Both transitions will induce thermal processes as electrons jump back to valence band. With higher laser light intensities, multi-photon absorption is favored, because the probability of nonlinear absorption increases strongly with laser intensity. The coherent multi-photon transitions would generate electron–hole pairs similar to inter-band transitions.

Thus, the initial electronic excitation is followed by complex secondary processes, which can be classified into thermal and photochemical processes. The type

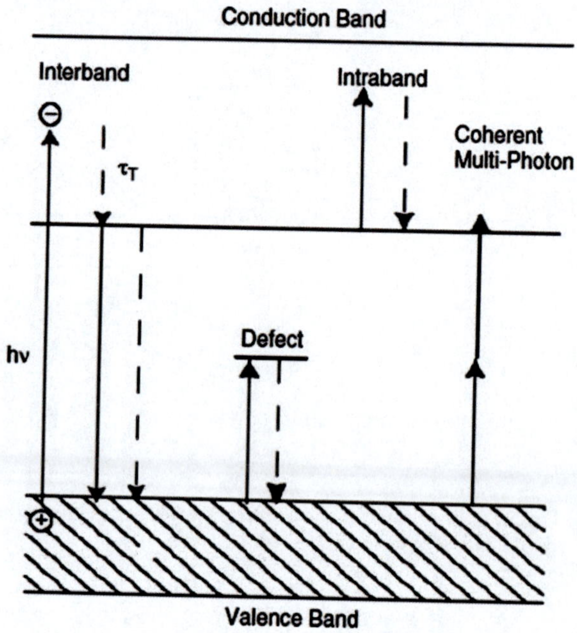

Fig. 5.8 Schematic of the different types of electronic excitation in a solid

of interaction between laser radiation and the material depends on laser parameters (wavelength, pulse duration, and fluence) and on the properties of the materials [26].

Laser ablation (material removal) can be analyzed on the basis of photo-thermal (purely photolytic), photochemical (purely photolytic), and photo-physical processes, in which both thermal and non-thermal mechanisms contribute to the overall ablation rate.

Confinement of deposited energy to desired regions on a material's surface can be achieved by controlling the laser's spatial intensity profile. The predominant methods for control include beam steering by fixed or galvanometric scanning mirrors, beam focusing through telescoping or converging optics, and beam shaping with homogenizers [12], amplitude masks, refractive elements [13], and diffractive optical elements [13]. However, one can also use more advanced optical devices such as spatial light modulators [14], deformable mirrors [15], and tunable acoustic gradient index (TAG) lenses [16] allowing real-time modulation of the beam's intensity profile on the surface. There has been extensive work in the area of beam shaping with a number of articles and books [17–18] as well as a chapter in this book devoted to the subject.

When light strikes the surface of a material, a portion will be reflected from the interface due to the discontinuity in the real index of refraction, and the rest will be transmitted into the material. The fraction of the incident power that is reflected from the surface R depends on the polarization and angle of incidence θ_i of the light as well as the index of refraction of the atmosphere n_1 and the material n_2. The reflection coefficients for the s-polarized and p-polarized components of the light can be calculated from the well-known Fresnel equations [19], which we provided as Eqs. 5.29 to 5.34 for normal incidents; now we re-write them in different forms [19]:

$$R_s = \left[\frac{E_r}{E_i}\right]^2 = \left[\frac{n_1 \cos(\theta_i) - n_2 \cos(\theta_t)}{n_1 \cos(\theta_i) + n_2 \cos(\theta_t)}\right]^2 \tag{5.35a}$$

$$R_p = \left[\frac{E_r}{E_i}\right]^2 = \left[\frac{n_1 \cos(\theta_t) - n_2 \cos(\theta_i)}{n_1 \cos(\theta_t) + n_2 \cos(\theta_i)}\right]^2 \tag{5.36a}$$

They are related to the transmission coefficients through $T_s = 1 - R_s$ and $T_p = 1 - R_p$.

For a perfectly flat surface, both Eqs. 5.35a and 5.36a produce the following forms of equation sets, 5.35b and 5.36b, assuming media 1 is a vacuum so $n_1 = 1$ and $n_2 = n$ for media or target index of refraction with φ being angle of incidence and parameters such as $\kappa =$ thermal diffusivity being target material property:

$$R_s = \frac{(n - \cos\varphi)^2 + \kappa^2}{(n + \cos\varphi)^2 + \kappa^2} \tag{5.35b}$$

$$R_p = \frac{\left(n - \frac{1}{\cos\varphi}\right)^2 + \kappa^2}{\left(n + \frac{1}{\cos\varphi}\right)^2 + \kappa^2} \tag{5.36b}$$

For the case of normally incident light on a flat surface, the above equations reduce to the more familiar form:

$$R = R_s = R_p = \left(\frac{n_1 - n_2}{n_1 + n_2}\right)^2 \tag{5.37}$$

which is analogous to Eq. 5.33.

A variation of the reflectivity with angle of incidence is shown in Fig. 5.9. At certain angles the surface electrons may be constrained from vibrating. Otherwise electrons would have to leave the surface, and they would be unable to do that (collective vibrational modes) without disturbing the matrix, i.e., absorbing the photon. Thus, if the electric vector is in the plane of incidence, the vibration of the electron is inclined to interfere with the surface, and absorption is thus high.

While if the plane is at right angles to the plane of incidence, then the vibration can proceed without reference to the surface, and reflection is preferred. There is a

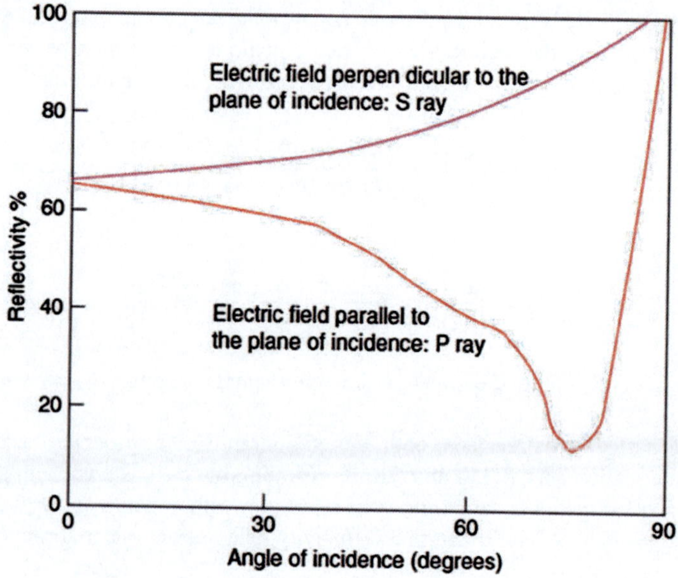

Fig. 5.9 Reflectivity of steel to polarized 1.064 μm radiations

particular angle—the "Brewster's" angle—at which the angle of reflection is at a right angle to the angle of refraction. When this occurs, it is impossible for the electric vector in the plane of incidence to be reflected since there is no component at right angles to itself. Thus, the reflected ray will have an electric vector only in the plane at right angles to the plane of incidence. At this angle the angle of refraction = 90°, which is the angle of incidence and hence by Snell's law the refractive index, $n = \tan$ (Brewster's angle). Any beam, which has only, or principally, one plane for the electric vector is called a "polarized" beam.

The reflectivity of a given material will depend on the frequency of the light source through the dispersion relation of its index of refraction. For instance, in the case of normal incidence, values for reflectivity of metals in the near UV and visible spectral range are typically between 0.4 and 0.95 and between 0.9 and 0.99 for the IR [20]. In addition, the reflectivity of a surface will depend on the temperature of the material through changes in the permittivity, band structure, plasma oscillations, or material phase [21]. For instance, upon melting, the reflectivity of silicon increases by a factor of about 2 [21], while that of a metal such as Ni changes by only a few percent [22]. In the case of small scale or structured materials, additional optical resonances are possible, such as surface and bulk plasmons and polaritons, which can lead to enhanced absorption or reflection due to the details of the photon–electron interactions [23].

Once inside the material, absorption causes the intensity of light to decay with depth at a rate determined by the material's absorption coefficient α. In general, α is a

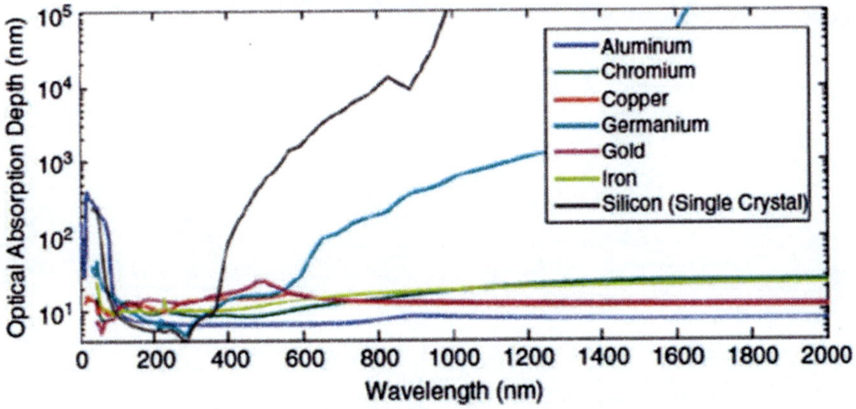

Fig. 5.10 Optical absorption depth for several materials over a range of wavelength [27]

function of wavelength and temperature, but for constant α, intensity I decays exponentially with depth z according to the Beer–Lambert law:

$$I(z) = I_0 e^{-\alpha z} \tag{5.38}$$

where I_0 is the intensity just inside the surface after considering reflection loss.

The magnitude of the gradient of intensity yields the volumetric energy deposition rate $\alpha I_0 e^{-\alpha z}$.

It is convenient to define the optical penetration or absorption depth, $\delta = 1/\alpha$, which is the depth at which the intensity of the transmitted light drops to $1/e$ of its initial value at the interface. Figure 5.10 shows optical absorption depths as a function of wavelength for a variety of metals and semiconductors. The important thing to note from Fig. 5.10 is that the absorption depths are short relative to bulk material dimensions. For instance, in the case of most metals undergoing UV illumination, the absorption depth is on the order of 10 nm. Although the interpretation of absorption depth was developed for a plane wave, the fact that energy absorption is approximately confined within the absorption depth still holds for more general beam profiles. Therefore, choosing wavelength with short absorption depths can allow local modification of surface properties without altering the bulk of the material.

The above treatments considered only linear optical phenomena; however, this is not necessarily the case in all materials, nor for all incident laser conditions. Some materials such as glasses exhibit strong nonlinearities in their index of refraction [24], which can lead to a number of interesting effects such as self-focusing, defocusing, or soliton propagation [25]. When dealing with CW or nanosecond duration laser pulses, it is typically assumed that most of the absorption is due to single photon interactions. However, for picosecond (ps) and femtosecond (fs) lasers, the extremely high instantaneous intensity enables phenomena such as

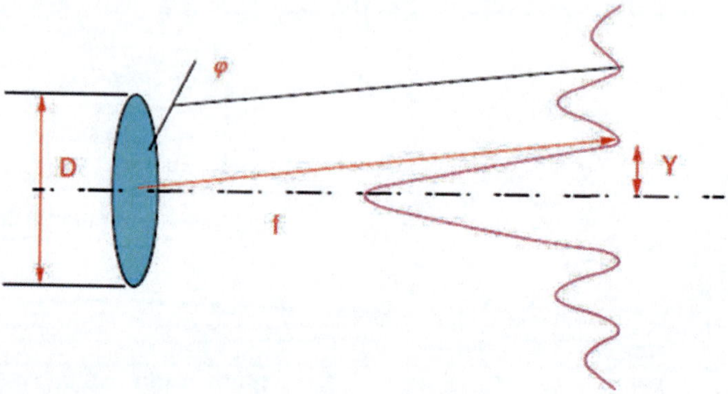

Fig. 5.11 Focus pattern of parallel laser light [32]

optical breakdown and multi-photon absorption which can significantly decrease absorption depths [26].

Another consideration of physics of high-power laser interaction with target materials is *focal spot size*, which determines the maximum energy density that can be achieved when the laser beam power is set to engage target surface or workpiece, and in the industry, this phenomenon also plays an important rule for materials processing.

In order to adjust the beam, guide it to the workpiece, and shape it, there are many devices. In nearly all of them the simple laws of geometric optics are sufficient to understand how they work. However, to calculate the precise spot size and depth of focus, one needs to refer to Gaussian optics and diffraction theory.

As we discussed in reference [69] by Zohuri (i.e., Sect. 5.2), a beam of finite diameter is focused by a lens onto a plane as shown in Fig. 5.11. When a beam of finite diameter D is focused by a lens onto a plane, the individual parts of the beam striking the lens can be imaged to be point radiators of new wavefront. The light rays passing through a lens will converge on the focal plane and interfere with each other, thus constructive and destructive superposition will take place, and light energy is distributed as shown in Fig. 5.11. The central maximum contains about 86% of the total power. The focusing diameter is measured between the points where the intensity has fallen to $1/e^2$ of the central peak value.

For a circular beam with a plane wavefront, the diffraction limited beam diameter, which is the smallest focal diameter ($\omega_{\min}$), is given by Eq. 5.39:

$$\omega_{\min} = 2.44\left(\frac{f \cdot \lambda}{D_L}\right) \tag{5.39}$$

where f is the lens focus length, λ is the light wavelength, and D_L is the unfocused beam diameter. Equation 5.30 is a representation of the upper limit of laser processing precision at a specific frequency, although in theory, laser should be free of these limitations, considering its coherency nature.

Equation 5.39 can be modified to the smallest possible focal spot size in this case:

$$\omega_0 = 2.44\left[\frac{f \cdot \lambda}{D}\right](2p + l + 1) \quad (5.40)$$

where f is the lens focal length, D is the beam diameter at the lens, λ is wavelength of the light, and p and l are the mode number.

The ideal laser should produce an intense, perfectly collimated beam of light that could be focused to a very small spot size. That tiny intense spot of light could be used to cut any shape in any materials. Unfortunately, such a well-focused high-intensity laser was hard to find before 1990, and even present ones are impractical for some laser cutting in industry.

The problem can be explained through the following equation:

$$D_F = M^2 \frac{4}{\pi} \cdot \lambda \cdot \frac{f}{D_L} \quad (5.41)$$

where D_F is the diameter of the laser beam in the focal plane of the focusing lens, M^2 is the beam quality factor, λ is wavelength of the laser light, f is the focal length of the focusing lens, and D_L is the diameter of the collimated laser beam on the focusing lens.

Further analysis of Eq. 5.41 indicates that if one wants a smaller diameter of the laser beam D_F, it requires either a smaller ration of f/D_L, or to make f very small and D_L very large. However, one should also consider the laser's Raleigh length R_L:

$$R_L = D_F \cdot \frac{f}{D_L} \quad (5.42)$$

which is the distance above and below the focal plane where diameter of the beam has increased by and the beam intensity has dropped by $\sqrt{2}$. In practice, one can cut very thick materials, which are as thick as Raleigh lengths. To cut thicker materials, one would need a large f/D_L ratio, so the only parameters left is the wavelength and beam quality M^2 to be adjusted to reduce the spot diameter D_F for the purpose of cutting very small objects with very small laser beam as a cutting tool or utilization of very-high-power laser as a directed-energy weapon.

Ultrafast laser oscillators generate pulse energy initially at nano-joule (i.e., 10^{-9}J) scale; amplification to milli-joule level is needed in micromachining, but one knows, when pulse lasting time is very small, peak energy intensity goes up far beyond the safe operation range of normal optical amplification systems. Ultrashort pulsed lasers successfully solved this difficulty using chirped pulse amplification (CPA) and pulse compression techniques. The energy of ultrafast lasers is highly concentrated in time domain, but their frequency distributions are much broader than normal laser systems. Light at different frequencies travel at different speeds through optical mediums, and this is not a very good thing at first glance, but contrary to one's intuition, this forms the base for the final solution. In free space, different components in a broadband laser pulse travel at nearly the same speed. When normal optical components are in the optical path, long wavelength light components travel through the medium faster than short wavelength components; thus pulse lasting time is stretched, and energy intensity is lowered. Special optical devices such as chirped mirrors or special prism pairs are used to compensate the pulse spreading; they allow short light components pass through faster than long wavelength components, so they compress pulse lasting time.

Further analysis of Eq. 5.40 clearly shows not only the influence of modes on the focal property, but other factors that also affect focal spot size, such as spherical aberration and thermal lens effects, should be taken under consideration as well.

Most lenses are made with a spherical shape, but they cannot be of perfect shape because there exists spherical aberration. Lenses in laser systems transmit or reflect high-power laser radiation. However, laser power variations can cause shape changes of the lenses, so the focal point will change when the radiation power changes, thus affecting the focal spot size.

According to Lasag[33], manufacturer of Nd:YAG laser system, the spot diameter of laser beam focused onto the workpiece and Fig. 5.12 can be approximated as:

$$2\omega \approx \frac{2f\theta}{M} \tag{5.43}$$

where f is the focal length of the lens, θ is the divergence of the laser beam before the expander, and M is the expansion factor of the beam expander.

5.5.2 *Depth of Focus*

The laser light is first converged at the lens focal plane and then diverged to a wider beam diameter again. The *depth of focus* (DOF) is the distance over which the focused beam has about the same intensity. This is also called beam waist (Fig. 5.13). It is defined as the distance over which the focal spot size changes ±5%. The equation for DOF is (Lasag 1997) [33]:

$$\text{DOF} = \left(\frac{8\lambda}{\pi}\right)\left[\frac{f}{D}\right]^2 = 2.44\lambda\left[\frac{f}{D}\right]^2 \tag{5.44}$$

where λ is the wavelength, f is the lens focal length, and D is the unfocussed beam diameter. Usually longer depth of focus is preferred, because equal energy density along the beam is preferred when using the laser to process materials.

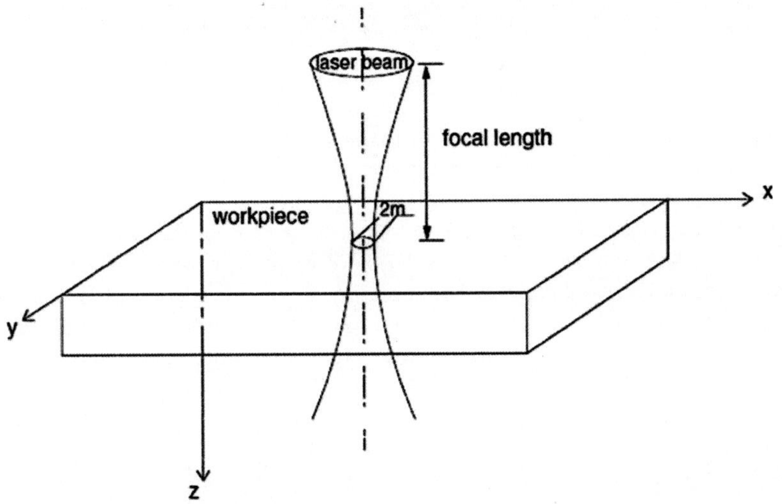

Fig. 5.12 Schematic of laser beam impinging on a workpiece

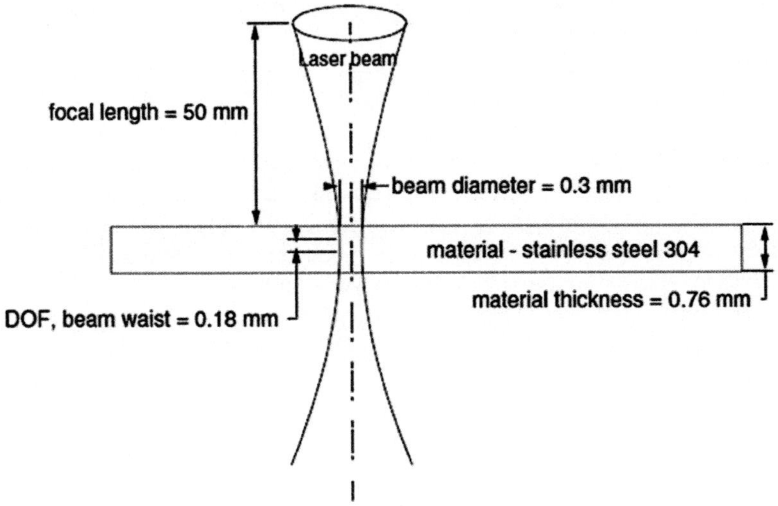

Fig. 5.13 Laser beam focusing onto the sample

Position of the focal plane is known to have an effect on the final shape of the hole as well as degree of penetration. As the laser focusing spot moves up and down, the laser material interaction area $A\ r_F$ varies too, according to expression (Semak and Matsunawa 1997) [34].

$$Ar_F = n\omega_0 \left[1 + \left(\frac{z - z_\omega}{2\omega_0 f/D} \right)^2 \right]^{\frac{1}{2}} \tag{5.45}$$

where $\omega = \omega_0$ is the laser beam radius at the beam waist equal to ω_0 which is the minimum radius of the laser beam. In our analytical and numerical calculations, Eq. 5.11 is proved to be the most accurate one to use.

While calculating the drilled-hole profiles, defocusing of the laser beam effects should be taken into account. As a rule, the divergence of the laser beam would increase the radial heat flux on the hole walls, by decreasing energy at the entrance of laser beam of the evolving hole. However, multiple reflections inside the hole will accumulate the energy at the bottom of the drilled hole, which was described before.

5.5.3 *Laser Beam Quality*

The concept of M^2 is important to describe actual propagation of laser beams. M^2 is a beam quality that measures the difference between the actual beam and the Gaussian beam. In order to find out the M^2 of a laser system, first we need to measure the spot size along the laser optical axis.

Edge method is used if beam profile meter is not available. Edge method uses a knife edge to block the laser beam, and a power meter measures the power after the blocking of the knife edge, by recording the 86% and 14% location of the full power. Subtracting the two values, one gets the beam spot size at that point. This method measures the $1/e^2$ radius of the laser beam. Because the laser spot size is rather small, the relative measurement error for such dimension using edge method can be large. Usually a beam expander, or a collimator, is used to expand and "parallel" the beam, and the spot size out of the collimator is several millimeters. For such dimension, knife-edge method can easily reduce the relative measurement error to less than 2%. As illustrated by Fig. 5.14, six measurements at three different distances from the collimator are taken, Z_n and D_n, where $n = 1, 2, 3$. The distance from any chosen point along the optical axis and the spot size at that location are recorded.

The beam size at location Z_n satisfies the following equation:

$$D_n^2 = D_0^2 + \left[\frac{4M^2\lambda}{\pi} \right]^2 \left[\frac{Z_n - Z_0}{D_0} \right] \quad \text{for} \quad n = 1, 2, 3 \tag{5.46}$$

where D_n is the beam size at location Z_n, D_0 is the beam waist, Z_0 is the beam waist location, λ is the wavelength, and M^2 is the beam quality parameter which is unknown. In this relation, λ is known, D_n and Z_n can be measured, and M^2, D_0, and Z_0 are also unknowns.

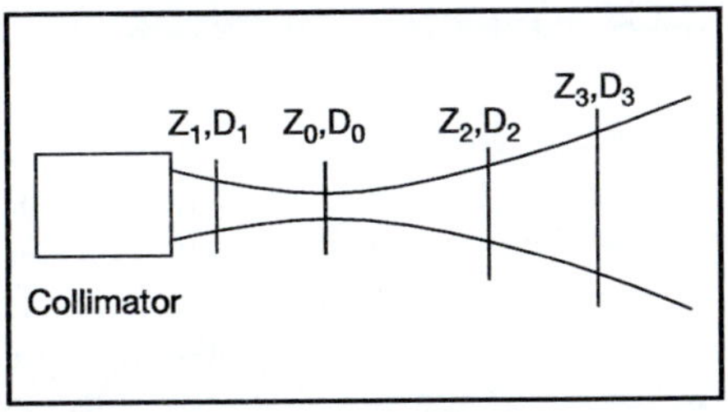

Fig. 5.14 Measurement of beam properties

Taking the measured data into Eq. 5.41, we get three highly nonlinear equations for three unknowns. One can solve these equations using, e.g., Mathcad. Knowing M^2, one can calculate the beam divergence, focused spot size, and depth of focus (DOF) as follows:

$$D_{\min} = \frac{4fM^2\lambda}{\pi D_L} \tag{5.47}$$

$$D_{\text{Gaussian}} = \left[\frac{Af}{\pi D_0}\right] \tag{5.48}$$

$$\theta_{\text{act}} = \frac{M^2\lambda}{\pi D_0} \qquad \theta_{\text{infinity}} = \left[\frac{D_3 - D_2}{2(Z_3 - Z_2)}\right] \tag{5.49}$$

$$\text{DOF} = \pm 0.08\lambda\left[\frac{D_{\min}}{M^2\lambda}\right] \tag{5.50}$$

where D_L is the laser beam size when it propagates to the front side of the focus objective lens, f is the focus length, $D_{\min}$ is the minimum beam diameter that can be achieved, and θ_{act} is the real beam divergence. It can be verified that the equations are solved with an error less than 10^{-10}. In practical cases, one should first measure the pulse energy or average power of the laser beam and then measure the beam spot size along the optical axis. Using Eqs. 5.47 and 5.48, one can calculate the beam waist and beam waist location, and one can find M^2 of the beam and then calculate the other indexes in Eq. 5.50. Knowing DOF, one can calculate M^2 value as:

$$M^2 = \left[0.08\frac{(D_{\min})^2}{\lambda(\text{DOF})}\right]^{1/2} \tag{5.51}$$

where $D_{\min}$ is the minimum beam diameter that can be achieved. Knowing the diameter at any location, one can determine the intensity of the laser beam at that location.

5.5.4 Spherical Aberration

There are two reasons why a lens will not focus to a theoretical point. First is the diffraction limited problem as discussed in Sect. 7.2, and the second is that lens is not of a perfect shape. Most lenses are made with a spherical shape since this can be accurately manufactured without too much cost and alignment of the beam is not so critical as with a perfect aspheric shape. The net result is that the outer ray entering the lens is brought to a shorter axial focal point than the rays nearer the center of the lens. This leaves a blur in the focal point location. The plane of best geometric focus is a little short of the plane wavefront (paraxial point). The size of the minimum spot $d_{\min}$ is given by:

$$d_{\min} = K(n,q,p)\left(\frac{D_L}{f}\right)^3 S_a = 2\Theta_a S_a \tag{5.52}$$

where Θ_a is the angular fault (half angle), S_a is distance from lens, D_L is diameter of top hat beam mode on lens, f is the focal length of the lens, $K(n, q, p)$ is the factor dependent on the refractive index n, q is the lens shape, and p is lens position:

$$K(n,q,p) = \frac{1}{128n(n-1)} \times \left[\frac{n+2}{n-1}q^2 + 4(n+1)pq + (3n+2)(n-1)p^2 + \frac{n^3}{n-1}\right] \tag{5.53}$$

where q is the lens shape factor $= (r_2 + r_1)/(r_2 - r_1)$, r_1, r_2 are the radii of curvature of the two faces of the lens, and p is position factor $= 1 - 2f/S_a$.

5.5.5 Thermal Lens Effect

In optical elements, which transmit or reflect high power radiation, there will be some heating of the component, which will alter its refractive index and shape (Bronski 2003) [35]. As the power, or the absorption, changes, so will the focal point and spot size.

The two main elements usually concerned are the output coupler and the focusing lens. The beam guidance mirrors could also be of concerned if adequate cooling is not supplied (water cooling or dry air cooling).

Thermal lensing is mainly caused by the rise in temperature increasing the refractive index (dn/dT) and thus shortening the focal length. A lesser effect is the thermal distortion (dl/dT). The focal length shift for thin lenses can be calculated to be:

$$\Delta F = \frac{2APF^2}{\pi k D_L^2}\left(\frac{dn}{dT}\right) \tag{5.54}$$

where A is the optic absorptance, k is the thermal conductivity, T is the temperature, and P is incident power of the laser beam.

Uneven heating causes further complications. The approximate Gaussian power distribution of the incident beam heats the middle more than the edge causing a radial temperature gradient (usually the edge of the lens is being cooled). A typical temperature difference would be 14° for a 1500 W beam of 38 mm diameter passing through an optic with 0.2% absorptance.

At this time let us mention another problem. During laser micromachining, from the fact that the absorption is on the surface of the lens, there is also going to be a temperature gradient in the depth direction. The thicker the optic, the more bowed will be the internal isotherms. Such an aberration will affect the M^2 value, transverse mode, and the spatial distribution of the laser beam as well.

5.6 Theoretical Discussion of Laser Absorption and Reflectivity

This subject was touched upon in Sect. 7.3, and we expand it further again. To consider the coupling of the laser energy to a material, we need first to know the optical reflectivity R and the transmissivity T for light incident on a surface which divides two semi-infinite media. The transmissivity plus the reflectivity equals unity at a *single surface*:

$$R + T = 1 \tag{5.55}$$

In most practical situations, we are dealing with more than one surface; typically, we have a slab of material with light impinging on one surface. Some light is reflected, and the rest either is absorbed or passes completely through the slab. In such a situation, we shall describe the net result of all the reflection, after multiple passes inside the slab and appropriate absorption have been accounted for, in terms of the reflectance R, the absorptance A, and the transmittance T:

$$R + T + A = 1 \tag{5.56}$$

The absorption of laser energy takes place through photon interaction with bound and free electrons in the material structure, which raises them to higher energy levels. Energy conversion takes place through various collision processes involving electrons, lattice phonons, ionized impurities, and defect structures.

If the surface being machined reflects too much light energy, the absorbed energy is decreased, the operation efficiency is lowered, and the reflected light may do harm to the optical systems. Reflection and absorption of laser beams are closely related to laser micromachining or dueling with target surface materials. The value of absorption and reflection is shown in Eq. 5.56.

What we really are interested in from the point of view of material response is A, the absorptance of the material. In most materials of interest from the practical aim of using lasers to melt, weld, etc., T is 0; thus Eq. 5.56 reduces to:

$$R + A = 1 \tag{5.57}$$

In metals, radiation is predominantly absorbed by free electrons in an "electron gas." These free electrons are free to oscillate and reradiate without disturbing the solid atomic structure. Both Eqs. 5.56 and 5.57 are analogous to Eqs. 5.23 and 5.24 for what we found out and discussed in Sect. 5.4. As an electromagnetic (*EM*) wavefront arrives at a surface of the target, then all the free electrons in the surface vibrate in phase generating an electric filed 180° out of phase with the incoming beam creating "electron gas." This "electron gas" within the metal structure means that the radiation is unable to penetrate metals to any significant depth, only 1–2 atomic diameters (or free paths); thus metals are opaque, and they appear shiny.

According to Fig. 5.15, reflectivity decreases as wavelength becomes shorter, while absorption increases when photon energy increases.

If sufficient energy is absorbed, then the vibration becomes so intense that the molecular bonding is stretched so far that it is no longer capable of exhibiting mechanical strength, and the material is said to have melted. On further heating, the bonding is further loosened due to the strong molecular vibrations, and the material is said to have evaporated. The vapor is still capable of absorbing the radiation, but only slightly since it will only have bound electrons. The exception occurs if the gas is sufficiently hot so that electrons are shaken free, and the gas is then said to be in plasma status.

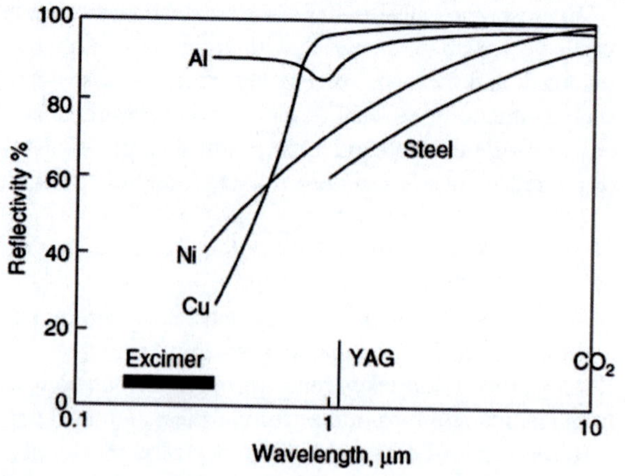

Fig. 5.15 Reflectivity as a function of wavelength for different metals

To understand reflectivity, we must use some general results from the theory of electromagnetic waves. Let us summarize these briefly at this point. The electric field of the electromagnetic wave, from Eq. 5.7, is:

$$\mathcal{E}(z,t) = \Re e\left[\mathcal{E}_0 e^{-2\pi kz/\lambda} e^{i\omega(t-nz/c)}\right] \qquad \text{(Eq.5.7 as before)}$$

The relationships we need are those among the index of refraction n, the extinction coefficient k, and the material properties. These relationships can be derived by substituting Eq. 5.7 in the wave equation as:

$$\frac{\partial^2 \mathcal{E}(z,t)}{\partial z^2} = \frac{1}{\mathrm{v}^2}\frac{\partial^2 \mathcal{E}(z,t)}{\partial t^2} \tag{5.58}$$

where V is the phase velocity of the wave, and for traveling wave in media of density, the phase velocity is given by:

$$\mathrm{v} = f\lambda = \nu\lambda \quad \text{and} \quad \omega = 2\pi f = 2\pi\nu = \frac{2\pi}{\lambda}$$

In case media as vacuum $\mathrm{v} = c$ speed of light in vacuum.

If we get twice derivation of Eq. 5.7 and substitute the result in Eq. 5.58, it results in:

$$\frac{\partial^2 \mathcal{E}}{\partial z^2} = \mu\varepsilon\frac{\partial^2 \mathcal{E}}{\partial t^2} + \mu\sigma\frac{\partial \mathcal{E}}{\partial t} \tag{5.59(a)}$$

Equation 5.59(a) results in the expression of Eq. 5.59(b) below as:

$$\left(\frac{2\pi k}{\lambda} + \frac{i\omega n}{c}\right)^2 = \mu\varepsilon\left(-\omega^2\right) + i\omega\mu\sigma \tag{5.59(b)}$$

Note that we are using rationalized MKS units throughout. The material properties enter through μ, ε, and σ, which are the magnetic permeability, the dielectric function, and the electric conductivity of the medium. Using the usual equations between the field vectors:

$$\vec{\boldsymbol{D}} = K_e\varepsilon_0\vec{\boldsymbol{\mathcal{E}}} \tag{5.60(a)}$$

$$\vec{\boldsymbol{B}} = K_m\mu_0\vec{\boldsymbol{H}} \tag{5.60(b)}$$

$$\vec{\boldsymbol{J}} = \sigma\vec{\boldsymbol{\mathcal{E}}} \tag{5.60(c)}$$

We have:

$$\varepsilon = K_e \varepsilon_0 \tag{5.60(d)}$$

$$\mu = K_m \mu_0 \tag{5.60(e)}$$

In Eqs. 5.60, ε_0 and μ_0 are the electric permittivity and magnetic permeability of a vacuum. K_e is the dielectric constant and K_m the magnetic permeability of the material. By substituting Eqs. 5.60(d) and 5.60(e) into Eq. 5.59(b) and using $(2\pi/\lambda) = (\omega/c)$, we obtain that:

$$(k + in)^2 = -K_m K_e \varepsilon_0 \mu_0 c^2 + iK_m \mu_0 \sigma \left(\frac{c^2}{\omega}\right) \tag{5.61}$$

Finally, if we introduce $c^2 = (\varepsilon_0 \mu_0)^{-1}$ and do some algebra, we get:

$$n - ik = \sqrt{K_m} \sqrt{K_e - i\frac{\sigma}{\varepsilon_0 \omega}} \tag{5.62}$$

This equation relates the material parameters K_e, K_m, and σ, which in general may be complex, to index of refraction n and extinction coefficient k. To describe the propagation of the light wave thus requires a knowledge of K_e, K_m, and σ.Before we describe these, let us look at two more general properties of our propagating electromagnetic wave.

The first of these is absorption. If the medium is absorbing, the intensity will fall off to $1/e$ of its initial value in a distance δ, obtained by setting $\mathcal{E}^2$ of Eq. 5.7 equal to $(1/e)\mathcal{E}^2_{\max}$, or:

$$\frac{4\pi k \delta}{\lambda} = 1 \tag{5.63(a)}$$

Or:

$$\delta = \frac{\lambda}{4\pi k} \tag{5.63(b)}$$

This shows why k is called the extinction coefficient, for it determines skin depth δ. Equation 5.63(b) is fairly general in that once k is known, δ can be calculated. As noted, a knowledge of the material properties is required to calculate k.

The second general property we wish to derive is the expression for reflectivity, in terms of n and k. To do this, consider light impinging normally onto an ideal solid surface, as shown in Fig. 5.16. Here we have illustrated the incident ($\vec{\mathcal{E}}_i$), reflected ($\vec{\mathcal{E}}_r$), and transmitted ($\vec{\mathcal{E}}_t$) electric waves at a vacuum–material interface. For the

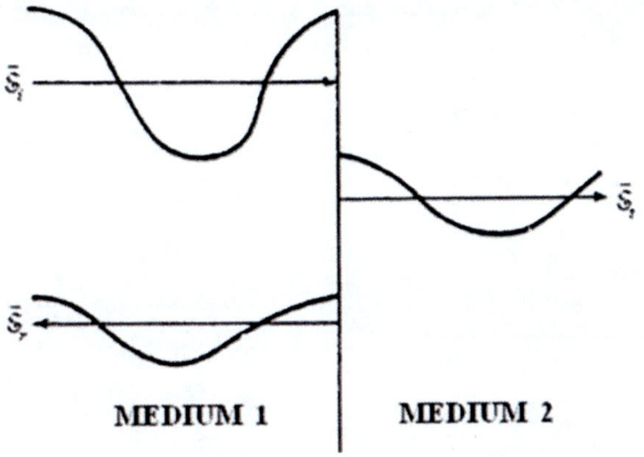

Fig. 5.16 Incident, transmitted, and reflected electric vectors at an interface

present, we limit our discussion to the case of normal incidence. We now consider the boundary condition. We electric field, we have:

$$\mathcal{E}_i + \mathcal{E}_r = \mathcal{E}_t \tag{5.64}$$

For the magnetic field $\vec{B}$, we can write:

$$B_i - B_r = B_t \tag{5.65}$$

The minus sign is before B_r because $\vec{\mathcal{E}} \times \vec{B}$ is positive in the direction of propagation of the wave. Now, the relationship between $\vec{B}$ and $\vec{\mathcal{E}}$, or, since $\vec{B} = \mu\vec{H}$, between $\vec{H}$ and $\vec{\mathcal{E}}$, is required in order to proceed further.

This follows directly from Maxwell's equations:

$$\nabla \times \vec{\mathcal{E}} = -\mu \frac{\partial \vec{H}}{\partial t} \tag{5.66}$$

$$\nabla \times \vec{H} = \sigma\mathcal{E} + \varepsilon \frac{\partial \mathcal{E}}{\partial t} \tag{5.67}$$

It is convenient to rewrite Eq. 5.7 and introduce $\omega\lambda = 2\pi c$, to have $\mathcal{E}$ explicitly in terms of ω instead of both ω and λ. Recall that $\vec{\mathcal{E}}$ is a vector, and take it as being along the x– direction. Thus, we can write:

$$\mathcal{E}_x = \mathcal{E}_0 e^{i\omega t} e^{-\frac{i\omega}{c} z(n-ik)} \tag{5.68}$$

Here we have dropped the "Re" notation and shall simply note that we always mean the real part when we write the wave in exponential form. We shall use unit vectors $\widehat{x}$, $\widehat{y}$, and $\widehat{z}$.

Now the curl expression reduces to:

$$\Delta \times \vec{\mathcal{E}} = \widehat{y}\frac{\partial \mathcal{E}_x}{\partial z} \tag{5.69}$$

which, with Eq. 5.66, tells us that $\vec{H}$ n has only a y–component, so, we have:

$$\Delta \times \vec{H} = -\widehat{x}\frac{\partial H_y}{\partial t} \tag{5.70}$$

Thus Eqs. 5.66 and 5.67 become:

$$\frac{\partial \mathcal{E}_x}{\partial z} = -\mu\frac{\partial H_y}{\partial t} \tag{5.71}$$

and

$$-\frac{\partial H_y}{\partial z} = \sigma \mathcal{E}_x + \varepsilon\frac{\partial \mathcal{E}_x}{\partial t} \tag{5.72}$$

and, of course, $\mathcal{E}_y = \mathcal{E}_z = H_x = H_z = 0$. Putting the expression for $\mathcal{E}_y$ from Eq. 5.68 to Eq. 5.71 leads to:

$$H_y = \frac{n - ik}{\mu c}\mathcal{E}_0 e^{-\frac{i\omega}{c}z(n-ik)}e^{i\omega t} \tag{5.73}$$

This is a desired relationship:

$$H_y = \left(\frac{n - ik}{\mu c}\right)\mathcal{E}_x \tag{5.74}$$

At this point we note in passing that Eq. 5.72 or 5.59(b) could be used to yield the relationship of n and k to μ, ε, and σ. If the reader is unfamiliar with these relationships, it is instructive to carry out the algebra.

Returning to our consideration of the reflected electric and magnetic fields, we rewrite Eqs. 5.64 and 5.65 with the help of the relationship between H and $\mathcal{E}$, from Eq. 5.74:

$$\mathcal{E}_i + \mathcal{E}_r = \mathcal{E}_t$$

and

$$\mu_1 H_i - \mu_1 H_r = \mu_2 H_t$$

become

$$\mathcal{E}_i - \mathcal{E}_r = \left(\frac{n_2 - ik_2}{n_1 - ik_1}\right)$$

Solving for $\mathcal{E}_r/\mathcal{E}_i$ by eliminating $\mathcal{E}_t$, we get:

$$\frac{\mathcal{E}_r}{\mathcal{E}_i} = \frac{n_1 - n_2 - i(k_1 - k_2)}{n_1 + n_2 - i(k_1 + k_2)}$$

Finally, the reflectivity R at the surface is:

$$R = \left|\frac{\mathcal{E}_r}{\mathcal{E}_i}\right|^2 = \frac{(n_1 - n_2)^2 + (k_1 - k_2)^2}{(n_1 + n_2)^2 + (k_1 + k_2)^2} \tag{5.75}$$

Take medium 1 as a vacuum and drop the subscript 2, since in a vacuum $n_1 = 1$ and $k_1 = 0$, then Eq. 5.75 reduces to:

$$R = \frac{(n-1)^2 + k^2}{(n+1)^2 + k^2} \tag{5.76}$$

Equation 5.76 is the second relationship we will find useful in discussing the coupling of optical radiation with metals. Note that it is derived for the special case of normal incidence and is applicable to a vacuum–material interface.

5.6.1 *Reflectivity of Materials at Infrared Wavelength*

We turn now to a derivation of the optical reflectivity of metals for infrared wavelengths, where experiment has shown that the free-electron theory (sometimes called the Drude–Lorentz theory) of metals is adequate. This theory rests on three assumptions:

1. The first is that electromagnetic radiation interacts only with the free electrons in the metal.
2. The second is that the free electrons obey Ohm's law, or, more specifically, that:

$$m^* \frac{dv}{dt} + \frac{m^*}{\tau} v = -e\mathcal{E} \tag{5.77}$$

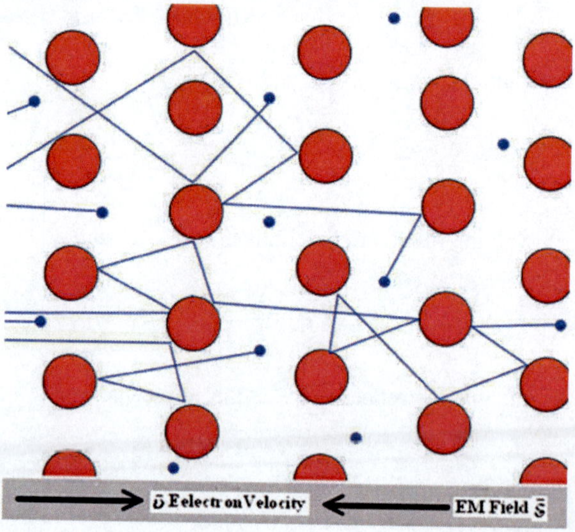

Fig. 5.17 Drude–Lorentz model. Electrons (shown here in blue or small dot) constantly bounce between heavier, stationary crystal ions (shown in red larger dot)

where m^* is the effective mass of the electron, υ the drift velocity of electro, τ the relaxation time due to collisions (i.e., mean free time between ionic collisions; see Fig. 5.17), and $-e\mathcal{E}$ the force on the electron due to the electromagnetic field.

3. The third assumption is that the free electrons of a metal can be described in terms of a single effective mass, carrier concentration, and relaxation time.

There has been a good deal of discussion about the validity of these assumptions in the literature. Recent work [10] indicates that, for wavelengths in the intermediate infrared (a few microns to many tens of microns) and beyond, the free-electron theory does a reasonable job of predicting the reflectivity of metals.

To derive the free-electron optical reflectivity, we try solutions to Eq. 5.77 of the form:

$$\upsilon \approx e^{i\omega t}$$

So that Eq. 5.77 with substitution of above relation results in:

$$\left[m * (i\omega) + \frac{m*}{\tau}\right]\upsilon = -e\mathcal{E} \tag{5.78}$$

Manipulating Eq. 5.78, we can find that:

$$\upsilon = -\frac{e\tau}{m * (1 + i\omega\tau)}\mathcal{E} \tag{5.79}$$

Now the current flow obeys:

$$\vec{J} = \sigma\vec{\mathcal{E}} = -Ne\vec{v} \tag{5.80}$$

where N is the electron concentration (number of electron per unit volume). By comparison of the last two equations:

$$\frac{\sigma}{Ne} = \frac{e\tau}{m^*(1 - i\omega\tau)} \tag{5.81}$$

Or:

$$\sigma = \frac{Ne^2\tau}{m^*(1 - i\omega\tau)} \tag{5.82}$$

Now the dc conductivity is:

$$\sigma_0 = \frac{Ne^2\tau}{m^*} \tag{5.83}$$

We see σ as a complex quantity and seek to write it as the sum of a real and imaginary part. Thus:

$$\sigma = \frac{Ne^2\tau(1 - i\omega\tau)}{m^*(1 + \omega^2\tau^2)} \tag{5.84}$$

Define:

$$\sigma = \sigma_1 - i\sigma_2 \tag{5.85}$$

The result is:

$$\sigma_1 = \frac{\sigma_0}{1 + \omega^2\tau^2} \tag{5.86(a)}$$

$$\sigma_2 = \frac{\sigma_0\omega\tau}{1 + \omega^2\tau^2} \tag{5.86(b)}$$

To proceed further, we need *to* use the general expression for electromagnetic waves developed in Sect. 5.6. Recall Eq. 5.62:

$$n - ik = \sqrt{K_m}\sqrt{K_e - i\frac{\sigma}{\varepsilon_0\omega}} \qquad \text{(previous 5.62)}$$

And from the complex σ of Eq. 5.85, we can write:

$$n - ik = \sqrt{K_m}\sqrt{K_e - i\frac{\sigma_1 - i\sigma_2}{\varepsilon_0\omega}} \tag{5.87}$$

If we assume only free-electron optical interactions, the metal does not polarize under the wave, and $K_e = 1$. In addition, for metals in the infrared, $K_m = 1$. Thus:

$$n - ik = \sqrt{K_m}\sqrt{K_e - i\frac{\sigma_1 - i\sigma_2}{\varepsilon_0\omega}} \tag{5.88(a)}$$

Or:

$$n - ik = \sqrt{K_m}\sqrt{K_e - \frac{\sigma_2}{\varepsilon_0\omega} - i\frac{\sigma_1}{\varepsilon_0\omega}} \tag{5.88(b)}$$

It remains only to separate the real and imaginary parts of Eq. 5.62, which will yield two equations in n and k and thus give n and k in terms of the dc conductivity and/or the relaxation time τ. Then we can use our expression for the reflectivity from Eq. 5.76 to generate R from n and k.

To carry out the algebra we use the identity:

$$\sqrt{A + iB} = \sqrt{\frac{R+A}{2}} + i\sqrt{\frac{R-A}{2}}$$

where

$R = \sqrt{A^2 + B^2}$

Letting:

$$A = 1 - \frac{\sigma_2}{\varepsilon_0\omega}$$

and

$$B = -\frac{\sigma_1}{\varepsilon_0\omega}$$

we have:

$$2n^2 = \left(i - \frac{\sigma_2}{\varepsilon_0\omega}\right) + \sqrt{\left(1 - \frac{\sigma_2}{\varepsilon_0\omega}\right)^2 + \left(\frac{\sigma_1}{\varepsilon_0\omega}\right)^2} \tag{5.89(a)}$$

$$2k^2 = -\left(1 - \frac{\sigma_2}{\varepsilon_0\omega}\right) + \sqrt{\left(1 - \frac{\sigma_2}{\varepsilon_0\omega}\right)^2 + \left(\frac{\sigma_1}{\varepsilon_0\omega}\right)^2} \qquad (5.89(b))$$

Equations 5.89(a) and 5.89(b), together with Eq. 5.76, give the reflectivity of R. Notice that R is a function solely of σ_1, σ_2, and ω. Look again at Eq. 5.86, and note that σ_0 can be used to replace τ in the expression for σ_1 and σ_2:

$$\sigma_1 = \frac{\sigma_0}{1 + \omega^2 \frac{m*^2 \sigma_0^2}{N^2 e^4}} \qquad (5.90(a))$$

$$\sigma_2 = \frac{\sigma_0^2 \frac{m*}{Ne^2} \omega}{1 + \omega^2 \frac{m*^2 \sigma_0^2}{N^2 e^4}} \qquad (5.90(b))$$

Equation 5.90 shows that σ_1, σ_2, and thus R depend on frequency ω, constant m^*/N, and dc conductivity σ_0. Thus:

$$R = f(\omega, \sigma_0, m*/N) \qquad (5.91)$$

This means that we can use the dc conductivity to predict the reflectivity. Furthermore, if we know the temperature variation of σ_0, we can use this method to calculate R as a function of temperature. This is a useful result, because it is difficult to measure optical reflectivity as a function of temperature, whereas it is fairly easy to measure σ_0 vs. temperature.

A wealth of data on electrical conductivity has been amassed for most metals and alloys. Thus, the free-electron model is currently enjoying a great deal of attention as a way of providing reflectivity-vs-temperature information in the study of laser effects.

There is, of course, one problem in using $\sigma_0(T)$ data to predict R, and that is the parameter $m*/N$. It turns out that R is fairly insensitive to this parameter at infrared wavelengths. To see this, we show here some numerical illustration. Define:

$$\beta = {}^{m*}/_{m_0} \qquad (5.92)$$

where m_0 is the free electron mass and then the parameter β/N is equivalent to $m*/N$.

Figure 5.18 shows a plot of $\partial R/\partial(\beta/N)$ as a function of β/N for $\lambda = 10$ μm and various values of σ_0. Here σ_0 is in units of reciprocal ohm-centimeters. Typical values are, for example, $\sigma_0 = 10^5 \Omega^{-1}\text{cm}^{-1}$ and $\beta/N = 10^{-23}\text{cm}^3$ for aluminum. Then the value of $\partial R/\partial(\beta/N)$ is about 7.2×10^{20} cm^{-3}. If we take a 10% error in β/N, we get:

Fig. 5.18 Sensitivity of R to the parameter β/N [10]

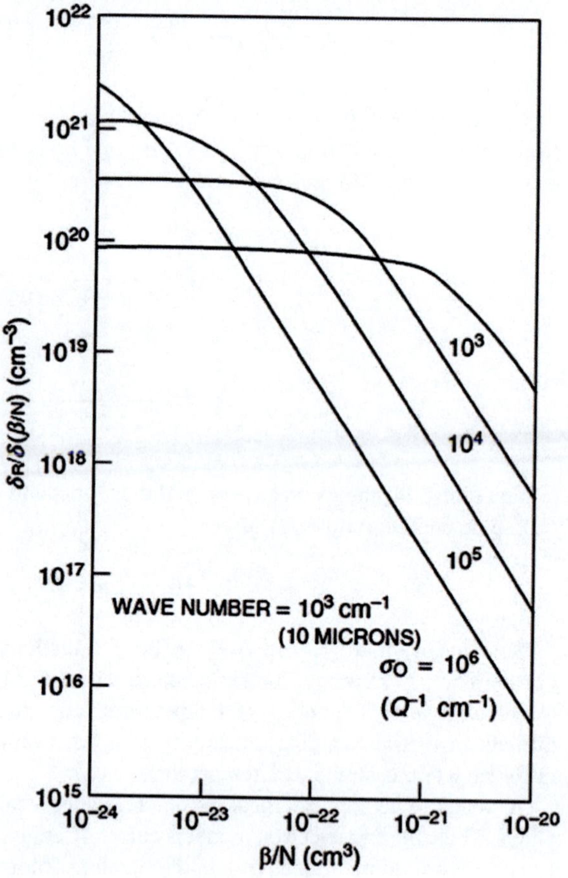

$$\frac{\partial R}{\partial(\beta/N)}\Delta(\beta/N) = 7.2 \times 10^0 \times 10^{-24}$$

$$\Delta R = 0.00072$$

Since for these values $R = 0.97366$, the change in R is only about 0.1%. We can obtain quite good predictions by the Drude–Lorentz model using the experimental values of σ_0 and the most simple choice for β/N, namely, one free electron for each valence electron per atom in the metal, and $\beta = 1$. For alloys, it is sufficient to choose the major constituent of the alloy. For example, with stainless steel, we choose iron, or two electrons per atom, to compute N and hence β/N.

Figures 5.19 and 5.20 do show the predictions of the free-electron theory for a variety of metals and some comparison to experimental data [10]. The abrupt change when the metal molts is caused by the abrupt change in the 0 conductivity. Notice in

Fig. 5.19 Temperature dependence of the absorptance at 10.6 μ for aluminum and stainless steel

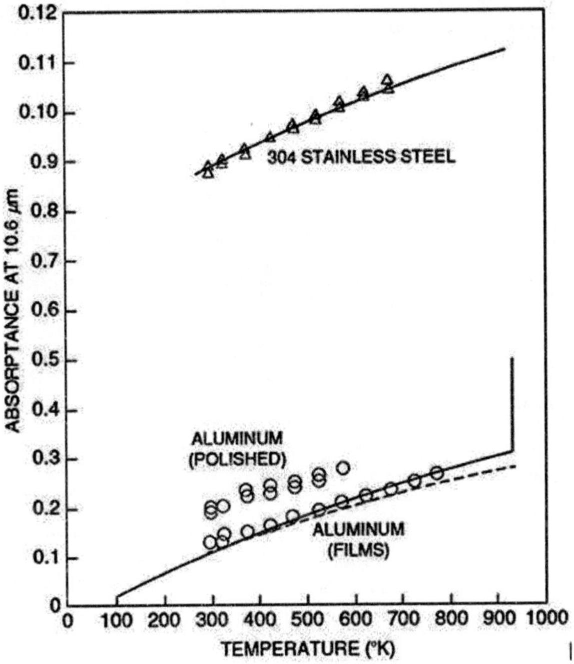

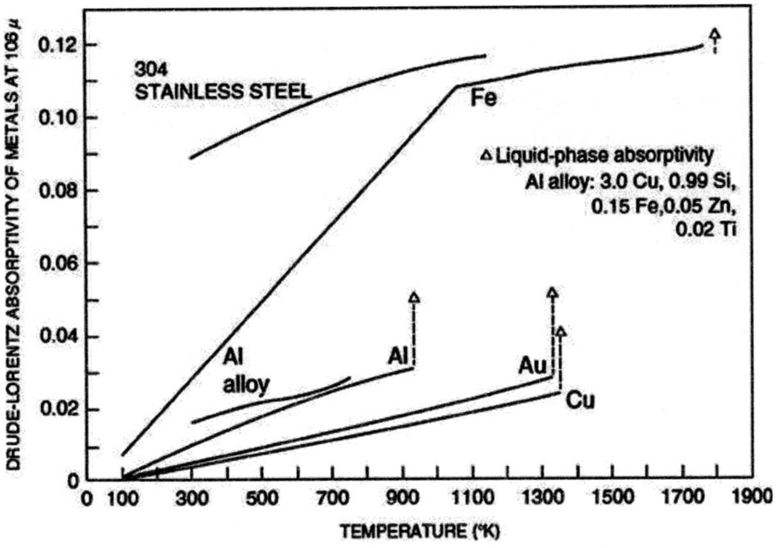

Fig. 5.20 Free-electron theory predictions of absorptivity of several metals at 10.6 μ. The open symbols indicate the molten state

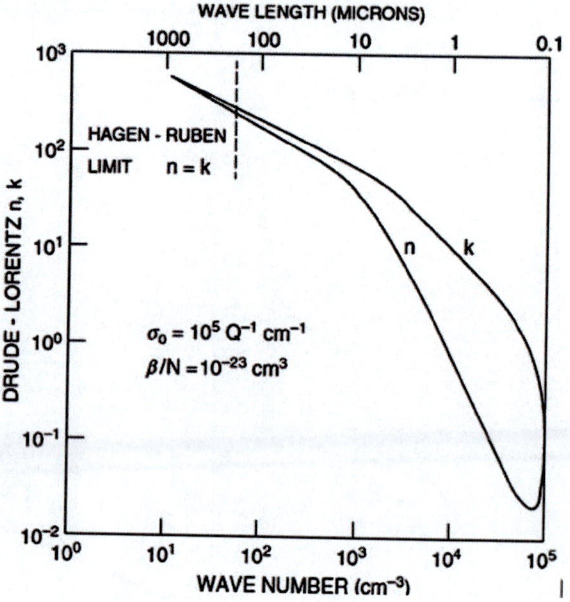

Fig. 5.21 *n* and *k* as functions of wavelength [10]

the comparison to data that aluminum films give values closet to the theory. This is probably because they prevent the best surfaces. Defects, oxide layers, etc. tend to trap the incident radiation and cause the real surface to absorb more radiation than the ideal surface. These graphs are in terms of absorptance, which is the experimentally measured quantity, and, since metals are opaque, $\mathcal{A} = 1 - \mathcal{R}$, which is correct for specular reflection at normal incidence from an opaque substance.

Let us return to the expressionism for *n* and *k* to look at some limiting forms and thus show how these complete expressions reduce to simple relationships. Remember that *R* (Eq. 5.76) is determined by *n* and *k* (Eq. 5.89), which are in turn obtained from the dc conductivity and *m*/*N* (Eq. 5.90). The variation of *n* and *k* with wavelength is shown in Fig. 5.21 for a typical good conductor like aluminum or copper at room temperature. Note that at long wavelengths, $n = k$. We can derive this by using Eqs. 5.86(a) and 5.86(b) for σ_1 and σ_2 and noting that as $\omega \to 0$, $\sigma_1 \to \sigma_0$ and $\sigma_2 \to \sigma_0\omega\tau$. By substituting these into Eqs. 5.89(a) and 5.89(b) for a and 1, we can readily show that:

$$n = k = \sqrt{\frac{\sigma_0}{2\varepsilon_0\omega}} \tag{5.93}$$

This is called the Hagen–Rubens limit. Note that *n* is very large. Under these conditions algebra can be used to reduce Eq. 5.76 to the following form:

$$R = \frac{n-1}{n+1} \tag{5.94(a)}$$

or

$$R = 1 - \frac{2}{n} \tag{5.94(b)}$$

and Eq. 5.93 can be substitute for n to get:

$$R = 1 - 2\sqrt{\frac{2\omega\varepsilon_0}{\sigma_0}} \tag{5.95}$$

This is the Hagen–Rubens reflectivity.

We can also comment on the skin depth. We have, at long wavelength ($\omega \to 0$):

$$\delta = \frac{\lambda}{2\pi}\sqrt{\frac{2\varepsilon_0\omega}{\sigma_0}} \tag{5.96}$$

This can be rewritten as:

$$\delta = \sqrt{\frac{2}{\mu_0\sigma_0\omega}} \tag{5.97}$$

Equation 5.97 is the common expression for skin depth used at long wavelengths.

Finally, we see from Fig. 5.22 that n and k re-converge at short wavelengths. This is called the plasma resonance. To see this, one must look at the behavior of n and k over a larger spectrum. We have already discussed the long-wavelength limiting behavior of n and k. This is the Hagen–Rubens region, where $n = k$. At short wavelengths, it is easy to show from Eqs. 5.90(a) and 5.90(b) that:

$$\sigma_1 \to \frac{N^2e^4}{m*^2\sigma_0\omega^2} \tag{5.98(a)}$$

$$\sigma_1 \to \frac{N^2e^2}{m*\omega} \tag{5.98(b)}$$

Thus Eqs. 5.89(a) and 5.98(b) can be written, for large ω, as:

$$n^2 = 1 - \frac{Ne^2}{\varepsilon_0 m*\omega^2} \tag{5.99(a)}$$

$$k^2 = 0 \tag{5.99(b)}$$

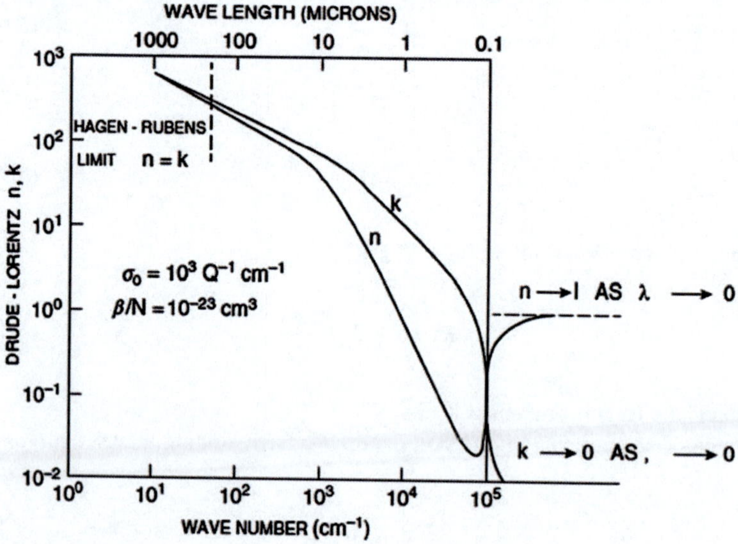

Fig. 5.22 *n* and *k* as functions of wavelength

Now the plasma frequency is usually defined from Eq. 5.108(a) by setting $n = 0$ to yield:

$$\omega_p^2 = \frac{Ne^2}{\varepsilon_0 m*} \tag{5.100(a)}$$

and thus:

$$n^2 = 1 - \frac{\omega_p^2}{\omega^2} \tag{5.100(b)}$$

We see, then, that at very high frequencies, the free-electron model predicts a transparent behavior ($k = 0$) and the index of refraction approaches that of a vacuum. The transition to this transparent behavior takes place at the plasma frequency, and it is a fairly abrupt transition, as Fig. 5.22 shows. In fact, some texts call this transition the "ultraviolet catastrophe."

Note that at ω near ω_p, Eqs. 5.99(a) and 5.99(b) are not valid. For these frequencies, we must use the full expression. If we use again the values of $\sigma_0 = 10^5 \Omega^{-1} \text{cm}^{-1}$ and $\beta/N = 10^{-23} \text{cm}^3$, which are appropriate to a good conductor like aluminum at room temperature, the reflectivity looks like Fig. 5.23. One can see that, in terms of the reflectivity, the transition is very abrupt, indeed.

The optical reflectivity of real metals is, as we have seen, in reasonable accord with the free-electron model at wavelengths in the infrared. The surface, however, must be nearly perfect for the predicted reflectivities to be achieved, and, of course,

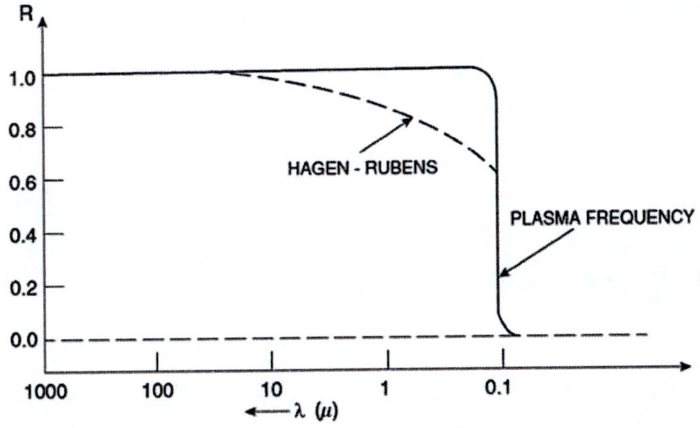

Fig. 5.23 *R* as a function of wavelength

Table 5.1 Room temperature absorbtaances of aerospace metal and alloys at 10.6 μm for various surface conditions and at normal incidence

Metal or alloy	Surface condition			
	Ideal	Polished	As-received	Sandblasted
Al	0.013	0.03	0.040 ±0.02	0.115 ±0.015
Au	0.006	0.01	0.02	0.14
Cu	0.011	0.016		0.06
Ag	0.005	0.011		
2014 Al		0.033	0.07 ±0.02	0.25
304 Stainless steel		0.11	0.4 ±0.2	
Ti Alloy (6Al, 4V)			0.65 ±0.2	
Mg Alloy Az-31B			0.06 ±0.03	

as the wavelengths approach the visible region band, effects become important, and the reflectivity shows rapid actuations with frequency. The absorptance of a practical metal surface is still largely an empirical matter. For high-power, continuous-wave radiation by a CO_2 laser, some data are available, but very little information on absorptivity as a function of surface temperature under these conditions is available.

Shown in Table 5.1 are room temperature absorptances for a few materials.

Data on the reflectivity of a metal during actual irradiation by a laser beam is quite difficult to obtain, although this information is central to the problem of laser–material interaction. One classic experiment along these lines was carried out by Bonch-Bruevich, Imas, Romanov, Libenson, and Maitsev in Russia in 1967

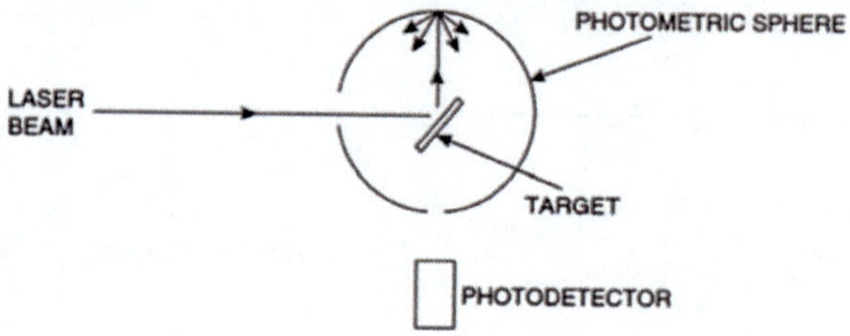

Fig. 5.24 Schematic representation of Bonch-Bruevich experiment

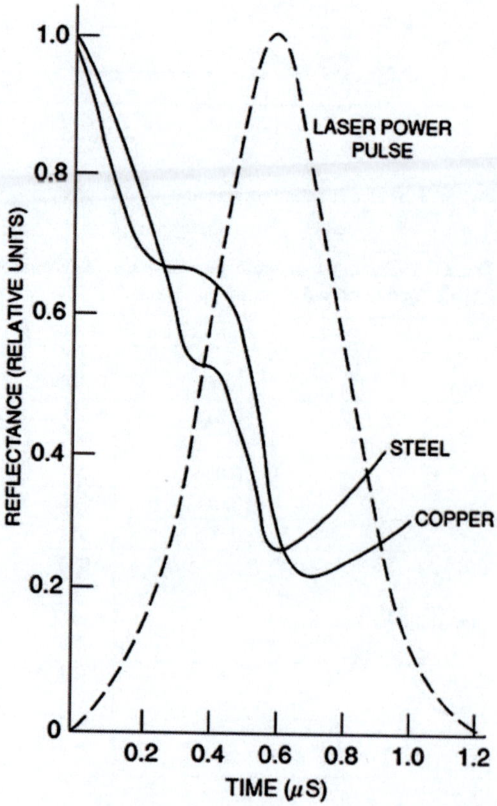

Fig. 5.25 Reflectance of steel and copper during irradiation by a laser beam [3]

[36]. They surrounded their specimens with a sphere to monitor the reflected radiation, as shown schematically in Fig. 5.24. The output of the photodetector is proportional to the reflectance of the specimen. Some of their results for steel and copper are shown in Fig. 5.25. The laser pulse (Nd:glass laser, 1.06 μm), with a peak power density of the order of 10^8 W/cm^2, is shown as a broken line. As time passes, of course, the laser pulse heats the surface, and the reflectance decreases. An especially interesting feature of these data is the shoulder. The author has suggested that this leveling off is associated with the surface reaching the melting point and pausing at that temperature while the thickness of the molten layer propagates into

the solid. In short, order, however, the molten layer begins to heat up, and the reflectance continues to decrease. As the power density of the laser pulse reaches its peak and begins to fall, the surface temperature can no longer be maintained, and as the surface cools, the reflectance begins to increase again.

In calculation of the laser interacting with metals, commonly used material properties are, for energy balance analysis, density, heat capacity, specific heat ratio, heat conductivity, heat diffusivity, latent heat, melting point, and vaporization point and, for stress and momentum analysis, viscosity, modulus of elasticity, shears modulus, Poison's ratio, and stress–strain constitutive relation.

5.7 Mathematical of Laser Absorption in Metals

A complete understanding of laser interaction with materials is still a matter of trials and adjustments. The real physical processes of laser beam interaction (drilling, cutting, welding, or being used as a directed-energy weapon application) with materials are very complex. The problem of laser interaction with materials presents many difficulties, both from modeling and from experimental sides. One would expect a reasonable description of the main phenomena occurring during laser interaction, but this is complicated because many of the physical processes equally contribute to the development of conservation equations, producing drawback because of a great complexity of the equations to be solved. In most instances, this leads to formulation of a model needed to be solved numerically. A lack of pertinent experimental data to compare with forces one to simplify some equations and use previous analytical and computational work done in this field [4].

The absorption coefficient, which can be derived from the material's dielectric function and conductivity, determines the absorption of light as a function of depth. However, the specific mechanisms by which the absorption occurs will depend on the type of material. In general, photons will couple into the available electronic or vibrational states in the material depending on the photon energy. In insulators and semiconductors, the absorption of laser light predominantly occurs through resonant excitations such as transitions of valence band electrons to the conduction band (interband transitions) or within bands (intersubband transitions) [20]. These excited electronic states can then transfer energy to lattice phonons. Photons with energy below the material's band gap will not be absorbed (unless there are other impurity or defect states to couple to or if there is multi-photon absorption). Such energies typically correspond to light frequencies below vacuum ultraviolet for insulators and below the visible to infrared spectrum for semiconductors. However, resonant coupling to high-frequency optical phonons in the near-infrared region is possible in some cases [31].

As we discussed in previous sections, light impinging on a material surface can be reflected, transmitted, or absorbed. In reality, all three occur to some degree. In order for laser machining to be practical, the laser light must be absorbed by the material. To yield an efficient process, it is necessary to couple as much of the incident

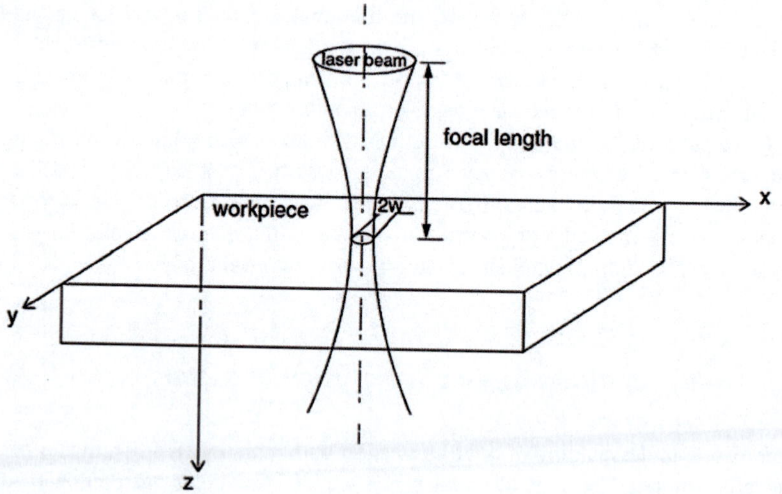

Fig. 5.26 Laser beam impinging on a finite size sample

intensity to the workpiece as possible. This coupling efficiency is described by the sample absorptivity *A* (in some parts of the book, this is also referred to as absorptance, absorption coefficient, or just absorption). The absorptivity is defined as the ratio between the absorbed energy and the incident energy. Absorptivity changes during the heating process and is a function of the sample's optical properties as well as the properties of the electromagnetic wave.

The mathematical model must be derived such that the parameters are easily handled. Once the model is verified by experiments, it can simulate the process and provide information such as heat-affected zone, transient temperature distribution, and cooling rates. Therefore, the model can reduce the experimentation by determining the effects of particular parameters beforehand.

This section presents the formulation of the general governing equation using the concept of the divergence of transport intensity as the net accumulation rate of energy per unit volume of the medium under consideration. To predict thermal effects produced by laser beam scanning, the surface of an absorbing sample requires that three-dimensional heat transfer equation be solved subject to finite size conditions of a sample. The geometry considered in this dissertation is illustrated in Fig. 5.26, which is finite in dimension slab irradiated by a laser beam impinging on its surface subject to convective and radiative losses

Dealing with interaction of high-power laser with materials and analysis of thermal response discussed in Sect. 5.8, we can define heat flux q_n as the magnitude of the *n*-direction (with *n* being *x*, *y*, and *z*), and one may introduce the vector sum, $\vec{q}$, of the directional fluxes as:

$$\vec{q} = q_x\hat{i} + q_y\hat{j} + q_z\hat{k} \tag{5.101}$$

where $\left(\hat{i}, \hat{j}, \hat{k}\right)$ are the unit vectors along each of the Cartesian coordinate directions.

Having defined a vector representation of the energy transport involved, one may write the expression for the net accumulation rate of thermal energy per unit volume as:

$$\nabla \cdot \vec{q} + Q = \frac{\partial u}{\partial t} \tag{5.102}$$

where Q is a volumetric term accounting for the internal generation of heat u. The right-hand side of Eq. 5.102 represents the total rate of change of the element's specific internal energy. Thus, Eq. 5.102 is a general statement of the first law of thermodynamics.

Using thermodynamic arguments, Eq. 5.102 can be expended further. If one considers a sample to be in a homogeneous continuum composition, with more than one phase, then one can uniquely determine the state of any property of the continuum using two independent properties, specific energy and specific enthalpy, and one can write (Nowak 1990) [37]:

$$du_m = \left(\frac{\partial u_m}{\partial v}\right)_T dv + c_v dT \tag{5.103}$$

$$dh_m = \left(\frac{\partial h_m}{\partial p}\right)_T dp + c_p dT \tag{5.104}$$

where:

u_m = specific internal energy.
h_m = specific enthalpy.
v = volume.
p = pressure.
c_v = specific heat at constant volume.
c_p = specific heat at constant pressure.
T = temperature distribution.

Both the specific heat at constant pressure and volume are defined as follows:

$$c_p = \left(\frac{\partial u}{\partial T}\right)_p \tag{5.105}$$

$$c_v = \left(\frac{\partial u}{\partial T}\right)_v \tag{5.106}$$

For solids and incompressible fluids, the specific volume is assumed constant. If one also neglects pressure changes in the given process, Eqs. 5.105 and 5.106 can be simplified to yield:

$$du_m = c_v dT \tag{5.107}$$

$$dh_m = c_p dT \tag{5.108}$$

Recalling the definition of specific enthalpy from classical thermodynamics:

$$h_m = u_m + pv \tag{5.109}$$

One can combine Eqs. 5.105 and 5.103 to find that:

$$c_p = c_v = c \text{for} dv \approx dp \approx 0 \tag{5.110}$$

If pressure is allowed to vary, Eq. 5.109 still holds approximately for solids and incompressible fluids. Taking the relations in Eqs. 5.107 and 5.109 and multiplying by the density of the medium result in expressions in terms relative to the unit volume as opposed to the unit mass. Using Eq. 5.109, these new volumetric terms can be expressed as:

$$du = \rho \cdot c(T) dT \tag{5.111}$$

$$dh = \rho \cdot c(T) dT \tag{5.112}$$

and

$$h = u + \rho dT \tag{5.113}$$

Integrating both sides of Eq. 5.111 over proper variables and differentiating with respect to time yields:

$$\frac{\partial u}{\partial t} = \rho \cdot \frac{\partial}{\partial t} \int_{T_0}^{T} c(T) dT \tag{5.114}$$

where T_0 is the temperature at the beginning and T is the temperature at the end of the infinitesimal time step, *dt*. For the particular very infinitesimal time interval *dt*, one can assume a constant specific heat, and then Eq. 5.114 simplifies to:

$$\frac{\partial u}{\partial t} = \rho \cdot c \frac{\partial T}{\partial t} \tag{5.115}$$

since the initial temperature is constant with respect to time. Substituting Eq. 5.124 into Eq. 5.102 yields the heat diffusion equation:

$$-\nabla q + Q = \rho \cdot c \frac{\partial T}{\partial t} \tag{5.116}$$

Solution of Eq. 5.116 for the time-dependent field, $T(x, y, z, t)$, requires the use of a constitutive equation relating the temperature to heat flow. For conductive heart transfer equation, this relation is Fourier's law of conduction. It states that the heat flux in a direction n is proportional to the temperature gradient in that direction. Mathematically this is expressed as:

$$q_n \hat{n} = -k \cdot \frac{\partial T}{\partial n} \hat{n} \tag{5.117}$$

where k is the thermal conductivity and q_n is the heat flux in the n-direction. Negative sign is necessary to satisfy the second law of thermodynamics.

In the analysis of the Fourier heat conduction model, the heat flux through a given plane is considered as a function of the spatial temperature gradient at that plane.

This depends upon the assumption that the temperature gradient remains almost constant between two successive and closely spaced planes. However, the distance between these planes is finite; therefore, error occurs when high-order terms, which are neglected, become important at high-power laser intensities. The heat flux through a given plane depends on the electron energy distribution through the material; therefore, the material cannot be considered as a homogeneous continuum when one is analyzing very short pulses (shorter than a picosecond) at intermolecular levels (distances less than 0.1 μm).

Substituting Eq. 5.126 into Eq. 5.125 yields:

$$-\nabla \cdot (k \nabla T) + Q = \rho \cdot c \frac{\partial T}{\partial t} \tag{5.118}$$

which is the general form of the governing differential equation for isobaric thermal conduction in a homogeneous solid or incompressible fluid.

In laser micromachining or its interaction with materials surface, the internal energy generation, Q, is commonly thought of as the rate of laser energy absorbed per unit volume in the irradiated medium. In case of metals, this absorption occurs beginning in a very thin layer at the surface of the workpiece or target of interest and, for many practical cases can be considered as a boundary condition to Eq. 5.118. This absorption process is calculated by Beer–Lambert's law, via the following relation:

$$I(z) = \varepsilon \cdot I_0(0)|_{\substack{z=0 \\ \text{material surface}}} \exp(-\mu \cdot z) \tag{5.119}$$

where $I(z)$ is the intensity of the incident radiation at a given distance, z, into the absorbing medium from the irradiated surface, ε is the surface emissivity of the medium, and μ is the absorption coefficient of the material, measure of the absorption of radiation propagating through the medium. Using the terminology used in Eq. 5.102, differentiating Eq. 5.117 with respect to the direction of propagation of the laser beam yields the volumetric extraction rate of energy absorbed by the differential element. Since the accumulation rate is negative of the extraction rate for a given volume, the necessary volumetric term for use in Eq. 5.119, is given by:

$$Q = -\frac{\partial I(z)}{\partial z} = (\varepsilon\mu) I_0(0)|_{\substack{z=0 \\ \text{material surface}}} \exp(-\mu \cdot z) \tag{5.120}$$

or for simplicity, we write Eq. 5.111 as:

$$Q = -\frac{\partial I(z)}{\partial z} = (\varepsilon\mu) I_0(0) \exp(-\mu \cdot z) \tag{5.121}$$

Using the Drude–Zener theory (Yilbas 1997) [38], Eq. 5.121 leads to the following expression for $Q(x, y, z, t)$:

$$Q(x, y, z, t) = A \exp(-\beta t)[\mu I_0(x, y, z, t) \exp(-\mu z)] \tag{5.122}$$

where A is the surface absorptivity, β is the pulse parameter, and $I_0(x, y, z, t)|_{z=0}$ is the laser radiation intensity at the material surface ($z = 0$).

In Eq. 5.122, absorption coefficient of the material ì measuring the absorption of radiation propagating through the medium is considered constant at this time, but based on a number of publications, ì should be a function of temperature and axial position of the irradiated laser beam in respect to the material under consideration. At this time let us modify Eq. 5.122 to be:

$$Q(x, y, z, t) = A \exp(-\beta t)[\mu(T, z) I_0(x, y, 0, t) \exp\{-\mu(T, z) \cdot z\}] \tag{5.123}$$

and function $\mu(T, z)$ will be defined as bulk absorption coefficient at distance z from the metal surface. Here for purpose of our analysis and discussion, we keep μ constant, but based on a number of publications, μ should be a function of temperature and axial position of the irradiated laser beam in respect to the material under consideration.

Incorporating Eq. 5.122 into Eq. 5.118, for laser pulse with a position-dependent intensity, for this form of pulse input, the Fourier differential Eq. 5.118 can be rewritten in very general form as:

$$\rho c(T)\frac{\partial T}{\partial t} = \nabla[k(T)\nabla T(x, y, z, t] + Q(x, y, z, t) \tag{5.124}$$

where *c(T)* is the density of the material of the workpiece, is the temperature-dependent specific heat of the material, *k(T)* is the temperature-dependent thermal conductivity, $T(x, y, z, t)$ is the resulting three-dimensional time-dependent temperature distribution in the material, and t is time; we can assume for the purpose of any boundary condition that T_0 is the initial temperature and x, y, z are the spatial Cartesian coordinates, while $Q(x, y, z, t)$ is the rate at which heat is supplied to the solid per unit time per unit volume, depending on the laser pulse parameter and physical and optical properties of materials irradiated. Note that both A and μ are functions of temperature and wavelength of the incident radiation as will be shown in latter sections of this chapter.

Solid or liquid evaporates at any temperature greater than 0 K. The evaporation rates strongly depend on the surface temperature T_s. Equation 5.124 considers heat diffusion into material only through conduction. Based on experimental and theoretical evidence, evaporation takes place, so one has to consider moving elements of vapor and liquid inside the material during laser beam interaction.

Considering this fact, the general governing differential equation allowing a phase change process can be written as:

$$\rho(T)c(T)\frac{\partial T}{\partial t} = \nabla(k\nabla T) + \rho(T)c(T)V_n(T)\nabla(T) + Q \tag{5.125}$$

where V_n is the normal component of the evaporation front or melt front velocity (recession velocity).

Let us define velocity, V_n. From the literature (Tokarev et al. 1995) [39], the interface moves into depth of the material at the speed defined as:

$$V(T_s) = \frac{1-c}{\rho}\sqrt{\frac{M}{2\pi k_B T_s}} p_{\text{sat}}(T_s) \tag{5.126}$$

where c is the condensation factor which is usually 0.18 and $p_{\text{sat}}(T_s)$ is the saturated vapor pressure defined from the Clapeyron–Clausius equation as:

$$p_{\text{sat}}(T_s) = \varphi_0 \exp\left(-\frac{E_a}{k_B T_s}\right) \tag{5.127}$$

and k_B is the Boltzmann constant, φ_0 is the pre-exponential factor, and E_a is an activation evaporation per atom.

Note that Eq. 5.126 is a pressure-dependent function, which has to be calculated simultaneously with temperature-dependent velocity. Thus, let us derive a velocity as a function of temperature and the latent heat of phase transition, which is also temperature-dependent.

The rate of change of latent heat with temperature can be expressed as (Yilbas et al. 1996) [40]:

$$\frac{\partial H}{\partial T} = \frac{H}{T} + (c_{pv} - c_{pl}) - \frac{H}{v_v - v_l}\left[\left(\frac{\partial v_v}{\partial T}\right)_p - \left(\frac{\partial v_l}{\partial T}\right)_p\right] \tag{5.128}$$

where c_{pv} and c_{pl} are specific heats at constant pressure for vapor and liquid states, respectively, and v_v and v_l are specific volumes for vapor and liquid states.

One has to realize that integration of the latent heat over the temperature ranging from 0 to critical temperature T_c is difficult, because we do not have enough information about values for the latent heat especially at extreme value of the range under consideration. It is safe to assume (Yilbas 1996) [41] that little inaccuracy is involved in taking the room temperature latent heat as the latent heat at absolute zero, because, by knowing that $v_v < < v_l$ and $\left(\frac{\partial v_v}{\partial T}\right)_p >> \left(\frac{\partial v_l}{\partial T}\right)_p$, i.e., the specific volume of gas, v_v is much greater than the condensed liquid, v_l, and its rate of change with temperature at constant pressure is correspondingly greater. Thus, Δc_p is extremely small for temperature up to room temperature T_a.

According to Maxwell's law, the function of velocity distribution of molecules can be defined as (Tabor 1991) [42]:

$$f(V_n)dV_n = \sqrt{\frac{M}{2\pi k_B T_s}} \exp\left(\frac{MV_n^2}{2\pi k_B T_s}\right) dV_n \tag{5.129}$$

where V_n is the velocity in the direction normal to the surface, and the other parameters were defined earlier. Using vernacular terms, function $f(V_n)dV_n$ is the ratio of number of atoms with velocity V_n to $V_n + dV_n$ per unit volume to the total number of atoms per unit volume. Only those molecules whose velocity is greater than V_{min} obtained from the equation given by Tabor (1991) [42] as:

$$\frac{1}{2}MV_{\min}^2 = H(T) \tag{5.130}$$

will escape from the retaining potential, where V_{min} lies in the z direction. If n is the number of atoms per unit volume, then the number of atoms with velocities V_n to $V_n + dV_n$ per unit volume is $nf(V_n)dV_n$, and the number of atoms with these velocities passing a unit area per unit time is $nf(V_n)V_n dV_n$.

Assuming that all the atoms for which, do not return to their equilibrium position under evaporation circumstances and if $V_n > V_{\min}$ and if we define as the number of atoms evaporated per unit time per unit area, then we can write:

$$
\begin{aligned}
N_G &= \int_{V_{\min}}^{\infty} [nf(V_n)V_n]dV_n \\
&= n\sqrt{\frac{M}{2\pi k_B T_s}}\int_{V_{\min}}^{\infty}\left[\exp\left(\frac{MV_n^2}{2\pi k_B T_s}\right)V_n\right]dV_n
\end{aligned}
\tag{5.131}
$$

After integration and substitution of Eq. 5.130 into Eq. 5.131, we obtain:

$$
N_G = n\sqrt{\frac{M}{2\pi k_B T_s}}\exp\left(\frac{H(T)}{k_B T_s}\right)V_n \tag{5.132}
$$

If atoms are equally spaced within the lattice, a surface layer would consist of $n^{2/3}$ with an evaporation time $n^{2/3}N_G$. The average velocity of the surface V_n would be:

$$
V_n = \frac{1}{n^{1/3}}\frac{N_G}{n^{2/3}} \tag{5.133}
$$

$$
V_n = \sqrt{\frac{k_B T}{2\pi M}}\exp\left(-\frac{H(T)}{k_B T}\right) \tag{5.134}
$$

where $H(T)$ is the *latent heat.*

Considering Eq. 5.134, the general governing differential equation defined in Equation

Eq. 5.135, with phase change processes at the surface of the material, interaction with high-power laser irradiation could be written as:

$$
\rho(T)c(T)\frac{\partial T}{\partial t} = \nabla(k\nabla T) + \rho(T)c(T)\sqrt{\frac{k_B T}{2\pi M}}\exp\left(-\frac{H(T)}{k_B T}\right)\nabla T + Q \tag{5.135}
$$

Note that latent heat $H(T)$, in Eqs. 5.143 and 5.144, is based on some theoretical and experimental evidence and is temperature dependent; the final form of it is going to be described later in this chapter.

In order to solve Eq. 5.134, appropriate boundary conditions should be applied. Sect. 5.8.1 discusses formulation of the appropriate boundary conditions.

5.8 Material and Thermal Response

Metal processing with lasers has reached a high level of maturity and acceptance in the industry. It is used for cutting, drilling, welding, forming, engraving, marking, hardening, and various forms of surface treatment of metals in a broad spectrum of modern industries, including the automotive and aerospace industries, the shipbuilding industry, the microelectronics industry, and the medical instrument industry, to name a few, with its application now as a directed-energy weapon both based on ground and space usages.

The details of the material response will depend on the particular material system and the laser processing conditions. As mentioned earlier, if laser-induced excitation rates are slow compared to the thermalization time, then the process is denoted as photothermal, and one can consider the absorbed laser energy as being directly transformed into heat. In this case, the material response will be a function of the local material heating and cooling rates, maximum temperatures reached, and temperature gradients, all of which can be determined from the solution to the heat equation for the given irradiation conditions. Because material heating rates can be so extreme, reaching as high as 109 K/s for nanosecond (ns) pulses and even higher for femtosecond lasers, significant changes to the material can occur.

One of the most important effects of intense laser irradiation is the conversion of the optical energy in the beam into thermal energy in the material. This is the basis of many applications of lasers, such as welding and cutting. We shall summarize here this thermal response. It is basically a classical problem, namely, heat flow. In the usual manner, we shall seek solutions to the equation which governs the flow of heat, namely:

$$\rho c \frac{\partial T}{\partial t} = \frac{\partial}{\partial x}\left(K\frac{\partial T}{\partial x}\right) + \frac{\partial}{\partial y}\left(K\frac{\partial T}{\partial y}\right) + \frac{\partial}{\partial z}\left(K\frac{\partial T}{\partial z}\right) + A \tag{5.136}$$

We use here ρ for the density, c for the specific heat, T for temperature, I for time, and K for thermal conductivity. A is the heat produced per unit volume per unit of time. In Eq. 5.145, ρ, c, and K are considered functions of both position and temperature, and A is a function of both position and time. In effect, the equation is a simple statement that the rate at which heat accumulates in an elemental volume $dxdydz$ is equal to the net flow of heat across the faces of that volume plus the rate at which heat is produced within the volume.

Thus, thermal response studies consist essentially of two parts. First, one needs to know the rate and source of production of heat by the laser, which yields A. Then one solves Eq. 5.145 subject to the boundary conditions of the situation of interest. This can be a very elaborate task and frequently can be done only with the aid of a computer.

There is a great deal of effort among workers in the field of laser effects to develop an all-inclusive computer program to solve Eq. 5.136 for every possible situation. However, the solution to Eq. 5.136 can be no better than the knowledge of A, and, as we shall see in later sections, it is often very difficult to establish A with any precision in a laser material interaction situation.

5.8.1 Boundary Conditions

Boundary conditions are important part of a model. They influence the programming and calculation results greatly. Specifying the suitable boundary conditions is the basis for successful computation. Three kinds of boundary conditions are typically encountered in heat conduction analysis. These are:

I. Given the boundary temperature, T
II. Given the boundary heat flux, q
III. A boundary heat flux balance relation

If the boundary temperature is given, there is no particular difficulty in modeling, and one needs only to specify the value of the boundary grids to the specified temperature.

The magnitude of heat flux due to convection to the ambient from the sample surface is expressed using Newtonian law of cooling:

$$q_{\text{conv}} = h_c(T_s - T_{\text{amb}}) \tag{5.137}$$

where h_c is convection heat transfer coefficient, T_s is the surface temperature, and T_{amb} is the temperature of ambient or surrounding.

In order to determine h_c, the characteristic length L of the workpiece should be:

$$L = \frac{Area}{Ob} \tag{5.138}$$

where $Area$, is the area of the surface and Ob is the perimeter of the workpiece. Then, the Nusselt number for the horizontal plate is (Bejan 1993) [43]:

$$N_u = 0.27(R_a)^{0.25} \tag{5.139}$$

where the Rayleigh number R_a is:

$$R_a = \frac{g\beta}{\kappa\gamma}L^3(T - T_{amb}) \tag{5.140}$$

In Eq. 5.140, g is the gravitational acceleration, β is the coefficient of the volumetric thermal expansion, κ is the thermal diffusivity, γ is the kinematic viscosity, and other parameters are as defined for Eq. 5.137. The convective heat transfer coefficient h_c can be calculated as:

$$h_c = N_u\frac{k_{\text{air}}}{L} \tag{5.141}$$

where k_{air} is the thermal conductivity of the air surrounding the workpiece. Note that in case of high-power laser as directed-energy weapon and its application as ground-based laser (GBL) or airborne laser (ABL), we need to consider the above discussion. Radiation to the ambient is expressed using the relation:

$$q_{\mathrm{rad}} = \sigma\varepsilon\left(T_s^4 - T_{\mathrm{amb}}^4\right) \tag{5.142}$$

where q_{rad} is the magnitude of radiation flux, ε is the emissivity of the material, and σ is the Stefan–Boltzmann constant. Note that convection is a linear function of temperature, whereas radiation is nonlinear, due to its dependence on the difference of the fourth powers of the surface and ambient temperature.

The magnitude of convection and conduction in the overall transport of heat can be evaluated from the value of the Peclet number, P_e, which is defined by:

$$P_e = \frac{uc_pL_r}{k} \tag{5.143}$$

where u is velocity and L_r is the characteristic length, taken as the pool radius at the top surface of the weld pool, and the other parameters were defined earlier.

Heat transported by a combination of convection and conduction mechanism is observed in the weld pool in laser micro-welding applications. When Peclet number is less than 1, the heat transport within the weld pool occurs primary by conduction. When Peclet number is much Where, u is velocity, L_r is the characteristic length, take as the pool radius at the top surface of the weld pool and the other parameters were defined earlier.

Heat transported by a combination of convection and conduction mechanism is observed in the weld pool in laser micro-welding applications. When Peclet number is less than 1, the heat transport within the weld pool occurs primary by conduction. When Peclet number is much higher than 1, then the convective heat transport is the main mechanism of heat transfer in the material.

If one is using shielding gas in laser micromachining, then heat transfer coefficient h_c required in Eq. 5.141 is calculated from Mazumder and Steen (1980) [44] and Gordon and Cobonpue (1961) [45] in the case of a vertically impinging jet:

$$h_c = 13\,\mathrm{Re}^{0.3}\mathrm{Pr}^{0.35}k_g\frac{1}{B} \tag{5.144}$$

where B is the jet plate distance, is the Reynolds number at jet exit, is the Prandtl number for gas, and k_g is thermal conductivity of the gas.

Performing an energy balance on a boundary where both convection and radiation losses occur, one can relate the flux conducted to the interface to the convection and radiation losses by the expression:

$$q_n = q_{\mathrm{conv}} + q_{\mathrm{rad}} \tag{5.145}$$

Substituting for q_n from Eq. 5.117, for q_{conv} from Eq. 5.137, and finally for q_{rad} from Eq. 5.142, into Eq. 5.145, we get:

$$-k\left(\frac{\partial T}{\partial n}\right)_s = h_c(T_s - T_\infty) + \sigma\varepsilon\left(T_s^4 - T_\infty^4\right) \quad (5.146)$$

where the term on the left-hand side of the Eq. 5.146 represents the magnitude of the heat conducted normal to the boundary surface. Equation 5.146 should be applied to all exposed boundaries of the finite sample under consideration. Setting the right-hand side of Eq. 5.146 to 0 accommodates insulated boundaries.

In laser micro-drilling or interaction with materials, Eq. 5.146 should be applied in the region of the newly vaporized material (hole area), and an appropriate heat transfer coefficient should be used. Since the question of modeling of the micro-drilling phenomena is associated with creating the hole formation has not yet been solved, then to choose an appropriate heat transfer coefficient is an educated guess and open to debate.

When phase change transition takes place, the boundary condition at the liquid–vapor interface is:

$$-k\frac{\partial T}{\partial z}\bigg|_z + \rho v_{dv} H_v = (1 - R)I_0 \quad (5.147)$$

where k is thermal conductivity of the solid or liquid phase, $\frac{\partial T}{\partial z}\big|_z$ is the temperature gradient at the surface along the normal (z-axis), ρ is the density of the solid or liquid phase, v_{dv} is the component of boundary velocity due to evaporation, R is the reflectivity for the laser wavelength, and I_0 is the intensity of the laser beam at the surface. For surface temperatures less than approximately half of the critical temperature, the energy of evaporation per atom U can be assumed to be constant. Then the component of the boundary velocity due to evaporation v_{dv} was defined by Eq. 5.134 and also in similar fashion by Niedrig and Bostonjglo (1996) [46].

$$v_{dv} = V_0 \exp\left(\frac{-U}{T_s}\right) \quad (5.148)$$

where V_0 is a coefficient of the order of magnitude of the sound velocity, T_s is the surface temperature, and U is the energy of evaporation per atom defined as:

$$U = \frac{MH_v}{N_a k_B} \quad (5.149)$$

where H_v is the latent heat of vaporization (per unit mass), k_B is the Boltzmann constant, M is the atomic mass, and N_a is the Avogadro's number. Velocity V_n, defined in Eq. 5.138, is equal to v_{dv}.

The classic Stefan boundary condition is applied to the solid–liquid boundary (melting front $z = z_m$) as:

$$\rho H_f v_m = k_s \frac{\partial T_s}{\partial z}\bigg|_{z=z_m} - k_l \frac{\partial T_l}{\partial z}\bigg|_{z=z_m} \tag{5.150}$$

where H_f is the latent heat of fusion, v_m is the melting front velocity, and subscripts "s " and " l " represent "solid" and "liquid", respectively. The Stephan boundary condition assumes an instant transition from solid to liquid at the melting temperature Tm and does not allow superheating at the melting front. The approximation is adequate for the slow velocities of melt front propagation typical of laser welding and drilling, where the melting kinetics can be disregarded.

To understand how the boundary conditions were defined, let us first comprehend material removal process together with energy transport in multiple phase transitions, which is going to be discussed later on. To establish effects of convective heat transfer coefficient due to evaporation recoil generated melt flow, Semak et al. (1999) [47] performed two simulations. In the first one, recoil pressure and related melt flow were disregarded, and in the second case, recoil pressure and melt flow were accounted for. Those observations were supported by calculating the melt surface temperatures at the axis of the laser beam for different values of absorbed intensity I_0 presented in Fig. 5.27. When recoil pressure and melt flow were included, the steady-state values of temperature were reached faster, and the

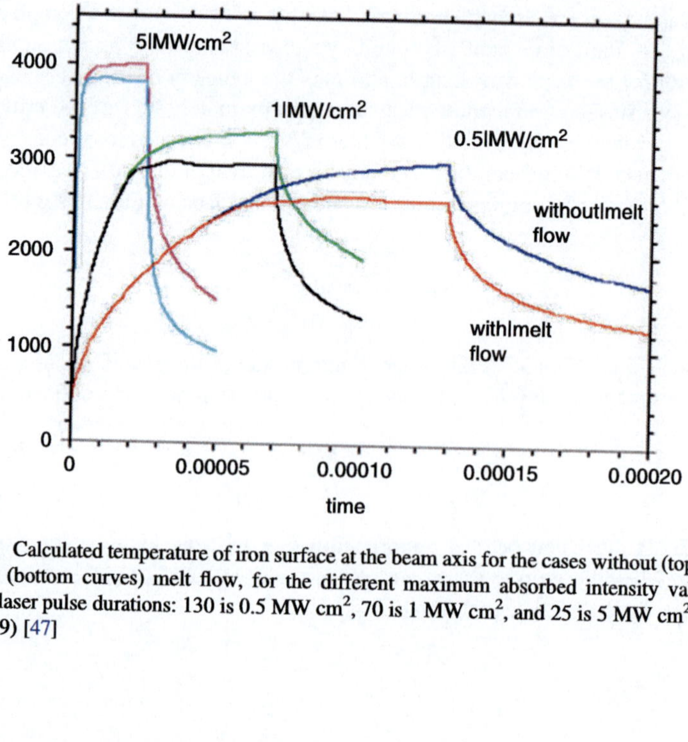

Fig. 5.27 Calculated temperature of iron surface at the beam axis for the cases without (top curves) and with (bottom curves) melt flow, for the different maximum absorbed intensity values and different laser pulse durations: 130 is 0.5 MW cm^2, 70 is 1 MW cm^2, and 25 is 5 MW cm^2 (Semak et al. 1999) [47]

maximum temperature was lower and the cooling rates much higher, than for the case where recoil pressure and flow were neglected. Analyzing Fig. 5.27, one can calculate that ignoring recoil pressure and associated convective heat transfer results in 1–5% error in calculations of surface temperature in the center of the laser beam.

5.9 Solutions of Governing Equation

This section considers the solution of the general governing Eq. 5.136 subject to its related boundary conditions, by means of exact and numerical methods. This section consists of a discussion of the approximate solution of the problem using mathematical tools and is divided into four sub-section parts. First four sections, Sects. 5.9.1.1, 5.9.1.2, 5.9.1.3, and 5.9.1.4, present an analytical method of the solution of the general governing equation using the Fourier theory. Section 5.9.1.5 is considering interaction of laser energy with materials using very short laser pulses and introduces electron–phonon theory approach to solve the heat transfer problem of the interaction of ultrashort pulses with matter. Section 5.10 consists of a discussion of the approximate solution of the problem using the Finite Difference Method (FDM) and Finite Element Method (FEM) and presents the computer solutions developed

5.9.1 Analytical Methods

There have been many analytical solution published for various special cases of the Fourier heat transfer equation. Section 5.9.1.1 illustrates an analytical solution for heating case without phase change in case of simple infinite solid, by laser defined as a step function (constant heating). Section 5.9.1.2 solves the problem similar as Sect. 5.9.1.1 but considering the slab of finite thickness. Section 5.9.1.3 deals with the analytical solution of heat transfer equation with spatial-dependent laser pulse heating. Next Sect. 5.9.1.4 considers analytical solution to the governing heat transfer equation for laser pulse with a time-dependent Gaussian pulse heating. Section 5.9.1.5 considers analytical solution of heat transfer equation with time-dependent Gaussian laser pulse heating with convective boundary conditions. Section 5.9.1.6 describes heating analysis with time-dependent pulse intensity and where evaporation is considered as the exclusive phenomenon taking place during the ablation process. Section 5.9.1.7 presents the heating analysis with pulsed laser heating process by considering both Fourier conduction and electron–phonon kinetic theory approaches.

5.9.1.1 No Phase Change, Simple Infinite Solid

Before we go through such details, let us consider the simplest situation of high-power laser interaction with metal where we have no phase change and semi-infinite solid case. Let the laser beam be perfectly uniform over an extremely large area, so that we have a one-dimensional situation. Assume also that the material parameters are temperature-independent and that the solid is uniform and isotropic and of semi-infinite extent (see Fig. 5.28). Finally, assume that there is no phase change; the rate at which energy enters the material is not sufficient to induce melting or vaporization.

First rewrite Eq. 5.136 one-dimensionally and in z-direction, using the fact that ρ, c , and K are constant:

$$\frac{\partial^2 T(z,t)}{\partial z^2} - \frac{1}{\kappa}\frac{\partial T(z,t)}{\partial t} = -\frac{A}{K} \tag{5.151}$$

Here we have introduced $\kappa = K/(\rho c)$, which is the thermal diffusivity. Let us adopt the convention that T is measured with respect to the initial (or ambient) temperature of the material. This is possible because Eqs. 5.145 and 5.160 define T only to within an additive constant. Then we have as a boundary condition that $T \rightarrow 0$ as $z \rightarrow \infty$. The boundary condition on the front face ($z = 0$) depends on what we assume for radiative and convective losses. It can be shown that, for most cases of interest, the rate at which the laser creates heat at the interface is overwhelmingly larger than convective and radiation losses, so we ignore them for the present calculation. Thus, the boundary condition is that there is no heat flux at $z = 0$, that is:

$$K \left.\frac{\partial T}{\partial z}\right|_{z=0} = 0$$

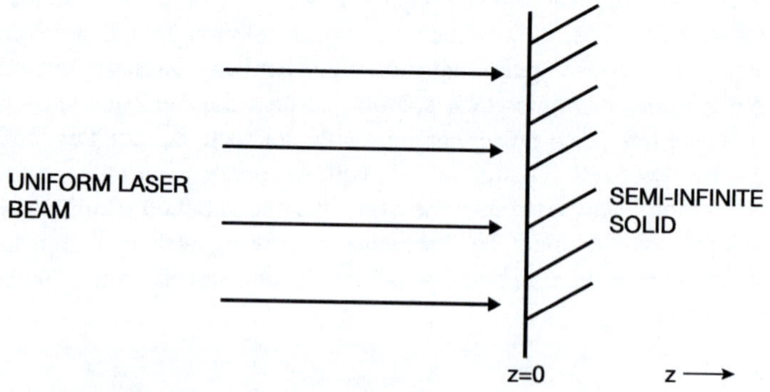

Fig. 5.28 Uniform irradiation of a semi-infinite solid

Now consider A. Denote by I the power density of the laser radiation at the surface; the dimensions of I are power per unit area. The power density of the radiation transmitted to the surface is $I(1 - R)$. Then the power density as a function of z is:

$$F = (1 - R)Ie^{-4\pi kz/\lambda} \tag{5.152}$$

This follows from the fact that the energy in the electromagnetic wave goes as $\mathcal{E}^2$. Now to get the power transferred per unit volume, consider elemental volumes of length dz and unit area:

$$A = -\frac{\partial F}{\partial z} = (1 - R)Ie^{-4\pi kz/\lambda} \tag{5.153}$$

The minus sign appears because $\partial F/\partial z$ is the power per unit volume lost by the radiation and A is the power per unit volume absorbed by the material. Finally, we define the absorption coefficient:

$$\frac{1}{\delta} = \alpha = \frac{4\pi k}{\lambda} \tag{5.154}$$

which is, of course $1/\delta$, the inverse of skin depth, and k is called the extinction coefficient. Thus Eq. 5.153 takes the following form:

$$A = (1 - R)I(t)\alpha e^{\alpha z} \tag{5.155}$$

where we have included the possibility of $1/\delta$ varying with time. Therefore, the equation which needs to be solved is:

$$A = (1 - R)I(t)\alpha e^{\alpha z} \tag{5.156}$$

In keeping with our assumption of temperature-independent thermal parameters, we assume further that R is independent of temperature. Equation 5.155 is valid for temperature dependent R and can be used to give $A(z, t, T)$.

For metals, α is a fairly large number. As we saw in Sect. 5.6, k is of the order of 100 at $\lambda = 10$ μm, so that α is of the order of 10^6 cm^{-1}. Hence the absorption occurs in a very narrow layer at the surface. It then becomes more convenient to seek solutions of the homogeneous part of Eq. 5.156, as:

$$\frac{\partial^2 T}{\partial z^2} - \frac{1}{\kappa}\frac{\partial T}{\partial t} = 0 \tag{5.157}$$

subject to the boundary conditions that $T = 0$ at $z = \infty$, but with a specific flux into the surface at $z = \infty$, i.e.:

$$-K\frac{\partial T}{\partial z}\bigg|_{z=0} = (1-R)I(t)$$

or, with the definition:

$$F(t) = (1-R)I(t) \tag{5.158}$$

$$-K\frac{\partial T}{\partial z}\bigg|_{z=0} = F(t) \tag{5.159}$$

First examine the case of $F = F_0$, a constant. This is appropriate to irradiation by a continuous laser, given temperature-independent material properties. We note here only the solution, for many excellent texts on heat conduction can be consulted for the details [3]. See also Example 5 in Appendix H of Reference [69]. Also, here we will show a detailed solution of this problem.

Temperature in a Semi-Infinite Solid

Before we solve this problem, we do couple of simple problem to establish a foundation and refresher for those readers that have out of these fundamental mathematic problem as well.

Problem-1: Let Γ denote the curved problem *BJPKQLA* of the Bromwich contour Fig. (a) with equation $s = Re^{i\theta}$, $\theta_0 \leq \theta \leq 2\pi - \theta_0$, i.e., Γ is the arc of a circle of radius R with center at O. Suppose that on Γ we have:

$$|f(s)| < \frac{M}{R^k}$$

where $k > 0$ and M are constant. Show that:

$$\lim_{R\to\infty}\int_\Gamma e^{st}f(s)ds = 0$$

If Γ_1, Γ_2, Γ_3, and Γ_4 represent arcs *BJ*, *JPK*, *KQL*, and *LA*, respectively, we have:

$$\int_\Gamma e^{st}f(s)ds = \int_{\Gamma_1} e^{st}f(s)ds + \int_{\Gamma_2} e^{st}f(s)ds + \int_{\Gamma_3} e^{st}f(s)ds + \int_{\Gamma_4} e^{st}f(s)ds$$

Then if we can show that each of the integrals on the right-hand side approach 0 as $R \to \infty$, we will have proved the required result. To do this we consider these four integrals.

(continued)

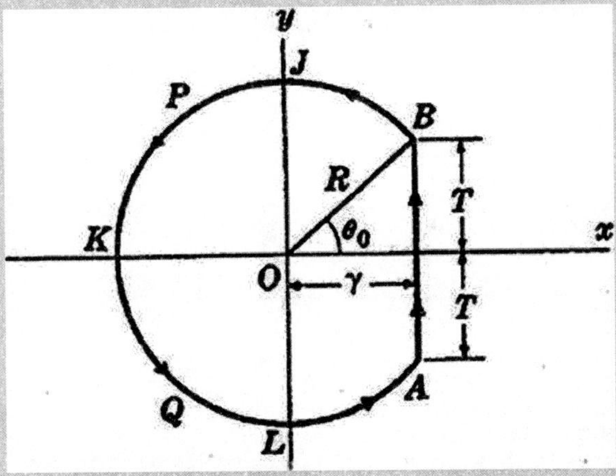

Fig. (a) Residue roots circle

Case I.
Integral over Γ_1 or BJ
Along Γ_1 we have, since $s = Re^{i\theta}$, $\theta_0 \leq \theta \leq \pi/2$:

$$I_1 = \int_{\Gamma_1} e^{st} f(s) ds = \int_{\theta_0}^{\pi/2} e^{Re^{i\theta}t} f\left(Re^{i\theta}\right) iRe^{i\theta} d\theta$$

Then:

$$|I_1| \leq \int_{\theta_0}^{\pi/2} \left|e^{(R\cos\theta)t}\right| \left|e^{i(R\sin\theta)t}\right| \left|f\left(Re^{i\theta}\right)\right| \left|iRe^{i\theta}\right| d\theta$$
$$\leq \int_{\theta_0}^{\pi/2} e^{(R\cos\theta)t} \left|f\left(Re^{i\theta}\right)\right| R d\theta$$
$$\leq \frac{M}{R^{k-1}} \int_{\theta_0}^{\pi/2} e^{(R\cos\theta)t} d\theta = \frac{M}{R^{k-1}} \int_0^{\phi_0} e^{(R\sin\phi)t} d\phi$$

where we have used the given condition $|f(s)| \leq M/R^k$ on Γ_1 and the transformation $\theta = \pi/2 - \phi$ where $\phi_0 = \pi/2 - \theta_0 =$ arcsin $(\gamma/R) = \sin^{-1}(\gamma/R)$.
Since $\sin\phi \leq \sin\phi_0 \leq \cos\theta_0 = \gamma/R$, this last integral is less than or equal to:

$$\frac{M}{R^{k-1}} \int_0^{\phi_0} e^{\gamma t} d\phi = \frac{Me^{\gamma t}\phi_0}{R^{k-1}} = \frac{Me^{\gamma t}}{R^{k-1}} \sin^{-1}\frac{\gamma}{R}$$

(continued)

But as $R \to \infty$, this last quantity approached zero as can be seen by noting, for example, that $\sin^{-1}(\gamma/R) \approx \gamma/R$ for large R. Thus $\lim_{R\to\infty} I_1 = 0$.

Case II.

Integral over Γ_2 or JPK

Along Γ_2 we have, since $s = Re^{i\theta}$, $\pi/2 \leq \theta \leq \pi$:

$I_2 = \int_{\Gamma_2} e^{st} f(s)ds = \int_{\pi/2}^{\pi} e^{Re^{i\theta}t} f\left(Re^{i\theta}\right) iRe^{i\theta} d\theta$

Then, as in Case 1, we have:

$$|I_2| \leq \frac{M}{R^{k-1}} \int_{\pi/2}^{\pi} e^{(R\cos\theta)t} d\theta = \frac{M}{R^{k-1}} \int_0^{\pi/2} e^{-(R\sin\phi)t} d\phi$$

Upon letting $\theta = \pi/2 + \phi$

Now $\sin\phi \geq 2\phi/\pi$ for $0 \leq \varphi \leq \pi/2$ so that the last integral is less than or equal to:

$$\frac{M}{R^{k-1}} \int_0^{\pi/2} e^{-2r\phi t/\pi} d\phi = \frac{\pi M}{2tR^k}\left(1 - e^{Rt}\right)$$

which approaches 0 as $R \to \infty$. Thus $\lim_{R\to\infty} I_2 = 0$.

Case III.

Integral over Γ_3 or KQL

This case can be treated in a manner similar to Case II.

Case IV.

Integral over Γ_4 or QL

This case can be treated in a manner similar to Case I.

Problem-2: Find the inverse Laplace transform of the following function with branch point by use of complex inversion formula.

$$\pounds^{-1}\left\{\frac{e^{-a\sqrt{s}}}{s}\right\} = ?$$

Solution: By the complex inversion formula and using Fig. (b), the required inverse Laplace transform is given by:

$$F(t) = \frac{1}{2\pi i} \int_{\gamma - i\infty}^{\gamma + i\infty} \frac{e^{st - a\sqrt{s}}}{s} ds \tag{1}$$

Since $s = 0$ is a branch point of the integrand, we consider:

(continued)

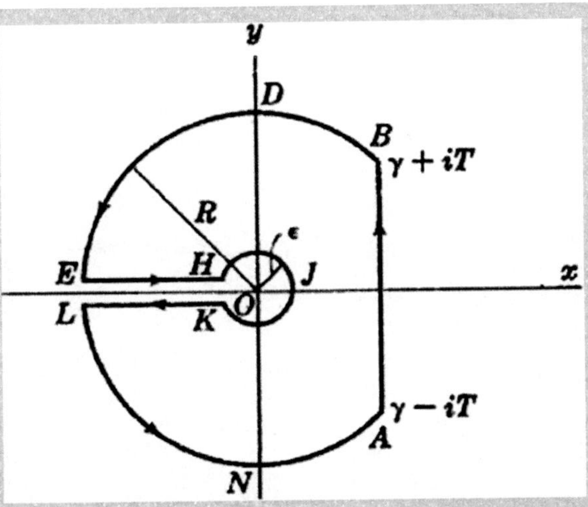

Fig. (b) Branch contour

$$\frac{1}{2\pi i}\oint_C \frac{e^{st-a\sqrt{s}}}{s}ds = \frac{1}{2\pi i}\int_{AB}\frac{e^{st-a\sqrt{s}}}{s}ds + \frac{1}{2\pi i}\int_{BDE}\frac{e^{st-a\sqrt{s}}}{s}ds$$
$$+\frac{1}{2\pi i}\int_{EH}\frac{e^{st-a\sqrt{s}}}{s}ds + \frac{1}{2\pi i}\int_{HJK}\frac{e^{st-a\sqrt{s}}}{s}ds$$
$$+\frac{1}{2\pi i}\int_{KL}\frac{e^{st-a\sqrt{s}}}{s}ds + \frac{1}{2\pi i}\int_{LNA}\frac{e^{st-a\sqrt{s}}}{s}ds$$

where C is the contour of Fig. (a) consisting of the line AB ($s = \gamma$), the arcs BDE and LNA of a circle of radius R and the center at origin O, and the arc HJK of a circle of radius ε with center at O.

Since the only singularity $s = 0$ of the integrand is not inside C, the integral on the left is 0 by Cauchy's theorem. Also, the integrand satisfies the condition of Problem-1, so that on taking the limit as $R \to \infty$, the integrals along BDE and LNA approach 0. It follows that:

$$F(t) = \lim_{\substack{R\to\infty \\ \varepsilon\to 0}} \frac{1}{2\pi i}\int_{AB}\frac{e^{st-a\sqrt{s}}}{s}ds = \frac{1}{2\pi i}\int_{\gamma-i\infty}^{\gamma+i\infty}\frac{e^{st-a\sqrt{s}}}{s}ds$$
$$= \lim_{\substack{R\to\infty \\ \varepsilon\to 0}} \frac{1}{2\pi i}\left\{\int_{EH}\frac{e^{st-a\sqrt{s}}}{s}ds + \int_{HJK}\frac{e^{st-a\sqrt{s}}}{s}ds + \int_{KL}\frac{e^{st-a\sqrt{s}}}{s}ds+\right\} \quad (2)$$

(continued)

Along EH, $s = xe^{\pi i}$, $\sqrt{s} = \sqrt{x}e^{\pi i/2} = i\sqrt{x}$, and as s goes from $-R$ to $-\varepsilon$, x goes from R to ε. Hence we have:

$$\int_{EH} \frac{e^{st-a\sqrt{s}}}{s} ds = \int_{-R}^{-\varepsilon} \frac{e^{st-a\sqrt{s}}}{s} ds = \int_{R}^{\varepsilon} \frac{e^{-xt-ai\sqrt{x}}}{x} dx$$

Similarly, along KL, $s = xe^{-\pi i}$, $\sqrt{s} = \sqrt{x}e^{-\pi i/2} = -i\sqrt{x}$, and as s goes from $-\varepsilon$ to $-R$, x goes from ε to R. Then, we have:

$$\int_{KL} \frac{e^{st-a\sqrt{s}}}{s} ds = \int_{-\varepsilon}^{-R} \frac{e^{st-a\sqrt{s}}}{s} ds = \int_{\varepsilon}^{R} \frac{e^{-xt+ai\sqrt{x}}}{x} dx$$

Along HJK, $s = \varepsilon e^{i\theta}$, and we have:

$$\begin{aligned}\int_{HJK} \frac{e^{st-a\sqrt{s}}}{s} ds &= \int_{\pi}^{-\pi} \frac{e^{\varepsilon e^{i\theta}t-a\sqrt{\varepsilon}e^{i\theta/2}}}{\varepsilon e^{i\theta}} i\varepsilon e^{i\theta} d\theta \\ &= i\int_{\pi}^{-\pi} e^{\varepsilon e^{i\theta}t-a\sqrt{\varepsilon}e^{i\theta/2}} d\theta\end{aligned}$$

Thus (2) becomes:

$$\begin{aligned}F(t) &= -\lim_{\substack{R\to\infty \\ \varepsilon\to 0}} \frac{1}{2\pi i}\left\{\int_{R}^{\varepsilon} \frac{e^{-xt-ai\sqrt{x}}}{x} dx + \int_{\varepsilon}^{R} \frac{e^{-xt+ai\sqrt{x}}}{x} dx + i\int_{\pi}^{-\pi} e^{\varepsilon e^{i\theta}t-a\sqrt{\varepsilon}e^{i\theta/2}} d\theta\right\} \\ &= -\lim_{\substack{R\to\infty \\ \varepsilon\to 0}} \frac{1}{2\pi i}\left\{\int_{\varepsilon}^{R} \frac{e^{xt}\left(e^{ai\sqrt{x}} - e^{-ai\sqrt{x}}\right)}{x} dx + i\int_{\pi}^{-\pi} e^{\varepsilon e^{i\theta}t-a\sqrt{\varepsilon}e^{i\theta/2}} d\theta\right\} \\ &= -\lim_{\substack{R\to\infty \\ \varepsilon\to 0}} \frac{1}{2\pi i}\left\{2i\int_{\varepsilon}^{R} \frac{e^{-xt}\sin a\sqrt{x}}{x} dx + i\int_{\pi}^{-\pi} e^{\varepsilon e^{i\theta}t-a\sqrt{\varepsilon}e^{i\theta/2}} d\theta\right\}\end{aligned}$$

Since the limit can be taken underneath the integral sign, we have:

$$\lim_{\varepsilon\to 0}\int_{\pi}^{-\pi} e^{\varepsilon e^{i\theta}t-a\sqrt{\varepsilon}e^{i\theta/2}} d\theta = \int_{\pi}^{-\pi} 1\ d\theta = -2\pi$$

and so we find that:

(continued)

$F(t) = 1 - \frac{1}{\pi}\int_0^{\infty} \frac{e^{-xt}\sin a\sqrt{x}}{x} dx$
and this result can be written as error function. See Problem-3:

$$F(t) = 1 - erf\left(a/2\sqrt{t}\right) = erfc\left(a/2\sqrt{t}\right)$$

Problem-3: Prove that:

$$\frac{1}{\pi}\int_0^{\infty} \frac{e^{-xt}\sin a\sqrt{x}}{x} dx = erf\left(a/2\sqrt{t}\right)$$

Letting $x = u^2$, the required integral becomes:

$$I = \frac{2}{\pi}\int_0^{\infty} \frac{e^{-u^2 t}\sin au}{u} du$$

Then differentiating with respect to a and using the following:

$$\frac{\partial I}{\partial a} = \frac{2}{\pi}\int_0^{\infty} e^{-u^2 t}\cos au du = \frac{2}{\pi}\left(\frac{\sqrt{\pi}}{2\sqrt{t}} e^{-a^2/4t}\right) = \frac{1}{\sqrt{\pi t}} e^{-a^2/4t}$$

Hence, using the fact that $I = 0$ when $a = 0$:

$$I = \int_0^{a} \frac{1}{\sqrt{\pi t}} e^{-p^2/4t} dp = \frac{2}{\sqrt{\pi}}\int_0^{a/2\sqrt{t}} e^{-u^2} du = erf\left(a/2\sqrt{t}\right)$$

and the required result is established.

Now that we learned all above knowledge and using Fig. (c) here, let us now derive the formula and solution for the temperatures $T(z, t)$ in a semi-infinite solid $z \geq 0$, initially at temperature 0, when a constant flux of heat (F_0) is maintained at the boundary $z = 0$.

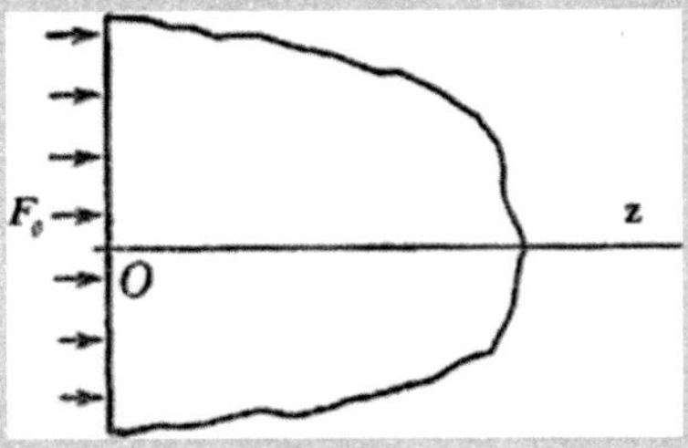

Fig. (c) Heat flux surface incident

(continued)

In this idealized case of a thick slab of material, we shall substitute for the thermal condition at the right-hand boundary, the condition that T tends to 0 as z tends to infinity. The boundary value problem then becomes:

$$\frac{\partial^2 T(z,t)}{\partial z^2} - \frac{1}{\kappa}\frac{\partial T(z,t)}{\partial t} = 0(z > 0,\ t > 0) \tag{1}$$

$$T(z,0) = 0(z > 0) \tag{2}$$

$$-K\frac{\partial T(z,t)}{\partial z}\bigg|_{z=0} = F_0 \lim_{z\to\infty} T(z,t) = 0t > 0 \tag{3}$$

Let $u(z, s)$ be the Laplace transform, with respect to time t, of the temperature function $T(z, t)$. Transforming the members of Eqs. (1) and (3), we have the following problem in ordinary differential equations which $u(z, s)$ must satisfy, so we can write:

$$su(z,s) - T(z,0) = \kappa\frac{d^2u(z,s)}{dz^2} \quad \text{or} \quad \frac{d^2u(z,s)}{dz^2} - \frac{s}{\kappa}u(z,s) = 0 \tag{4}$$

where:

$$u(0,s) = \pounds\{T(o,t)\} = \frac{F_0}{s} \tag{5}$$

and $u = \vec{u}(z, s)$ is required to be bounded. Now the boundary condition (3) in the new Laplace plane becomes:

$$-K\frac{\partial T(z,s)}{\partial z} = \frac{F_0}{s} \qquad \lim_{z\to\infty} T(z,s) = 0 \tag{6}$$

Solving Eq. (4), we find that:

$$u(z,s) = c_1 e^{\sqrt{s/\kappa z}} + c_2 e^{-\sqrt{s/\kappa z}} \tag{7}$$

Then we choose $c_1 = 0$ so that u is bounded as $z \to \infty$, and we have:

$$u(z,s) = c_2 e^{-\sqrt{s/\kappa z}} \tag{8}$$

From boundary condition of Eq. (6) and what is shown in Eq. (5), we have $c_2 = F_0\sqrt{\kappa}/Ks\sqrt{s}$, so that:

$$u(z,s) = \frac{F_0\sqrt{\kappa}}{Ks\sqrt{s}} e^{-\sqrt{s/\kappa z}}$$

Hence by problems 2 and 3, we know that:

(continued)

$$\pounds^{-1}\left\{\frac{1}{\sqrt{s}}e^{-z\sqrt{s/\kappa}}\right\} = \frac{1}{\sqrt{\pi t}}e^{-z^2/(4\kappa t)}$$

and view of the factor $1/s$ in our formula for $u(z, s)$, it follows:

$$T(z,t) = \frac{F_0}{K}\sqrt{\frac{\kappa}{\pi}}\int_{\tau=0}^{\tau=t} e^{-z^2/(4\kappa\tau)}\frac{d\tau}{\sqrt{\tau}}$$
$$= \frac{F_0 z}{K\sqrt{\pi}}\int_{z/\left(2\sqrt{\kappa t}\right.}^{\infty}\frac{1}{\lambda^2}e^{-\lambda^2}d\lambda$$

Where the second integral is obtained from the first by substitution $\lambda = z/\left(2\sqrt{\kappa t}\right)$. Upon integration of the last integral by parts, we find that:

$$T(z,t) = \frac{F_0}{K\sqrt{\pi}}\left(2\sqrt{\kappa t}e^{-z^2/(4\kappa t)} - 2z\int_{z/\left(2\sqrt{\kappa t}\right.}^{\infty} e^{-\lambda^2}d\lambda\right)$$

which is the result of Eq. 5.160.

The solution to this problem is given by:

$$T(z,t) = \frac{2F_0\sqrt{\kappa t}}{K} ierf\left[z/2\sqrt{\kappa t}\right] \tag{5.160}$$

or

$$T(z,t) = \frac{2F_0}{K}\left\{\sqrt{\frac{\kappa t}{\pi}}e^{-z/4\kappa t} - \frac{z}{2}erf\left[z/\left(2\sqrt{\kappa t}\right)\right]\right\}$$

The functions which appear here are error functions, and it is useful to summarize some of their properties and definitions. (See Carslaw and Jaeger, Chapter II) [3].

The error function is:

$$erf(x) = \frac{2}{\sqrt{\pi}}\int_0^x e^{-\ell^2}d\ell$$

$$erf(0) = 0 erf(\infty) = 1 erf(-x) = -erf(x)$$

The complementary error function is:

$$erfc(x) = 1 - erf(x) = \frac{2}{\sqrt{\pi}}\int_x^{\infty} e^{-\ell^2} d\ell erfc(0) = 1$$

The integral of the complementary error function is:

$$ierfc(x) = \int_x^{\infty} erfc(\ell)d\ell$$

or

$$ierfc(x) = \frac{1}{\sqrt{\pi}}e^{-x^2} - xerfc(x)$$

$$ierfc(x) = \frac{1}{\sqrt{\pi}}e^{-x^2} - x + xerf(x)$$

Some derivation are useful:

$$\frac{\partial erf(x)}{\partial x} = -\frac{\partial erfc(x)}{\partial x} = \frac{2}{\sqrt{\pi}}e^{-x^2}$$

$$\frac{\partial^2 erf(x)}{\partial x^2} = -\frac{\partial^2 erfc(x)}{\partial x^2} = -\frac{4}{\sqrt{\pi}}xe^{-x^2}$$

$$\frac{\partial ierfc(x)}{\partial x} = -erfc(x)$$

$$\frac{\partial^2 ierfc(x)}{\partial x^2} = \frac{2}{\sqrt{\pi}}e^{-x^2}$$

Now we can show that the boundary condition is satisfied. Usage of the first form of Eq. 5.169 yields:

$$\frac{\partial T}{\partial z} = \frac{2F_0\sqrt{\kappa t}}{K}\left\{-erfc\left[z/\left(2\sqrt{\kappa t}\right)\right]\right\}\frac{1}{2\sqrt{\kappa t}}$$

and since $erfc(0) = 1$, therefore:

$$\left.\frac{\partial T}{\partial z}\right|_{z=0} = -\frac{F_0}{K}$$

One can also show that Eq. 5.160 satisfies Eq. 5.157.

We can use Eq. 5.160 to show what the front surface temperature behavior is under constant irradiation, by setting $z = 0$, so that:

$$T(0,t) = \frac{2F_0}{K}\sqrt{\frac{\kappa t}{\pi}} \tag{5.161}$$

As an illustration, let us calculate the time required to raise aluminum to its melting point for a power density of 5 kW/cm ^{2}:

$$K = 2.3\text{W}/\text{cm}^2$$
$$\kappa \approx 0.9\ \text{cm}^2/\text{s}$$
$$T = T_{melt} - T_{room} \approx 600^\circ\text{C}$$
$$F_0 = (1-R)I$$
$$(1-R) = 0.04$$
$$F_0 = 0.04 \times 5 \times 10^3 = 200\ \text{W}/\text{cm}^2$$

Then $t = \left(\pi K^2 T^2\right)/\left(4F_0^2\kappa\right)$ yields $t \approx 42$ s. In practice, it is very difficult to melt extremely thick slabs of aluminum with even a high-power laser, as these calculations suggest.

Equation 5.160, although derived for a very simple case, describes many very important features of thermal response to lasers. First we shall define the diffusion length, which is useful in that it permits a wide variety of order-of-magnitude calculations to be made. The thermal diffusion length D is defined as:

$$D = 2\sqrt{\kappa t} \tag{5.162}$$

Strictly speaking, the thermal diffusion length is defined as the distance required for the temperature to drop to $1/e$ of its initial value and depends somewhat on the geometry and the boundary condition. For most purposes, it is sufficient simply to take it as defined by Eq. 5.162. Looking at our Eq. 5.160, for example, we see that:

$$T(D,t) = \frac{2F_0}{K}\left\{\sqrt{\frac{\kappa t}{\pi}}\frac{1}{e} - \frac{D}{2}erfc(1)\right\}$$

Using tarcble of erfc, we get erfc(1) = 0.1573; see Table 5.2 as well.

$$T(D,t) = \frac{2F_0}{K}\left\{\sqrt{\frac{\kappa t}{\pi}}\frac{1}{e} - \frac{D}{2}(0.1573)\right\}$$

Table 5.2 Table of error function and its complementary

Value x	Erf (x)	Erfc(x)l
0	0	1
0.05	0.0563720	0.9436280
0.1	0.1124629	0.8875371
0.15	0.1679960	0.8320040
0.2	0.2227026	0.7772974
0.25	0.2763264	0.7236736
0.3	0.3286268	0.6713732
0.35	0.3793821	0.6206179
0.4	0.4283924	0.5716076
0.45	0.4754817	0.5245183
0.5	0.5204999	0.4795001
0.55	0.5633234	0.4366766
0.6	0.6038561	0.3961439
0.65	0.6420293	0.3579707
0.7	0.6778012	0.3221988
0.75	0.7111556	0.2888444
0.8	0.7421010	0.2578990
0.85	0.7706681	0.2293319
0.9	0.7969082	0.2030918
0.95	0.8208908	0.1791092
1	0.8427008	0.1572992
1.1	0.8802051	0.1197949
1.2	0.9103140	0.0896860
1.3	0.9340079	0.0659921
1.4	0.9522851	0.0477149
1.5	0.9661051	0.0338949
1.6	0.9763484	0.0236516
1.7	0.9837905	0.0162095
1.8	0.9890905	0.0109095
1.9	0.9927904	0.0072096
2	0.9953223	0.0046777
2.1	0.9970205	0.0029795
2.2	0.9981372	0.0018628
2.3	0.9988568	0.0011432
2.4	0.9993115	0.0006885
2.5	0.9995930	0.0004070
2.6	0.9997640	0.0002360
2.7	0.9998657	0.0001343
2.8	0.9999250	0.0000750
2.9	0.9999589	0.0000411
3	0.9999779	0.0000221

(continued)

Table 5.2 (continued)

Value x	Erf (x)	Erfc(x)l
3.1	0.9999884	0.0000116
3.2	0.9999940	0.0000060
3.3	0.9999969	0.0000031
3.4	0.9999985	0.0000015
3.5	0.9999993	0.0000007

From Eq. 5.161, we have:

$$T(D,t) = T(0,t)\left\{\frac{1}{e} - 0.1573\sqrt{\pi}\right\}$$

Thus:

$$T(D,t) = 0.09T(0,t) \tag{5.163}$$

In this case, $(1/e)T(0,t) \approx 0.37T(0,t)$. Referring to our example of irradiating aluminum for 42 s to reach the melting point, we note that the diffusion length at that time is given by:

$$D = 2\sqrt{0.9 \times 42} \approx 12 \text{ cm}$$

and by Eq. 5.163, and the value of $T = 600^{\circ}$C, the temperature at distance D into the material is 0.09×600, or about 54 °C above ambient temperature given by above values.

Now let us illustrate the solution Eq. 5.160 graphically as demonstrated in Fig. 5.29. For convenience rewrite the equation by introducing $D = 2\sqrt{\kappa t}$ and by reducing it to the error function erf, so that:

$$T(z,t) = \frac{2F_0}{K}\left[\frac{D}{2\sqrt{\pi}}e^{-z^2/D^2} - \frac{z}{2} + \frac{z}{2}erf(z/D)\right]$$

Now let $\eta = z/D$, and:

$$T(z,t) = \frac{F_0 D}{K}\left[\frac{e^{-\eta^2}}{2\sqrt{\pi}} - \eta + \eta erf(\eta)\right] \tag{5.164(a)}$$

or equivalently in terms of ierfc, we can rewrite the above equation as follows:

$$T(z,t) = \frac{F_0 D}{K} ierfc(\eta) \tag{5.164(b)}$$

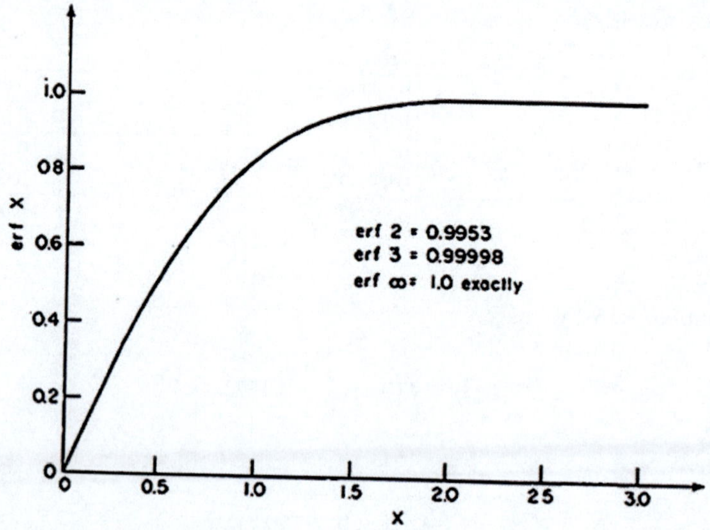

Fig. 5.29 The Error function plot

Finally, we define a dimensionless temperature $\theta = TK/(F_0D)$, that [48]:

$$\theta = ierfc(\eta) \tag{5.164(c)}$$

Thus, the plot of the integral of the complementary error function here is the graph of the solution to the problem of constant heat flux on the surface of a semi-infinite solid.

Now, although the graph of Eq. 5.164(c) (see Fig. 5.30) represents very succinctly the solution to our problem, it does not really snow how the temperature varies as a function of position and time. For this purpose, it is useful to look at the temperature profiles for various times and see how the profile changes with time. These curves can be generated quickly from $\theta = ierfc(\eta)$ by recalling the definitions of θ and η, and writing them in the following form:

$$T = \left(\frac{2F_0\sqrt{\kappa}}{K}\right)\sqrt{t}\theta \tag{5.165(a)}$$

$$z = 2\sqrt{\kappa}\sqrt{t}\eta \tag{5.165(b)}$$

Thus, at a given time, the $\theta = ierfc(\eta)$ curve scales according to Eq. 5.166; the basic shape of the curve is unchanged, but it is stretched one way or the other depending on the parameters, and this stretching progresses in time as $\sqrt{t}$. The case of aluminum is shown in Fig. 5.31.

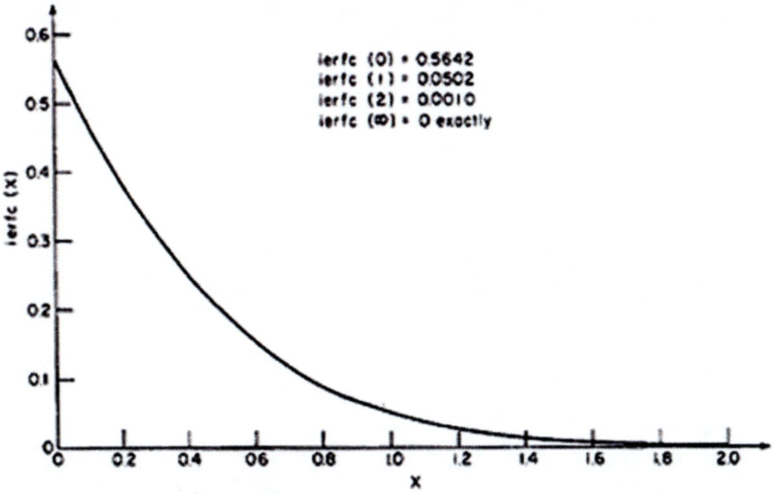

Fig. 5.30 The integral of the complimentary error function

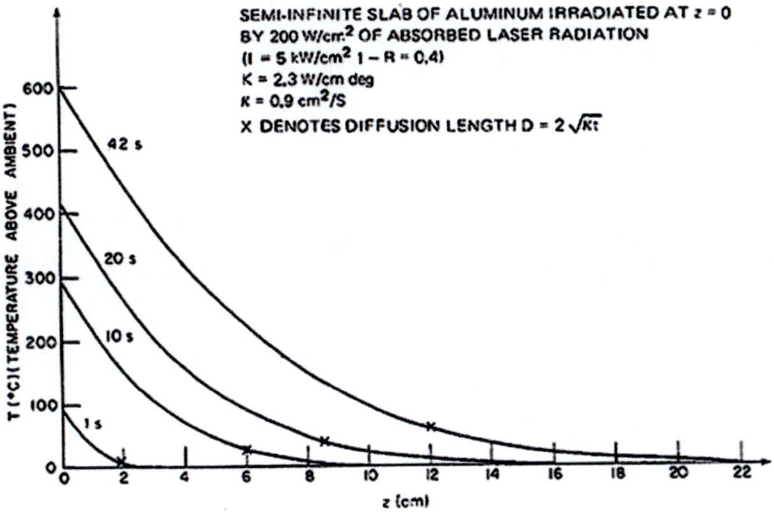

Fig. 5.31 Laser-induced temperature rise in aluminum as a function of depth

We can also look at the variation of temperature with time at a fixed position. The variation of the surface ($z = 0$) is simply $T(0, t) \approx \sqrt{t}$ as Eq. 5.170 shows. Wherever $z/\left(2\sqrt{\kappa t}\right)$ is very small, the temperature variation will approach $\sqrt{t}$. Thus, at any position $T \approx \sqrt{t}$ at sufficiently large t. The temperature vs. time profiles at fixed position for times such that $z/\left(2\sqrt{\kappa t}\right)$ is not small can be

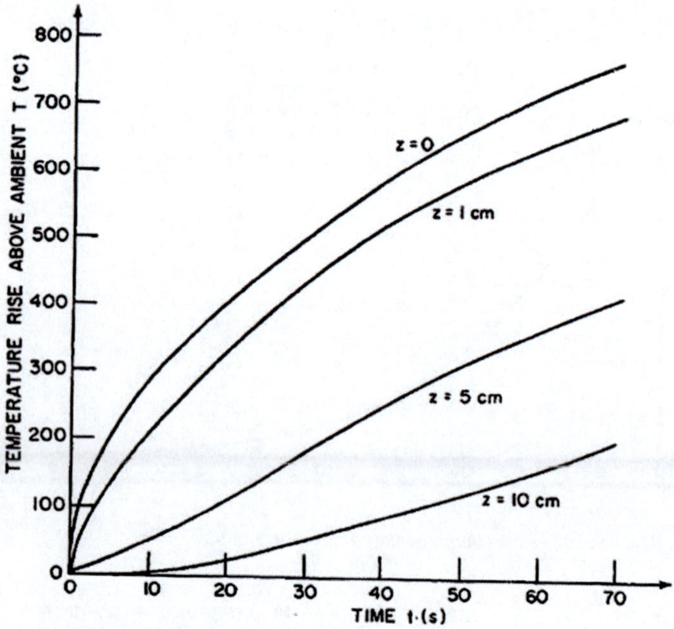

Fig. 5.32 Laser-induced temperature rise on front surface of aluminum as a function of time

calculated, of course, from Eq. 5.169. Some results for aluminum, with the parameters used above, are shown in Fig. 5.32. Notice that at z = 10 cm, the temperature profile is far from the "long-time," or $\sqrt{t}$, behavior even at 40 or 50 s, whereas the surface has already begun to melt.

We now turn to some order-of-magnitude arguments. One such argument can be used to estimate the power–pulse length combination which might be expected to yield surface vaporization.

Consider a laser pulse that has a simple time behavior shown in Fig. 5.33 and uniformly irradiates the surface of the material. The pulse length is t_p and the intensity is such that, combined with the reflectance, the absorbed power density is F_0. Again, we assume that the optical energy is absorbed in a very thin layer at the surface. Let D_p be the diffusion length associated with the time t_p.

The question is whether a significant amount of surface vaporization will occur before the pulse ends. One approach would be to use Eq. 5.161 to calculate the surface temperature at the time t_p and compare it to this vaporization temperature. However, this would ignore the influence of the latent heats of melting and of vaporization, which have an important influence. We shall discuss thermal flow with phase changes later. For the present purpose, we can include them by considering the energy required to melt and vaporize a portion of the material. The key is to estimate what thickness of the material is involved, and in this order-of-magnitude

Fig. 5.33 Irradiation of a semi-infinite solid by a pulse

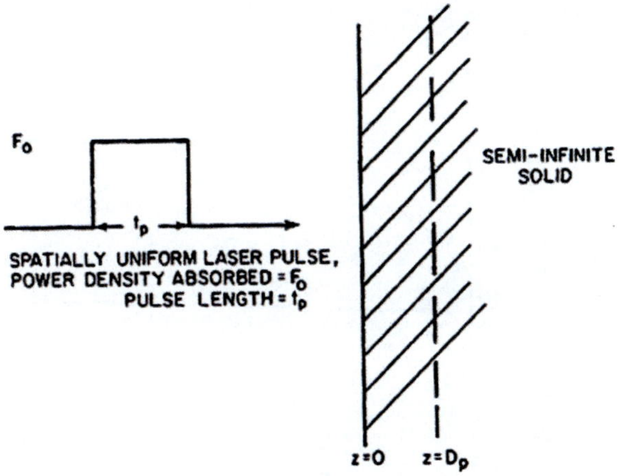

argument, we simply use the thermal diffusion length for this thickness. Thus, we set the criterion for vaporization as:

$$\frac{F_0 t_p}{D_p} \geq \rho[c_s(T_m - T_0) + L_m + c_\ell(T_b - T_m) + L_v]$$

where ρ is density of the material; c_s and c_ℓ are the specific heats of the solid and liquid, respectively; Tm is the melting point; T_b is the boiling point; and L_m and L_v are the heats of melting and vaporization. Notice we are ignoring differences between the solid and liquid for density and conductivity, as is appropriate in this crude argument. If numerical values are checked, L_v dominates the expression on the right side of the inequality. For example, for aluminum $L_v = 10{,}875$ J/g, whereas all the other terms contribute to a total of 3.046 J/g. Since the argument is crude, then, one usually takes:

$$\frac{F_0 t_p}{D_p} \geq \rho L_v$$

as the criterion for vaporization by a pulse. Since $D_p = 2\sqrt{\kappa t_p}$, we have:

$$F_0 \geq \text{aprrox } \frac{2\sqrt{\kappa} L_v \rho}{\sqrt{t_p}} \tag{5.166}$$

Some calculations based on Eqs. 5.165(a) and 5.165(b) are shown in Fig. 5.34. Most metals fall in the band indicated. For a given pulse time, at power densities greater than the band indicates, vaporization effects would be expected to be important. Some useful thermal constants are included in Table 5.3.

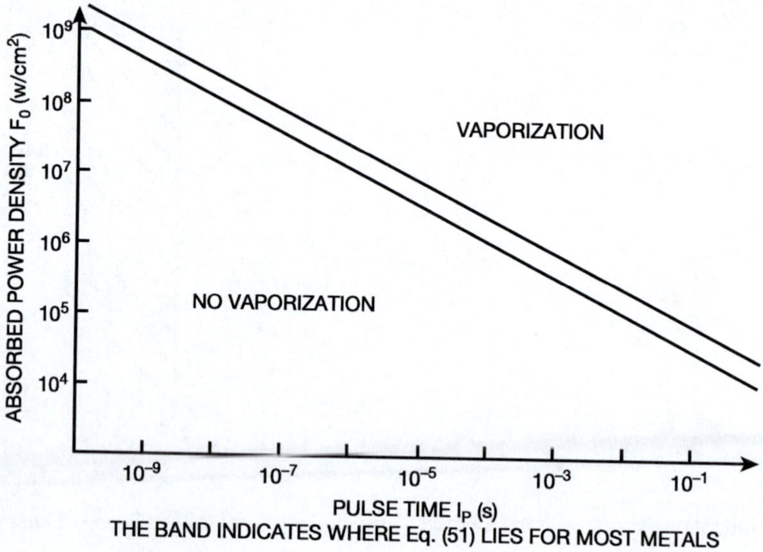

Fig. 5.34 Power density pulse time criterion for vaporization

In deriving Eq. 5.166, we have been seeking the power density required, at a given pulse length, for a thermal layer to be vaporized. The same expression, of course, tells us the pulse time at which vaporization becomes important for a fixed power density. Rewriting Eq. 5.166 gives for this time:

$$t_p > \text{aprrox} \frac{K^2 T_{\text{vap}}^2 \pi}{4F_0^2 \kappa} \tag{5.167}$$

Let us compare this to the time required for surface vaporization to begin. We do this by using our solution for heat flow in the semi-infinite solid for the surface (Eq. 5.161) and solving for the time at which the front surface reaches the vaporization temperature:

$$T_{\text{vap}} = \frac{2F_0}{K}\sqrt{\frac{\kappa t_{\text{vap}}}{\pi}}$$

or

$$t_{\text{vap}} = \frac{K^2 T_{\text{vap}}^2 \pi}{4F_0^2 \kappa}$$

Thus, at $t_p = t_{\text{vap}}$, this calculation would predict that vaporization at the surface begins. For example, at $F_0 = 10^6 \text{W/cm}^2$, vaporization begins at $t_p \approx 10^{-5}$ to 10^{-6} s, depending on the metal. On the other hand, for a thermal layer to be evaporated, it

Table 5.3 Thermo-chemical data for metals[a]

Metal or alloy	Diffusivity[b] (cm^2/s)	Conductivity[b] (W/cm K)	(cal/s cm K)	Specific heat[b,c] (J/g K)	(cal/g K)	Density (g/cm^3)	Solidus temperature (°C)	Liquid us temperature (°C)	Heat of melting (J/B)	(cal/g)	Vaporization temperature (°C)	Heat of vaporization (J/g)	(cal/g)
Al	0.85	2.40	0.57	1.05	0.25	2.7		660	400	95.6	2520	10,875	2600
2024 Al	0.60	1.732	0.414	1.0	0.24	2.77	502	638					
Ti·	0.0636	0.216	0.0516	0.753	0.18	4.51		1670	324	77.3	3289	8790	2100
Ti (6Al, 4V)	0.051	0.19	0.0455	0.837	0.20	4.47	1537	1649					
Fe						7.86		1536	247	59.1	2862	6260	1496
304 Stainless steel	0.0523	0.259	0.062	0.628	0.15	8.0	1399	1454					

[a]Data compiled by R. L. Stegman of the Naval Research Laboratory, Washington, DC
[b]These values represent an average over the temperature between room' temperature and the melting point
[c]Of the solid

requires, according to Eq. 5.167, $t_p \approx 10^{-3}$s. It turns out that both estimates are useful. In a later section, we shall discuss features of a more correct treatment, which accounts for both the heat of melting and the heat of vaporization in the dynamic situation of propagating solid–liquid and liquid–vapor interfaces.

5.9.1.2 No Phase Change, Slab of Finite Thickness

Let us turn now to a treatment of another geometry which can be useful in practical cases, namely, irradiation of one surface of a sheet or slab of finite thickness. Let the slab be taken as infinite in extent in the x and y direction, and let the laser irradiation be uniform over the entire surface $z = 0$. Thus, we again have a one-dimensional situation, as shown in Fig. 5.35. The thickness of the sheet is taken as ℓ, and the absorbed power density as a function of time is $F(t)$; we again assume that the radiation is absorbed in a very narrow layer at the front surface. The equation we wish to solve is then:

$$\frac{\partial^2 T(z,t)}{\partial z^2} - \frac{1}{\kappa}\frac{\partial T(z,t)}{\partial t} = 0$$

with boundary conditions:

$$-K\frac{\partial T(z,t)}{\partial z}\bigg|_{z=0} = F(t)$$

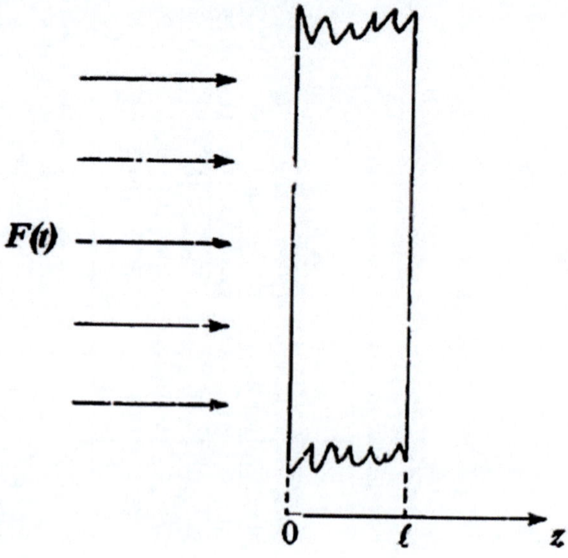

Fig. 5.35 Irradiation of slab of finite thickness

$$-K\frac{\partial T(z,t)}{\partial z}\bigg|_{z=\ell} = 0$$

The second boundary condition states that "the rear surface is insulated." We shall look at the consequences of this assumption a little later.

As we showed for the semi-infinite slab, the solutions turn out to be elaborate. Turning to the special case of $F(t) = F_0$, a constant, the solution is as follows and easily can be obtained very similar to the methods in Case I by using the Laplace transform or separation of variables methods by assuming $T(z,t) = Z(z)U(t)$ and given boundary conditions to obtain a general solution. In case Laplace Transformation approach the Ordinary Differential Equation (ODE) in Laplace time and space plane is given as:

$$su(z,s) - T(x,0) = \kappa\frac{d^2u(z,s)}{dz^2} \text{ or } \frac{d^2u(z,s)}{dz^2} - \frac{su(z,s)}{\kappa} = -\frac{F_0}{\kappa} \tag{5.168}$$

Taking Laplace transform over the boundary conditions provides the new form of them in Laplace domain as:

$$£\left\{-K\frac{\partial T(z,t)}{\partial z}\bigg|_{z=0}\right\} = \frac{\partial u(z,s)}{\partial z}\bigg|_{z=0} = \frac{F_0}{Ks}$$

$$£\left\{-K\frac{\partial T(z,t)}{\partial z}\bigg|_{z=\ell}\right\} = \frac{\partial u(z,s)}{\partial z}\bigg|_{z=\ell} = 0$$

The general solution of Eq. 5.168 is written as:

$$u(z,s) = c_1 \cosh\sqrt{s/\kappa}z + c_2 \sinh\sqrt{s/\kappa}z$$

Applying the new sets of boundary conditions to solve for c_1 and c_2 and taking reverse Laplace transform to transfer from coordinates of (z, s) to (z, t) as well as dealing with finding the residues at given s , and then calculating the sum of residues and some mathematical manipulations, we will end up with the following solution (Eq. 5.169) (readers should go through this exercise on their own):

$$T(z,t) = \frac{F_0\kappa}{K\ell}t + \frac{F_0\ell}{K} \times \left\{\frac{3(\ell-z)^2 - \ell^2}{6\ell^2} - \frac{2}{\pi^2}\sum_{n=1}^{\infty}\frac{(-1)^n}{n^2}e^{-\kappa n^2\pi^2 t/\ell^2}\cos\left[\frac{n\pi(\ell-z)}{\ell}\right]\right\} \tag{5.169}$$

We can check if this satisfies the boundary conditions:

$$\frac{\partial T(z,t)}{\partial z} = \frac{F_0 \ell}{K} \left\{ \frac{-(\ell - z)}{\ell^2} - \frac{2}{\pi^2} \sum_{n=1}^{\infty} \frac{(-1)^n}{n^2} e^{-\kappa n^2 \pi^2 t/\ell^2} \frac{n\pi}{\ell} \sin\left[\frac{n\pi(\ell - z)}{\ell}\right] \right\}$$

Now $\sin(n\pi) = 0$ and $\sin(0) = 0$, so the $\sum$ term vanishes at both $z = 0$ and $z = \ell$, and:

$$\left.\frac{\partial T(z,t)}{\partial z}\right|_{z=0} = -\frac{F_0}{K}$$

$$\left.\frac{\partial T(z,t)}{\partial z}\right|_{z=\ell} = 0$$

Similarly, the thermal diffusion equation is satisfied, as the reader can verify.

Let us look briefly at this solution. It consists of a linear term in t, together with a "correcting term," which can be plotted as shown in Fig. 5.36. In other words, what is plotted is the term:

$$\mathcal{D} = \frac{3(\ell - z)^2 - \ell^2}{6\ell^2} - \frac{2}{\pi^2} \sum_{n=1}^{\infty} \frac{(-1)^n}{n^2} e^{-\kappa n^2 \pi^2 t/\ell^2} \cos\left[\frac{n\pi(\ell - z)}{\ell}\right]$$

Let us examine some special cases. At $z = 0$, for example, we have:

$$T(0,t) = \frac{F_0 \kappa}{K\ell} t + \frac{F_0 \ell}{K} \left[\frac{1}{3} - \frac{2}{\pi^2} \sum_{n=1}^{\infty} \frac{1}{n^2} e^{-\kappa n^2 \pi^2 t/\ell^2}\right] \tag{5.170}$$

We can rewrite this as:

$$T(0,t) = \frac{F_0 \kappa}{K\ell} t + \frac{F_0 \ell}{K} \mathcal{D}_{z=0}$$

Now $\mathcal{D}$ at a fixed value of z is a function of $\kappa t/\ell^2$. If $\eta = \kappa t/\ell^2$, we can write:

$$T(0,t) = \frac{F_0 \ell}{K} [\eta + \mathcal{D}_{z=0}(\eta)] \tag{5.171}$$

Figure 5.37 shows how $\mathcal{D}$ depends on η, for $z = 0$, and was taken from the previous graph of $\mathcal{D}$, vs. $(1 - z/\ell)$ as per Fig. 5.36. Note that at small η, i.e., at $\kappa t \ll \ell^2$, $\mathcal{D} \approx 0$, so that the front surface initially heats up linearly with time, as:

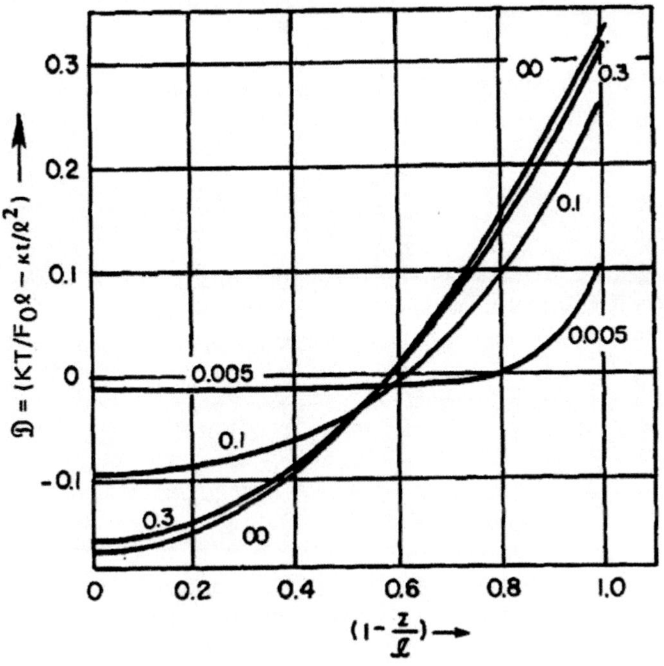

Fig. 5.36 $\mathcal{D}$ as a function of $1 - z/\ell$. (Reference [3], Page 113)

Fig. 5.37 $\mathcal{D}$ at $z = 0$ as a function of η

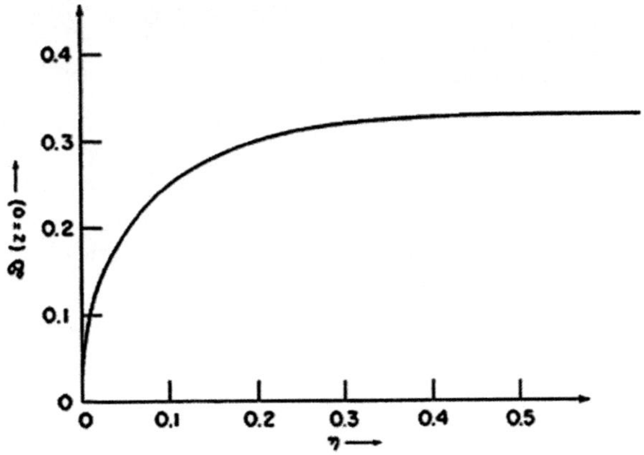

$$T(0,t) = \frac{F_0\kappa}{K\ell}t \tag{5.172}$$

For large t values, or $\kappa t \gg \ell^2$, $\mathcal{D}_{z=0}$ approaches a limiting value of about 0.33. Thus, at long times, we have:

$$T(0,t) = \frac{F_0\ell}{K}\left[\frac{\kappa t}{\ell^2} + 0.33\right] \kappa t \gg \ell^2 \tag{5.173}$$

Here again we see linear behavior, but this time there is an additive constant. If we have a very thick slab, we should get the same result as our previous solution for the time to reach 600 °C on the front surface of aluminum with an absorbed power density of 200 W/cm ^{2}. It turns out that the limiting form of Eq. 5.172 is not correct because it ignores the behavior of $\mathcal{D}_{z=0}(\eta)$ at small η. It. is necessary to use the full expression. Thus:

$$\eta + \mathcal{D}_{z=0}(\eta) = \frac{KT}{F_0\ell}$$

Assume that $\ell = 100$ cm, since we know from our infinite slab solution that the diffusion distance is 12 cm at $T = 600$°C on the front. Thus:

$$\eta + \mathcal{D}_{z=0}(\eta) = \frac{2.3 \times 600}{200 \times 100} = 0.069$$

Reading very roughly from the graph of $\mathcal{D}_{z=0}$ vs. η gives:

$$\eta \approx 0.004 \text{at} \mathcal{D} \approx 0.065$$

Thus, our solution is:

$$\eta \approx 0.004 = \frac{\kappa t}{\ell^2}$$

which gives, since $\kappa \approx 0.9$ cm 2/s, a time of about 44 s, in reasonable agreement with the semi-infinite slab solution.

Now let us turn to a consideration of the rear surface temperature. For this case $z = \ell$, so Eq. 5.169 reduces to:

$$T(z,t) = \frac{F_0 t\kappa}{K\ell} + \frac{F_0\ell}{K}\left\{-\frac{1}{6} - \frac{2}{\pi^2}\sum_{n=1}^{\infty}\frac{(-1)^n}{n^2}e^{-\kappa n^2\pi^2 t/\ell^2}\right\}$$

or, introducing η as before, we have:

$$T(\ell,t) = \frac{F_0\ell}{K}[\eta + \mathcal{D}_{z=\ell}(\eta)]$$

Comments could be made here for the rear surface temperature, and they would be similar to those we made for the front surface temperature. It is interesting to compare the front surface temperature to the back surface temperature. This has a simple form for thin sheets, where $\kappa t/\ell^2 \gg 1$. By referring to the graph, Fig. 5.36 of $\mathcal{D}$, vs. $(1 - z/\ell)$, one can read off values for $\mathcal{D}_{z=\ell}(\infty)$ and $\mathcal{D}_{z=0}(\infty)$, and thus:

$$T(0,t) - T(\ell,t) \approx 0.5\frac{F_0\ell}{K}$$

for

$$\frac{\kappa t}{\ell^2} \gg 1$$

Notice in Fig. 5.36 that the limiting values are approached rapidly; they are nearly realized by the time $\kappa t/\ell^2 \gg 1$. As a numerical illustration, if we have 0.3-cm-thick aluminum:

$$T(0,t) - T(\ell,t) \approx 13^{\circ}\mathrm{C}$$

For the same numbers, we used above. This situation, with the two surfaces heating at the same rate but separated by 13 °C, would start at a time of the order of $t \approx \ell^2/\kappa \approx 0.1$ s. At this time the front surface temperature is about 35 °C.

Let us turn now to a different sort of heat input. So far we have been discussing continuous irradiation. Another simple case, which is a reasonable approximation under certain conditions, is that of a laser pulse which is short enough to be treated as a delta function. Take again a slab of thickness ℓ, and assume that the energy is deposited in a very thin layer near the surface. F refers, as before, to the fraction absorbed by the material. The laser power density, that is, must be multiplied by the optical absorptance. In this case we solve the thermal diffusion equation subject to the boundary condition that:

$$-K\frac{\partial T(z,t)}{\partial z}\bigg|_{z=0} = -K\frac{\partial T(z,t)}{\partial z}\bigg|_{z=\ell} = 0$$

with the stipulation that there is an instantaneous release of E0 units of energy per unit area in the plane z = 0 at time 0. This type of problem is discussed by Carslaw and Jaeger [3] and is most easily solved by the Laplace transform methods. For our present purpose, we quote the solution:

$$T(z,t) = \frac{F_0\kappa}{K\ell}\left\{1 + 2\sum_{n=1}^{\infty}\cos\left(\frac{n\pi z}{\ell}\right)e^{-\kappa n^2\pi^2 t/\ell^2}\right\} \tag{5.174}$$

In Eq. 5.174, we have introduced E_0, the energy per unit area in the pulse. Thus:

$$E_0 = \int_0^{\infty} F(t)dt$$

For the case under consideration, $F(t)$ is considered to be a delta function.

Equation 5.174 is the basis for a scheme used quite frequently for the measurement of thermal parameters [49]. This scheme consists of using a thin sheet of the material to be studied and irradiating uniformly one surface with a very short laser pulse while monitoring the temperature rise induced on the back surface. If one knows E_0, and if the assumptions of no heat loss are valid, the experiment can yield values of both specific heat and thermal conductivity. One adjusts the pulse energy, and hence E_0, so that the induced temperature rise is small. In this way the values of specific heat and thermal conductivity are representative of essentially the ambient temperature of the material.

To see how this is applied, rewrite Eq. 5.174 for the back surface, $z = \ell$ as:

$$T(z,t) = \frac{F_0\kappa}{K\ell}\left\{1 + 2\sum_{n=1}^{\infty}(-1)^n e^{-\kappa n^2\pi^2 t/\ell^2}\right\} \tag{5.175}$$

If we introduce a characteristic time $t_c = \ell^2/\kappa\pi^2$, Eq. 5.184 looks approximately like the curve shown in Fig. 5.38. Here we have also introduced a characteristic temperature $t_c = E_0\kappa/K\ell^2$ and plotted T/T_c, vs. t/t_c, or:

$$\frac{T}{T_c} = 1 + 2\sum_{n=1}^{\infty}(-1)^n e^{-n^2 t/t_c} \tag{5.176}$$

Essentially the experiment consists of monitoring the temperature as a function of time and fitting it to Eq. 5.177. This can be done quite readily. First, the long-term temperature rise T_∞ yields the specific heat because:

$$\frac{T_\infty}{T_c} = 1$$

and, on substituting for T_c:

$$T_\infty = \frac{E_0\kappa}{K\ell}$$

or, since $\kappa = K/\rho c_p$:

$$c_p = \frac{E_0}{T_\infty \rho \ell} \tag{5.177}$$

This technique of measuring specific heat is, of course, not unique, to pulsed lasers. It is sometimes referred to as the slat, calorimeter. The accuracy of the method depends on knowing E_0, which is frequently difficult to ascertain with laser radiation. In some application, Eq. 5.181 is used to calculate E_0, the energy actually absorbed from the laser pulse, by using materials of known specific heat.

The pulsed laser measurement technique is especially suited to determining thermal diffusivity. The magnitude of the back surface temperature rise depends, as we saw, on the energy which is coupled into the material, and this may be difficult to know with any precision. However, the time dependence of the hack surface temperature is independent of the energy input and is controlled only by the diffusivity κ. A simple way to derive κ from a temperature-time profile can lie seen from Fig. 5.38. One measures the time required for the measured temperature response to reach some fraction, say one-half, of its limiting value. Let us call this time Il. It can lie shown numerically [49] from Eq. 5.185 that:

$$\frac{T}{T_c} = \frac{1}{2} = 1 + 2\sum_{n=1}^{\infty} (-1)^n e^{-n^2 t_{1/2}/t_c}$$

is satisfied when:

$$\frac{t_{1/2}}{t_c} = 1.37$$

Using $t_c = \ell^2/(\kappa\pi^2)$ yields:

$$\kappa = \frac{1.37}{\pi^2} \frac{\ell^2}{t_{1/2}} \tag{5.178}$$

Thus, a measurement of $t_{1/2}$, together with the thickness of the specimen, immediately gives thermal diffusivity. If one knows E_0 this experiment gives values of

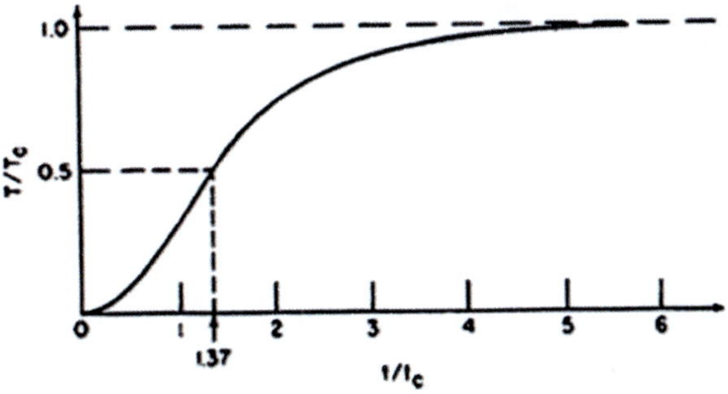

Fig. 5.38 Normalized back surface temperature response to a delta function heat pulse

both the specific heat and the diffusivity, and hence if one knows the density ρ, the experiment gives the thermal conductivity.

This technique has been applied often, at rather high temperatures, usually in the 1000 °C range and above. At these temperatures, steady-state methods of measuring the thermal conductivity are difficult to apply because radiation losses are so large. In the laser flash technique, the radiation loss goes like $T_\infty^4 - T_0^4$ where T_0 is the starting or ambient temperature, established by, say, a furnace and T_∞ is defined above. This radiation loss can be made quite small by adjusting E_0 so that T_∞ is only a few degrees larger than T_0. Since another precise value of E_0 is difficult to establish, these experiments typically measure only thermal diffusivity, not the conductivity.

A final remark on the criterion for the applicability of Eq. 5.186 for slab heating concerns the limits on the laser pulse duration. No laser pulse is, of course, a true δ function. Our solution would be expected to be correct for laser pulse times which are short compared to the time it takes the back surface to respond. The response times are of the order of t_c, so we have the criterion as:

$$t_p \ll t_c$$

or

$$t_p \ll \ell^2 / \left(\kappa \pi^2\right)$$

Calculations that include explicitly the time dependence of the laser pulse [50] indicate that our δ function solution is in error by less than about 2% provided that t_p is less than or equal to about 4% of t_c. Some typical value of t_c with specimens 1 mm thick is given below as:

	κ (cm²/s)	$t_p = \ell^2/(\kappa\pi^2)$ (ms)
Aluminum	0.85	1.2
Stainless steel	0.05234	19

Typical laser pulse lengths are about ½ to 1 ms for the so called “normal mode” lasers, and thus with 1-mm-thick specimens, the technique would be fairly accurate for stainless steel but not very good for aluminum. Thicker specimens would help, but this would make the rear surface temperature rise number. If our laser pulse has, say, 20 J/cm² in it and we use the 10.6 μm absorptance quoted earlier for as-received surfaces, the anticipated temperature rises at the back of the 1 mm specimens would be as shown below:

	A	E_0 (J/cm²)	ρ (g/cm²)	c_p (J/g °C)	$T_\infty = E_0/\rho c \ell$ (°C)
Aluminum	0.04	0.8	2.7	1.05	208
Stainless steel	0.4	8.0	8.0	0.628	16

We see the need to use shorter laser pulses with the same amount of energy. Another solution would be to carry out a more detailed heat flow calculation. Both tailoring of the pulse shape and more detailed calculations are usually employed in current applications of laser flash techniques [51].

5.9.1.3 Analytical Solution of Heat Transfer Equation with Spatial-Dependent Laser Pulse Heating

This section presents analysis of the conduction heating process introduced for a practical Nd:YAG Gaussian laser pulse with a position-dependent intensity. An analytical solution to the problem is obtained with appropriate boundary conditions for a typical continuous wave laser (CW) was developed in Sect. 5.9.1.1, and now we pay our attention to pulse laser irradiation. When the laser intensity is rather low, no phase transition occurs, and the only effect of laser absorption is heating of the material. In metals the laser radiation is absorbed by "free electron," and the energy transfer in metals is also due to electron heat conduction. The temperature field is described by the standard Fourier heat conduction, and for laser pulse with a position-dependent intensity, for this form of pulse input, the Fourier differential equation is described by Eq. 5.124. $Q(x, y, z, t)$ in this equation is defined by the Drude–Zener theory which leads to the following expression (Yilbas et al. 1996). [40]:

$$Q(x,y,z,t) = Ae^{-\beta t}\mu I_{\max}(x,y,t)e^{-\mu z} \tag{5.179}$$

where A is the surface absorptivity, β is the time pulse parameter, μ is the absorption coefficient of the material, and $I_{\max}(x, y, t)|_{z\,=\,0}$ is the maximum laser radiation intensity at the material surface ($z = 0$). According to Sparks (1975) [52], μ is independent of temperature, while surface absorption coefficient of the material, $A = 1 - R$, where, R R is the surface reflectivity, is a linear function of the surface temperature:

$$A = A_0 + A_1(T - T_0) \tag{5.180}$$

where A_0 is the surface absorptivity at room or ambient temperature, T_0, and for most of engineering materials, the absorption factor A_1 is almost unity for Nd:YAG laser wavelength. This temperature dependence of surface absorptivity results from the fact that A is proportional to the electron–phonon collision frequency which, in turn, is proportional to the crystal lattice temperature.

For metals and certain applications, at temperatures above the Debye temperature, it can be assumed that $k(T)$ and $c(T)$ do not change dramatically with temperature, where again $c(T)$ is the density of the material of the workpiece, is the temperature-dependent specific heat of the material, $k(T)$ is the temperature-dependent thermal conductivity. Details of these analysis were discussed in Sect. 5.7.

Therefore, assuming constant specific heat and thermal conductivity for a particular time interval, Eq. 5.124 can be simplified to:

$$k\left(\frac{\partial^2 T}{\partial z^2}\right) + \mu I_{\max}(x, y, t)e^{-\mu z} = \rho c \frac{\partial T}{\partial t} \tag{5.181}$$

In many practical cases, the transverse dimensions of laser focusing spot are large compared to the thickness of the heated layer and heat conduction problem. Equation 5.181 can be considered one-dimensional and can be solved using standard methods Carslaw and Jaeger (1969) [3].

It is unnecessary to solve for the complete pulse since the complete solution may be obtained by summation of the solutions for the individual parts of time exponential, and then equation is linear. Rearrangement of Eq. 5.181 gives:

$$\frac{\partial^2 T(z,t)}{\partial z^2} + \frac{\mu I_{\max}}{k} \exp(-\mu z) = \frac{1}{\kappa}\frac{\partial T(z,t)}{\partial t} \tag{5.182}$$

where thermal diffusivity, $\kappa = k/(\rho c)$, with boundary conditions and ($z > 0$, $t > 0$):

$$\left.\frac{\partial T(z,t)}{\partial z}\right|_{z=0} = 0 \quad \text{for} \lim_{z\to\infty} T(z,t) = 0 t > 0$$

$$T(z,t)|_{z\to\infty} = 0_1$$

$$T(z,0) = 0 (z > 0)$$

Laplace transform can be used to solve certain types of partial differential equations with two or more independent variables. To solve Eq. 5.182, let us show how to solve in detail differential equation using Laplace transformation of one-dimensional heat equation and homogeneous aspect of Eq. 5.182, given the boundary condition (B.C.) and initial condition (I.C.) as before in Sect. 5.9. They are established below again as:

$$\frac{\partial^2 T}{\partial z^2} = \frac{1}{\kappa}\frac{\partial T}{\partial t} \tag{5.183}$$

$$\text{B.C.} \quad T(0,t) = F_0 \quad t > 0$$

$$\text{I.C.} \quad T(z,0) = 0 \quad z > 0$$

This was done for a CW laser interaction with target materials in Sect. 5.9.1.1.

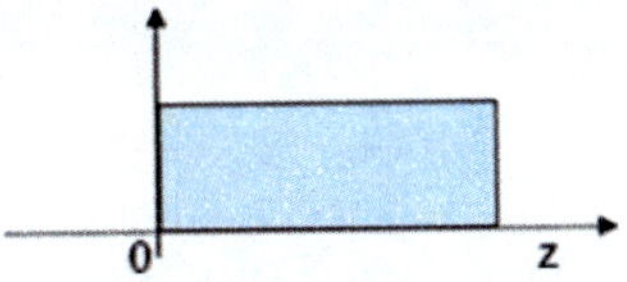

Fig. 5.39 Semi-infinite metal slab

The physical model of Eq. 5.183 is a semi-infinite slab of metal with a plane face on which the origin of the z-axis is located with the positive half of the axis directed into slab. This situation is illustrated in Fig. 5.39.

The approach will be to take Laplace transform of the dependent variable $T(z, t)$ in the heat equation with respect to the time t; as a result of which, an ordinary differential equation with z as its independent variable will be obtained for the transformed variable that will then depend on both the Laplace transform variables s and z. After the ordinary differential equation has been solved for the transformed variable, the inverse Laplace transform ($£^{-1}$) will be used to recover the time variation and so to arrive at the required solution as a function of z and t. Please note that if the Laplace transform is applied to the independent variable t in the function of two variables $T(z, t)$, the variable z will behave like a constant. Consequently, the rules for transforming derivatives of functions of a single independent variable also apply to a function of two independent variables.

Use the notation $\overline{T}(z,s) = £\{T(z,t)\}$ to denote the Laplace transform of $T(z, t)$ with respect to the time.

The formula for the transform of first derivative is:

$$£\left\{\frac{\partial T(z,t)}{\partial t}\right\} = s\overline{T}(z,s) - T(z,0) \tag{5.184}$$

To proceed further we must now use the condition that at time $t = 0$, the material of the slab is at 0 temperature, so $T(z, 0) = 0$; as a result of which:

$$£\left\{\frac{\partial T(z,t)}{\partial t}\right\} = s\overline{T}(z,s) \tag{5.185}$$

Next, since z is regarded as a constant, one has:

$$£\left\{\frac{\partial^2 T(z,t)}{\partial z^2}\right\} = \frac{\partial^2 \overline{T}(z,s)}{\partial z^2} \tag{5.186}$$

Using Eq. 5.186 when taking the Laplace transform of the heat conduction equation with respect to t, and using the linearity property of the transform, one obtains:

$$s\overline{T}(z,s) = \kappa\left[\frac{d^2\overline{T}(z,s)}{dz^2}\right] \text{for} \begin{matrix} z > 0 \\ t > 0 \end{matrix} \tag{5.187}$$

where now one can use an ordinary linear derivative with respect to z, so z can be considered to be the only independent variable. Therefore Eq. 5.187 can be rewritten as:

$$\frac{d^2\overline{T}(z,s)}{dz^2} - \frac{s}{\kappa}\overline{T}(z,s) = 0 \tag{5.188}$$

Equation 5.188 has the general solution of the following with two distinctive roots for its auxiliary or characteristic equation of:

$$r^2 - \frac{s}{\kappa} = 0$$

$$r = \pm\sqrt{\frac{s}{\kappa}}$$

and

$$\overline{T}(z,s) = c_1 e^{\left(\sqrt{s/\kappa}\right)z} + c_2 e^{-\left(\sqrt{s/\kappa}\right)z} \tag{5.189}$$

Since temperature has to be finite for $z > 0$ and $t > 0$, so for $s \to +\infty$, we have $c_1 = 0$, and then the Laplace transform of temperature is seen to be given by:

$$\overline{T}(z,s) = c_2 e^{-\sqrt{s/\kappa}} = c_2 \exp\left[-\left(\sqrt{s/\kappa}\right)z\right] \tag{5.190}$$

To determine c_2, we can use the boundary condition on the plane face of the slab that required $T(0,t) = F_0$ from which it follows that $£\{T(0,t)\} = F_0/s$. Thus, the Laplace transform of the solution with respect to the time t is

So, the coefficient $c_2 = F_0/s$. To recover the time variation from Eq. 5.190, it is necessary to find the inverse of Laplace function ($£^{-1}$), i.e.:

$$£^{-1}\left\{\overline{T}(z,s)\right\} = T(z,t) \tag{5.191}$$

Result of such analysis was shown in Sect. 5.9 in more detail, and we just use the results here that such inverse transformation yields:

$$T(z,t) = F_0 erfc\left(\frac{z}{2\sqrt{\kappa t}}\right) \tag{5.192}$$

where the error function, *erf*, is defined again as:

$$erf(z) = \frac{2}{\pi}\int_0^s e^{-u^2} du \tag{5.193}$$

and the complementary error function *erfc* is defined by Carslaw and Jaeger (1969) [3] as:

$$erfc(z) = 1 - erf(z) \tag{5.194}$$

where z is the independent variable and u is the dummy variable.

Continuing the forgoing mathematical calculations, the Laplace transformation of Eq. 5.182 and its related B.C. and I.C. with respect to t and substitution of boundary and initial conditions using z– direction for a one-dimensional problem yields:

$$\frac{\partial^2 \overline{T}(z,s)}{\partial t^2} - g^2\overline{T}(z,s) = -\frac{I_{\max}(z,t)\exp[-\mu z]}{ks} \tag{5.195}$$

where $g^2 = s/\kappa$ and $\overline{T}(z,s)$ the transform variable with complimentary and particular solutions, and we will take steps of solving them here. Equation 5.195 is a second-order linear but non-homogeneous differential equation and so we assume has a general solution of the following type:

$$\overline{T}(z,s) = y_c + \overline{T}_p(z,s) \tag{5.196}$$

where y_c is a general solution of the homogenous part of solution to Eq. 5.196, while $\overline{T}_p(z,s)$ is any specific function that satisfies the non-homogeneous part of Eq. 5.195.

Solution of y_c is similar to what we found in the case of Eq. 5.188, and as result of Eq. 5.189, therefore the general solution to y_c with the help of its characteristic or auxiliary equation is presented as:

$$\begin{aligned} &r^2 - g^2 = 0 \\ &r = \pm g \\ &y_c = c_1\exp(-gz) + c_2\exp(+gz) \end{aligned} \tag{5.197}$$

and to find the particular solution of non-homogeneous part of Eq. 5.195, we assume the following:

$$\overline{T}_p(z,s) = Ae^{-\mu z} \tag{5.198(a)}$$

then:

$$\frac{\partial \overline{T}_p(z,s)}{\partial z} = -A\mu e^{-\mu z} \tag{5.198(b)}$$

and:

$$\frac{\partial^2 \overline{T}_p(z,s)}{\partial z^2} = -A\mu^2 e^{-\mu z} \tag{5.198(c)}$$

By substituting Eq. 5.198(c) into Eq. 5.195, we find that:

$$\begin{aligned} -A\mu^2 e^{-\mu z} - g^2 A e^{-\mu z} &= -\frac{I_{\max}\mu}{ks} e^{-\mu z} \\ A &= \frac{I_{\max}\mu}{ks} \frac{1}{(g^2 - \mu^2)} \end{aligned} \tag{5.199}$$

Thus, the particular solution of $\overline{T}_p(z,s)$ is given as:

$$\overline{T}(z,s) = \frac{I_{\max\mu}}{ks} \exp(-\mu z) \tag{5.200}$$

Now the summation of the last term of Eqs. 5.197 and 5.200 is a general solution of Eq. 5.195 and is written as:

$$\begin{aligned} \overline{T}(z,s) &= y_c + \overline{T}_p(z,s) \\ &= c_1 \exp(-gz) + c_2 \exp(+gz) + \left(\frac{I_{\max\mu}}{ks}\right)\left(\frac{1}{(\mu^2 - g^2)}\right) \exp(-\mu z) \end{aligned} \tag{5.201}$$

where c_1 and c_2 are arbitrary constants. Substituting the boundary conditions related to Eq. 5.182 in Laplace transform plane indicates that:

$$c_1 = \frac{I_{\max}\mu^2}{ksg(\mu^2 - g^2)} \text{and} c_2 = 0$$

Therefore, the complex solution of Eq. 5.195 in the transform plane is:

$$\overline{T}(z,s) = -\frac{I_{\max}\mu}{ks}\left[\frac{\mu \exp[-gz]}{g(\mu^2 - g^2)} - \frac{\exp[-\mu z]}{(\mu^2 - g^2)}\right] \tag{5.202}$$

The problem now is to invert the solution of Eq. 5.202 which is a product of two s-functions. There are two ways to do this. The first is a convolution integral method, and the second one is to entail expansion of the functions into partial fractions. Using the second method, the full solution obtained by inverse Laplace transformation of Eq. 5.202 is:

$$
\begin{aligned}
T(z,t) = \frac{I_{\max}\mu}{ks}\Bigg\{ & -\frac{4}{\mu}\sqrt{\frac{\kappa t}{\pi}}\exp\left(-\frac{z^2}{4\kappa t}\right) \\
& -\left(\frac{1-\mu z}{\mu^2}-\frac{1+\mu z}{\mu^2}\right) erfc\left(\frac{z}{2\sqrt{\kappa t}}\right) \\
& -\frac{1}{\mu^2}\left[\exp\left(\kappa\mu^2 - t - \mu z\right) erfc\left(\frac{z}{2\sqrt{\kappa t}} - \mu\sqrt{\kappa t}\right)\right] \\
& -\frac{1}{\mu^2}\left[\exp\left(\kappa\mu^2 - t - \mu z\right) erfc\left(\frac{z}{2\sqrt{\kappa t}} + \mu\sqrt{\kappa t}\right)\right] \\
& -\frac{2}{\mu^2}\exp(-\mu z)\left[1-\exp\left(\kappa\mu^2 t\right)\right]\Bigg\}
\end{aligned}
\tag{5.203}
$$

where:

$erfc(z) = 1 - erf(z) = \frac{2}{\pi}\int_z^\infty e^{-u^2}du$, defined by Eqs. 5.193 and 5.194 and knowing that:

$$
ierfc(z) = \int_z^\infty erfc\xi d\xi = \frac{1}{\sqrt{\tau}}\exp\left(-z^2\right) erfc(z) \tag{5.204}
$$

The final solution of Eq. 5.203 is:

$$
\begin{aligned}
T(z,t) = & \frac{2I_{\max}}{k}\sqrt{\kappa t}\, ierfc\left(\frac{z}{2\sqrt{\kappa t}}\right) \\
& -\frac{I_{\max}}{k\mu}\exp(-\mu z) \\
& +\frac{I_{\max}}{2k\mu}\left[\exp\left(\kappa\mu^2 - t - \mu z\right) erfc\left(\mu\sqrt{\kappa t} - \frac{z}{2\sqrt{\kappa t}}\right)\right] \\
& +\frac{I_{\max}}{2k\mu}\left[\exp\left(\kappa\mu^2 - t + \mu z\right) erfc\left(\mu\sqrt{\kappa t} + \frac{z}{2\sqrt{\kappa t}}\right)\right]
\end{aligned}
\tag{5.205}
$$

Equation 5.205 gives the temperature profile inside the material for a given beam power intensity which is shown in Fig. 5.40.

It should be noted that as the time tends to infinity in Eq. 5.205, i.e.:

$$
\lim_{t\to\infty} T(z,t) = \infty \tag{5.206}
$$

no steady-state solution exists for the temperature distribution.

In the majority of calculations regarding the laser heating of solids surfaces, the temperature dependence of the surface reflectivity usually is neglected. If we set $A = 0$, growth of the temperature at the surface in the center of laser beam with time may be obtained by setting $z = 0$ into Eq. 5.205, i.e.:

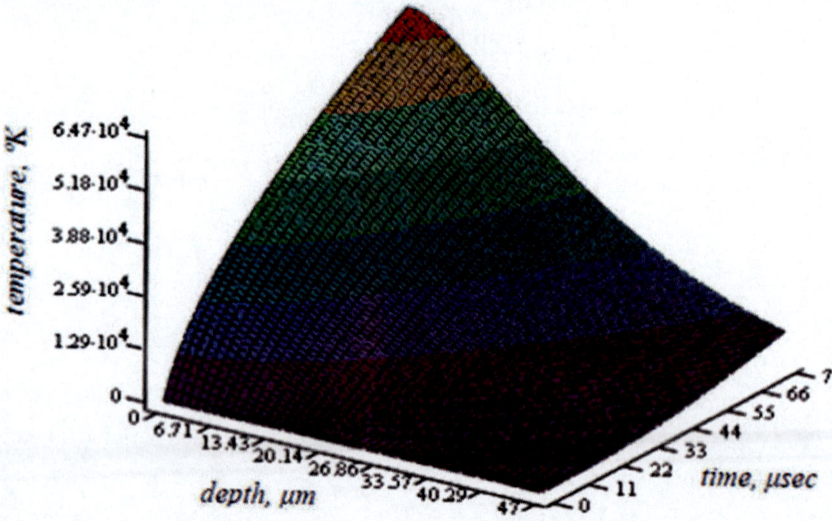

Fig. 5.40 Temperature distribution inside a material

$$T(0,t) = \frac{I_{\max}}{k\mu}\left[2\frac{\sqrt{\kappa\mu^2 t}}{\pi} + \exp\left(\kappa\mu^2 t\right)erfc\left(\mu\sqrt{\kappa t}\right) - 1\right] \tag{5.207}$$

A graphical representation of Eq. 5.216 is shown in Fig. 5.41.

Distribution in Fig. 5.41 represents the surface temperature of the workpiece during the laser interaction, and roughly one can estimate if there is going to be a drilling or melting on that surface. Writing in Mathcad parametric program as a function of thermal properties for different materials, thermal conductivity of lead $k = 19.66$, $c_p = 0.151$; titanium, $k = 20.5$, $c_p = 0.782$; and 304 stainless steel $k = 41.84$, $c_p = 0.418$. A typical result of such a parametric study is illustrated in Fig. 5.42.

Using very complicated function, programming in Mathcad is not always easy. Thus, to obtain a similar result, one can simply substitute values of thermo-physical parameters for solid first and then for liquid, and by displaying those on two different graphs, one can compare the results. This method was used on Eq. 5.213, because equation is very complex.

Further examinating Eq. 5.205 and assuming maximum absorption coefficient, let us say that μ goes to infinity, and one obtains the following result:

$$T(z,t) = \frac{2I_{\max}}{k}\sqrt{\kappa t}ierfc\left(\frac{z}{2\sqrt{\kappa t}}\right) \tag{5.208}$$

A graphical representation of Eq. 5.208 is shown in Fig. 5.43.

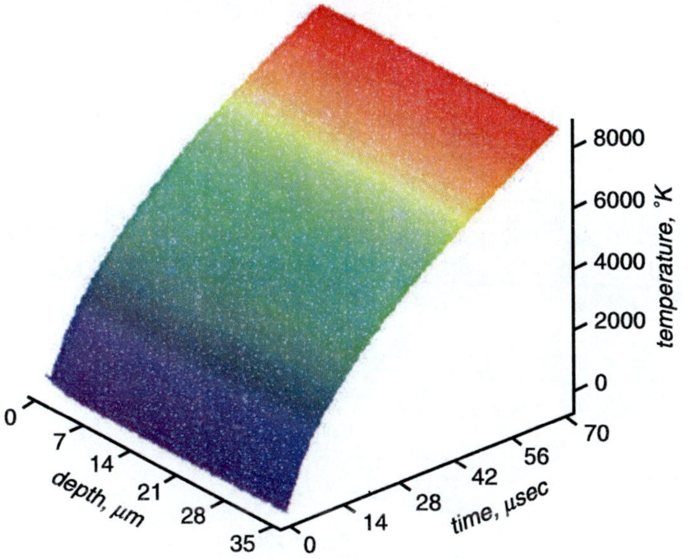

Fig. 5.41 Temperature distribution on the surface of the material

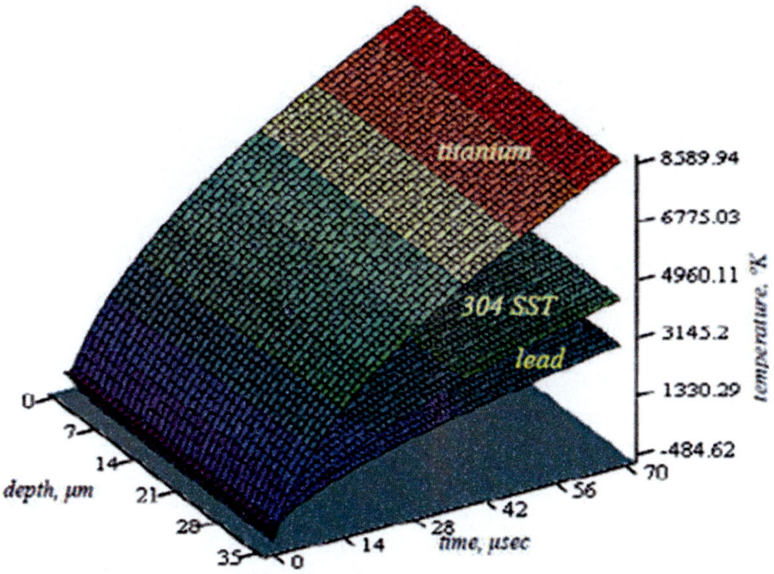

Fig. 5.42 Parametric study of thermal properties

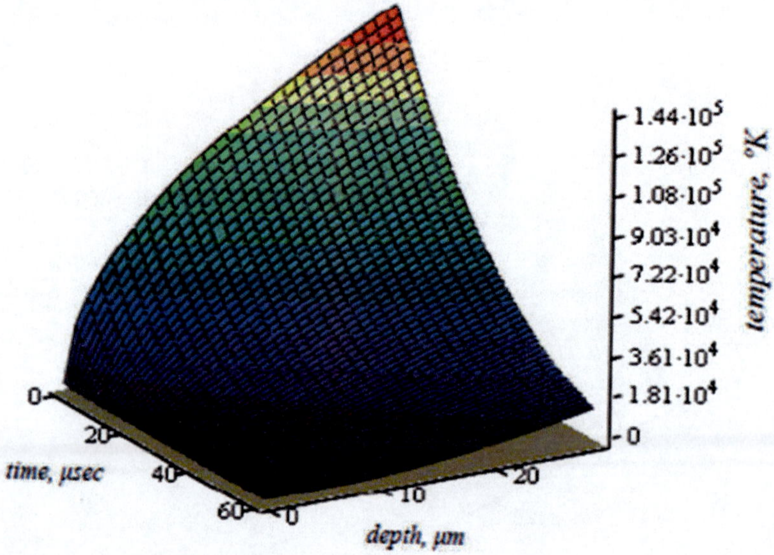

Fig. 5.43 Temperature distribution of the material with respect to time

Differentiation of Eq. 5.205 with respect to gives the temperature gradient inside the material, i.e.:

$$\begin{aligned}\frac{d}{dz}T(z,t) = & \frac{I_{\max}}{k}\exp\left(-\mu z\right) - \frac{I_{\max}}{k}erf\left(\frac{z}{2\sqrt{\kappa t}}\right) \\ & + \frac{I_{\max}}{2k}\exp\left(\kappa\mu^2 t - \mu z\right)erfc\left(\mu\sqrt{\kappa t} - \frac{z}{2\sqrt{\kappa t}}\right) \\ & + \frac{I_{\max}}{2k}\exp\left(\kappa\mu^2 t + \mu z\right)erfc\left(\mu\sqrt{\kappa t} - \frac{z}{2\sqrt{\kappa t}}\right)\end{aligned} \tag{5.209}$$

It is evident that $dT(z,t)/dz$ (Fig. 5.44) will only be zero at the surface, i.e., maximum temperature will occur at the surface.

The variation of $dT(z,t)/dz$ with distance z, i.e., (depth, expressed in μm) as a function of three different time instances, is shown in Fig. 5.45.

As illustrated in Fig. 5.45, the slope of the curves decreases reaching the minimum and then increases to attain almost 0 as the temperature profile becomes almost asymptotic with z. In this case, the behavior of $dT(z,t)/dz$ with z may be divided into three regions, which are indicated in Fig. 5.45.

In the first region, the heat gain due to laser irradiation dominates the conduction losses, i.e., the internal energy increase is considerably high as compared to conduction losses. In the second region, the slope has a $z =$ minimum value; in this case, the energy gain due to incident laser beam balances the conduction losses, i.e., the

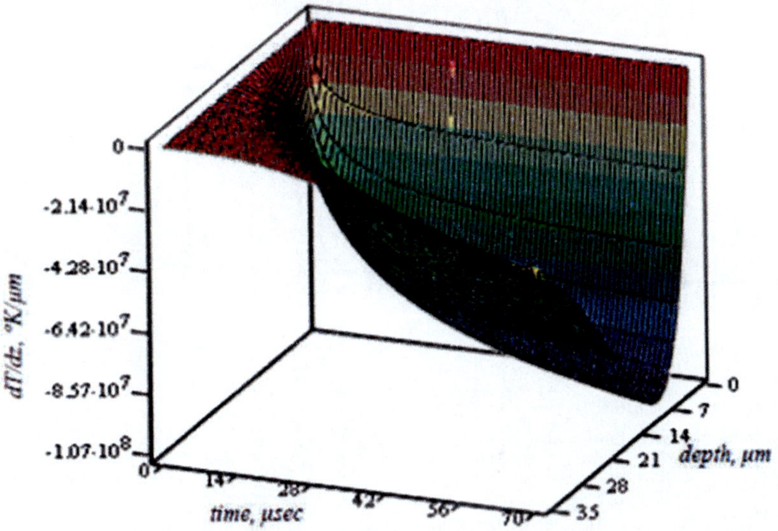

Fig. 5.44 Temperature gradient distribution inside the material with respect to time

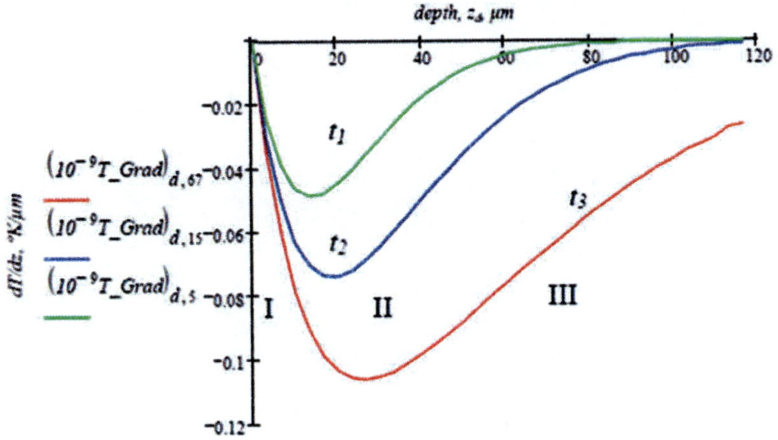

Fig. 5.45 Temperature gradient distribution inside the material for a different time

internal energy of the material remains almost constant. In this case, the distance z corresponding to this point may be defined as the equilibrium distance $(z)_{eq}$ (Fig. 5.46), and $dT(z, t)/dz$ becomes $(dT(z, t)/dz)_{\min}$.

In the third region, the slope increases to reach almost 0. In this region, conduction losses are dominant, and the energy gain to the external field is insignificant, i.e., the internal energy decreases as the distance increases.

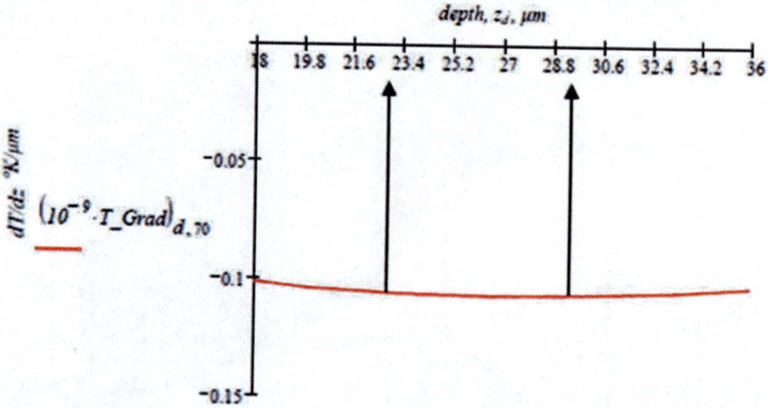

Fig. 5.46 Equilibrium distance

Variation of the equilibrium distance shown in Fig. 5.46 can be dimensionalized by:

$$(z\mu)_{eq} = C\left(\kappa\mu^2 t\right)_q^m \tag{5.210}$$

and is expected that increase in heating time increases the dimensionless equilibrium distance, which in turn increases the dimensionless equilibrium temperature, which can be defined as:

$$\left(\frac{T(z,t)}{I_{\max}/k\mu}\right)_{eq} = C(z\mu)_q^m \tag{5.211}$$

where C is the constant and m is the power. In other words, the equilibrium temperature is defined as the temperature where $dT(z,t)/dz$ is minimum.

Based on research done by Yilbas and Sami (1997) [53], the relationship between equilibrium temperature and equilibrium distance on the logarithmic scale is linear for all materials and pulse lengths.

If the pulse length is approximately 10^{-9} s, the analysis of the heat transfer process using the Fourier equation becomes invalid. In this case the heating process would be non-equilibrium.

In semi-infinite medium as shown in Fig. 5.39, at $T = 0$ with no heat flow at time $t = 0$ and with the heat source defined in Eq. 5.212 below,

$$I(z) = \varepsilon I_0(0)e^{-\mu z} \tag{5.212}$$

The solution to this same problem is:

$$\begin{aligned} T(z,t) = & \frac{2I_{\max}\sqrt{\kappa t}}{\mu k} ierfc\left(\frac{z}{2\sqrt{\kappa t}}\right) \\ & - \frac{I_{\max}}{k\mu^2} \exp\left(-\mu z\right) \\ & + \frac{I_{\max}}{2k\mu^2} \exp\left[\kappa\mu^2 t - \mu z\right] erfc\left(\mu\sqrt{\kappa t} - \frac{z}{2\sqrt{\kappa t}}\right) \\ & + \frac{I_{\max}}{2k\mu^2} \exp\left[\kappa\mu^2 t + \mu z\right] erfc\left(\mu\sqrt{\kappa t} + \frac{z}{2\sqrt{\kappa t}}\right) \end{aligned} \quad (5.213)$$

and is illustrated in Fig. 5.47. Solution in Eq. 5.213 is identical to Sparks' (1975) [54] solution for infinite thickness slab $\ell \gg \delta$.

Let us consider variation of the temperature distribution with respect to the difference from the material properties' variations with temperature. Analytically one can determine that based on extreme material properties for maximum and minimum values of thermal properties, known from the literature. Thermal conductivity varies between 14.9 and 16.2, for ambient temperature, T_a, and 32 for vaporization temperature, T_v, heat capacity cp varies from 500 to 824, and density ρ varies from 7870 to 8000 g/m^3. Discrepancies related to the thermo-physical properties of the material could be determined by substituting the lowest and the highest values to the solution given by Eq. 5.213. Examining graphs in Figs. 5.47 and 5.48, one can conclude that melting temperature dependencies of the material may invoke an estimated error about 50%, and we will talk more about melting process in Sect. 5.9.2 of this chapter.

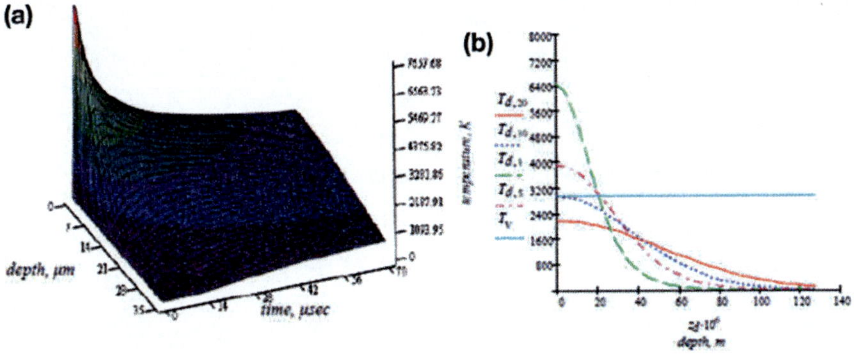

Fig. 5.47 Temperature distribution of laser irradiated material with the lowest values of thermo-physical parameters, (**a**) temperature as a function of depth and time, (**b**) temperature or melting of the material with depth calculated using Eq. 5.222

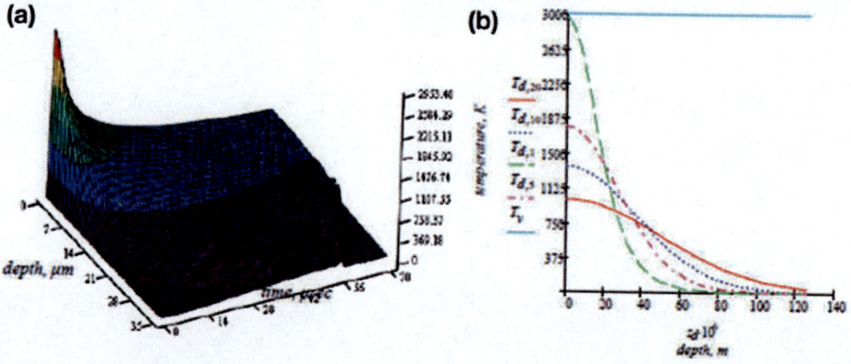

Fig. 5.48 Temperature distribution of laser irradiated material with the pulse can be thermo-physical parameters, (**a**) temperature as a function of depth and time, (**b**) temperature or melting of the material with depth

Based on Anisimov and Khokhlov (1995) [55], the rate of cooling after the end of the pulse can be estimated by Eq. 5.214 and graphically represented as in Fig. 5.49.

$$V_{imp}C(z,t) = \frac{2AI_{\max}}{k}\left[\sqrt{\frac{\kappa}{t\ell_t}}ierfc\left(\frac{z}{2\sqrt{\kappa t}}\right) + \frac{zerfc\left(\frac{z}{2\sqrt{\kappa t}}\right)}{2t\ell_t} + \sqrt{\frac{\kappa}{t\ell_t - t_p}}ierfc\left(\mu\sqrt{\kappa t} - \frac{z}{2\sqrt{\kappa t}}\right) - \frac{z}{2(t\ell_t - t_p)}erfc\left(\mu\sqrt{\kappa t} + \frac{z}{2\sqrt{\kappa t}}\right)\right] \quad (5.214)$$

5.9.1.4 Analytical Solution of Heat Transfer with Time-Dependent Gaussian Laser Pulse Heating

This section presents time-unsteady analysis of the conduction limited heating process introduced for a practical Nd:YAG laser pulse with a time-dependent intensity. An analytical solution to the problem is obtained with appropriate boundary conditions.

The output from a pulsed Nd:YAG laser is described by approximating the form of the true output by the subtraction of two exponential functions. This analytical form is given by:

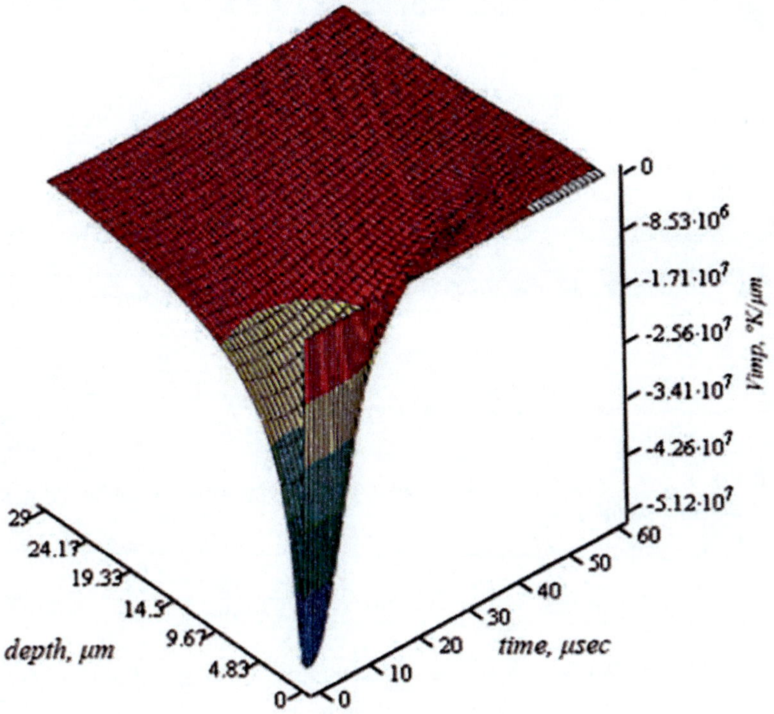

Fig. 5.49 Rate of cooling of the workpiece after the end of laser pulse

$$I_0 = I_{\max}\left[e^{(-\beta t)} - e^{(-\gamma t)}\right] \tag{5.215}$$

For metals and certain applications, at temperatures above the Debye temperature, it can be assumed that *k*(*T*) and *c*(*T*) do not change with temperature. Therefore, assuming constant specific heat and thermal conductivity for a particular time interval, Eq. 5.124 can be simplified to:

$$k\left(\frac{\partial^2 T}{\partial z^2}\right) + \exp\left[-(\beta t + \gamma t)\right] \cdot \mu I_{\max}(x, y, t)_{z=0} \exp(-\mu z) = \rho c \frac{\partial T}{\partial t} \tag{5.216}$$

It is unnecessary to solve for the complete pulse, since the complete solution may be obtained by summation of the solutions for the individual parts of time exponential, and then equation is linear. Rearrangement of Eq. 5.216 gives:

$$\frac{\partial^2 T}{\partial z^2} + \frac{\mu I_{\max}}{k} \exp(-\mu z) \exp\left[-(\beta t + \gamma t)\right] = \frac{1}{\kappa} \frac{\partial T}{\partial t} \tag{5.217}$$

with boundary conditions as:

$$\left.\frac{\partial T}{\partial t}\right|_{z=0} = 0 \tag{5.218}$$

$$T(\infty, t) = 0 \tag{5.219}$$

$$T(z, 0) = 0 \tag{5.220}$$

Laplace transformation of Eq. 5.217 with respect to t and substitution of boundary condition from Eq. 5.219 yield:

$$\frac{\partial^2 \overline{T}(z,s)}{\partial z^2} - g^2 \overline{T}(z,s) = \frac{-I_{\max}\mu e^{-\mu z}}{k(s+\beta)} \tag{5.221}$$

with $\overline{T}(z,s)$, $g^2 = s/\kappa$, and s as the transform variable, which has a complimentary and particular solution as:

$$\overline{T}(z,s) = c_1 \exp(-gz) + c_2 \exp(gz) + \frac{I_{\max}\mu e^{-\mu z}}{k(s+\beta)(\mu^2 - g^2)} \tag{5.222}$$

where c_1 and c_2 are arbitrary constants. Substitution of boundary Equation of 5.219 and 5.221 gives these constants as $c_2 = 0$ and:

$$c_1 = \frac{I_{\max}\mu^2}{kg(s+\beta)(\mu^2 - g^2)} \tag{5.223}$$

Therefore, the complex solution in the transform plane is:

$$\overline{T}(z,s) = -\frac{I_{\max}\mu}{k(s+\beta)}\left[\frac{\mu \exp(-gz)}{g(g^2-\mu^2)} - \frac{\exp(-\mu z)}{(g^2-\mu^2)}\right] \tag{5.224}$$

The problem now is to invert this solution, which is a product of two p-functions. There are two ways to do this. The first is a convolution integral method, and the second method is simpler, entailing expansion of the functions into partial fractions. Using the second method, the full solution, obtained by the inverse Laplace transformation of Eq. 5.225, is:

$$
\begin{aligned}
T(z,t) = {} & \frac{I_{\max}\mu}{2k}\frac{\kappa}{(\beta+\kappa\mu^2 0} \\
& \left\{ i\mu\sqrt{\frac{\alpha}{\beta}}\exp(-\beta t)\left[\exp\left(iz\sqrt{\frac{\beta}{\kappa}}\right)erfc\left(\frac{z}{2\sqrt{\kappa t}}+i\sqrt{\beta t}\right)\right]\right. \\
& -\left[\exp\left(-iz\frac{\sqrt{\beta}}{\kappa}\right)erfc\left(\frac{z}{2\sqrt{\kappa t}}-i\sqrt{\beta t}\right)\right] \\
& +\exp(\alpha\mu^2 t)\left[\exp(\mu z)erfc\left(\frac{z}{2\sqrt{\kappa t}}+\mu\sqrt{\kappa t}\right)\right] \\
& -\left[\exp(-\mu z)erfc\left(\frac{z}{2\sqrt{\kappa t}}-\mu\sqrt{\alpha t}\right)\right] \\
& \left.-2\exp(-\mu z)[\exp(\kappa\mu^2 t)-\exp(-\beta t)]\right\}
\end{aligned}
\tag{5.225}
$$

Using the relationship $erfc(-z) = 2 - erfc(z)$, give the solution in following form:

$$
\begin{aligned}
T(z,t) = {} & \frac{I_{\max}\mu}{2k}\frac{\kappa}{(\beta+\kappa\mu^2 0} \\
& \left\{ i\mu\sqrt{\frac{\alpha}{\beta}}\exp(-\beta t)\left[\exp\left(iz\sqrt{\frac{\beta}{\kappa}}\right)erfc\left(\frac{z}{2\sqrt{\kappa t}}+i\sqrt{\beta t}\right)\right.\right. \\
& \left.-\exp\left(-iz\frac{\sqrt{\beta}}{\kappa}\right)erfc\left(\frac{z}{2\sqrt{\kappa t}}-i\sqrt{\beta t}\right)\right] \\
& +\exp(\alpha\mu^2 t)\left[\left(\exp(\mu z)erfc\left(\frac{z}{2\sqrt{\kappa t}}+\mu\sqrt{\kappa t}\right)\right)\right. \\
& \qquad -\exp(-\mu z)erfc\left(\frac{z}{2\sqrt{\kappa t}}-\mu\sqrt{\alpha t}\right) \\
& \left.\left.-2\exp[-(\beta t+\mu z)]\right]\right\}
\end{aligned}
\tag{5.226}
$$

Note that:

$$
i^{2m}erfc(-z) = -i^{2m}erfc(z) + \sum_{q=0}^{m}\frac{z^{2q}}{2^{2(m-q)-1}(2q)!(m-q)!}
$$

$$
i^{2m+1}erfc(-z) = i^{2m+1}erfc(z) + \sum_{q=0}^{m}\frac{z^{2q+1}}{2^{2(m-q)-1}(2q+1)!(m-q)!}
$$

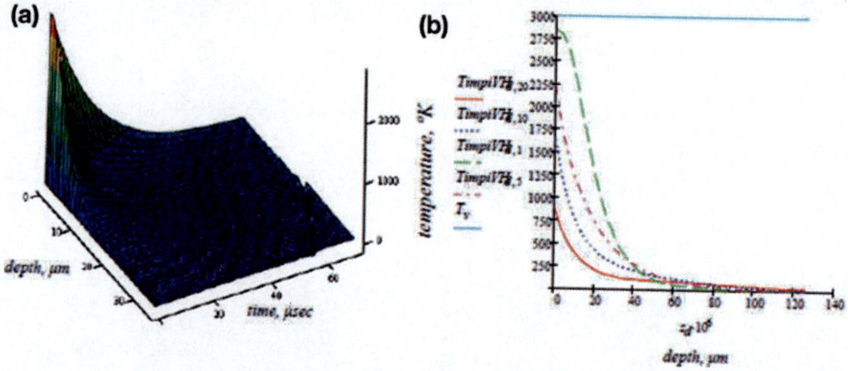

Fig. 5.50 (a) Temperature distribution inside the material (z-axis) with respect to time: (b) temperature vs. depth of the workpiece, at few different times, forms the beginning of the pulse, with the highest values of thermo-physical parameters calculated using Eq. 5.226

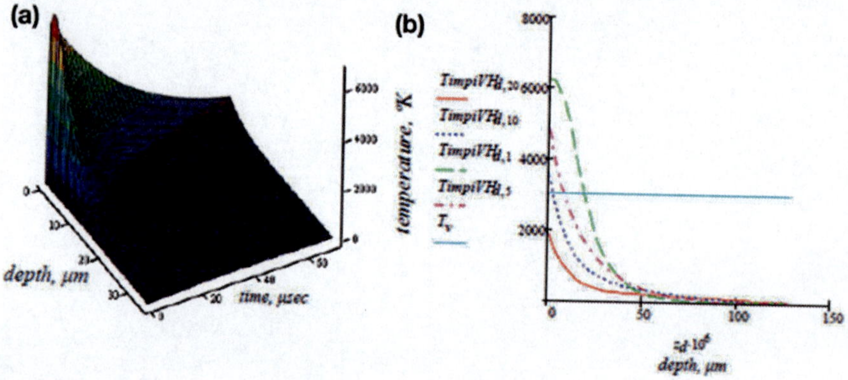

Fig. 5.51 (a) Temperature distribution inside the material (z-axis) with respect to time: (b) temperature vs. depth of the workpiece, at few different times, form the beginning of the pulse, with the highest values of thermo-physical parameters calculated using Eq. 5.226

The temperature distribution calculated in Eq. 5.226 is illustrated in Figs. 5.50 and 5.51.

The time at which the maximum value of the pulse occurs can be obtained by differentiating Eq. 5.226 with respect to t for the pulse. Mathematically speaking one has to find t in the condition that the first derivative of this function should be equal to 0, which gives the condition for the maximum time t_{max}.

$$t_{max} = \frac{\ln\left(\frac{\gamma}{\beta}\right)}{\gamma - \beta} \tag{5.227}$$

or

$$\sqrt{\beta t_{max}} = \sqrt{\frac{\ln\left(\frac{\gamma}{\beta}\right)}{\frac{\gamma}{\beta} - 1}} \tag{5.228}$$

Insertion of Eq. 5.228 into Eq. 5.226 gives the maximum temperature when the maximum pulse occurs. Solving Eqs. 5.226 and 5.228 gives the condition that the maximum temperature may coincide with the maximum pulse amplitude, i.e., at this point:

$$\sqrt{\beta t_{max}} = 1.282 \quad \text{and} \quad \sqrt{\gamma/\beta} = 1.596$$

The time at which the maximum temperature occurs is a function of pulse parameters, provided that the pulse rise times are much greater than the equilibrium time $C/\kappa\mu^2$, described in Sect. 5.9.1.1. The surface temperature depends on absorption depth in the initial stages of the pulse, and ability to absorb the laser depends on the shape of the pulse (Yilbas et al. 1996 [41]),

The knowledge of pulse parameters β and γ allows at least theoretically to adjust the pulse for maximum desired effect based on deductions from the physics of lasing system.

5.9.1.5 Analytical Solution of Heat Transfer with Time-Dependent Gaussian Laser Pulse Heating with Convective Boundary Conditions

This section presents time and unsteady analysis of the conduction with limited heating process introduced for a practical Nd:YAG laser pulse with a time-dependent intensity. An analytical solution to the problem is obtained with appropriate boundary conditions.

The output from a pulsed Nd:YAG laser is described by approximating the form of the true output by the subtraction of two exponential functions. This analytical form is given by:

$$I = AI_{max}[\exp(-\beta t) - \exp(-\gamma t)] \tag{5.229}$$

For metals and certain applications, at temperatures above the Debye temperature, it can be assumed that $k(T)$ and $c(T)$ do not change with temperature. Therefore, assuming constant specific heat and thermal conductivity for a particular time interval, Eq. 5.134 can be simplified to:

$$\rho c \frac{\partial T(z,t)}{\partial t} = k\left(\frac{\partial^2 T(z,t)}{\partial z^2}\right) + \exp(-\beta t) - \exp(-\gamma t) \times \left[A\mu I_{max}(x,y,t)_{z=0} \exp(-\mu z)\right] \tag{5.230}$$

It is unnecessary to solve for the complete pulse, since the complete solution may be obtained by summation of the solutions for the individual parts of time exponential, and then equation is linear. Rearrangement of Eq. 5.230 gives:

$$\frac{\partial^2 T}{\partial z^2} + A\frac{\mu I_{\max}}{k} \exp(-\beta t) - \exp(-\gamma t)\exp(\mu z) = \frac{1}{\kappa}\frac{\partial T}{\partial t} \tag{5.231}$$

with boundary conditions:

$$\left.\frac{\partial T}{\partial t}\right|_{z=0} = \frac{h}{k}[T(0,t) - T_0] \tag{5.232}$$

$$T(\infty, t) = 0 \tag{5.233}$$

$$T(z, 0) = 0 \tag{5.234}$$

A complete solution can be obtained by subtraction of solutions for the individual parts of the time exponential pulse. It should be noted that for the solution of a complete pulse, the ambient temperature is considered as 0 (T0 = 0). Hence, the heat transfer for the half-pulse becomes:

$$\frac{\partial^2 T}{\partial z^2} + A\frac{\mu I_{\max}}{k} \exp\left[-(\beta t + \mu z]\right. = \frac{1}{\kappa}\frac{\partial T}{\partial t} \tag{5.235}$$

The solution of Eq. 5.235 can be obtained through the Laplace transformation method, with respect to t as:

$$\frac{\partial^2 \overline{T}(z,s)}{\partial z^2} + \frac{AI_{\max}\mu \exp(-\mu z)}{k(s+\beta)} = \frac{1}{\kappa}\left[s\overline{T}(z,s) - T(z,0)\right] \tag{5.236}$$

with $\overline{T}(z,s)$ and $T(z,0) = 0$, where s is the transform variable, which has a complimentary and particular solution as before.

The full solution, obtained by the inverse Laplace transformation of Eq. 5.227, is:

$$T(z,t) = a_{10}\left\{\begin{array}{l} i\mu\sqrt{\frac{\kappa}{\beta}}\exp(-\beta t)\left[\begin{array}{l} l_1 \exp\left(iz\sqrt{\frac{\beta}{\kappa}}\right) erfc\left(\frac{z}{2\sqrt{\kappa t}} + i\sqrt{\beta t}\right) \\ l_2 \exp\left(-iz\sqrt{\frac{\beta}{\kappa}} erfc\left(\frac{z}{2\sqrt{\kappa t}} - i\sqrt{\beta t}\right)\right) \end{array}\right] \\ +\mu\sqrt{\kappa}\exp\left(\alpha\mu^2 t\right)\left[\begin{array}{l} l_3 \exp(-\mu z) erfc\left(\frac{z}{2\sqrt{\kappa t}} - \mu\sqrt{\beta t}\right) \\ -l_4 \exp(\mu z) erfc\left(\frac{z}{2\sqrt{\kappa t}} + \mu\sqrt{\beta t}\right) \end{array}\right] \\ -w_1 l_5 \exp\left(w_1 \frac{z}{\sqrt{\kappa}}\right)\exp\left(w_1^2\right) erf\left(\frac{z}{2\sqrt{\kappa t}} + w_1\sqrt{t}\right) \end{array}\right\}$$
$$+\left\{\begin{array}{l} \frac{a_{20}}{\beta + \kappa\mu^2}\exp\left(\alpha\mu^2 t\right) - \exp(-\beta t)\exp(-\mu z) \\ -\frac{a_{30}}{w_1}\left[\begin{array}{l} -\exp\left(w_1\frac{z}{\sqrt{\kappa}}\right)\exp\left(w_1^2 t\right) erf\left(\frac{z}{2\sqrt{\kappa t}} + w_1\sqrt{t}\right) \\ erfc\left(\frac{z}{2\sqrt{\kappa t}}\right) \end{array}\right] \end{array}\right\} \tag{5.237}$$

where:

$$a_{10} = \frac{-\mu I_{\max}\kappa(h + k\mu)}{k^2} \tag{5.238}$$

$$a_{20} = \frac{I_{\max}\kappa\mu}{k} \tag{5.239}$$

$$a_{30} = \frac{\sqrt{\kappa}h}{k}T_0 \tag{5.240}$$

The temperature distribution calculated in Eq. 5.228 is illustrated in Figs. 5.52, 5.53, 5.54, and 5.55.

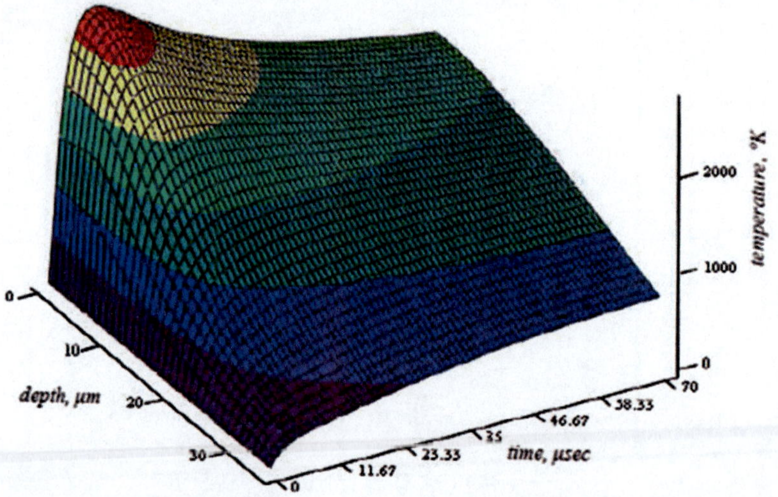

Fig. 5.52 (**a**) Temperature distribution inside the material (z-axis) with respect to time; (**b**) temperature vs. depth of the workpiece, at few different times, form the beginning of the pulse, with the highest values of thermo-physical parameters calculated using Eq. 5.228

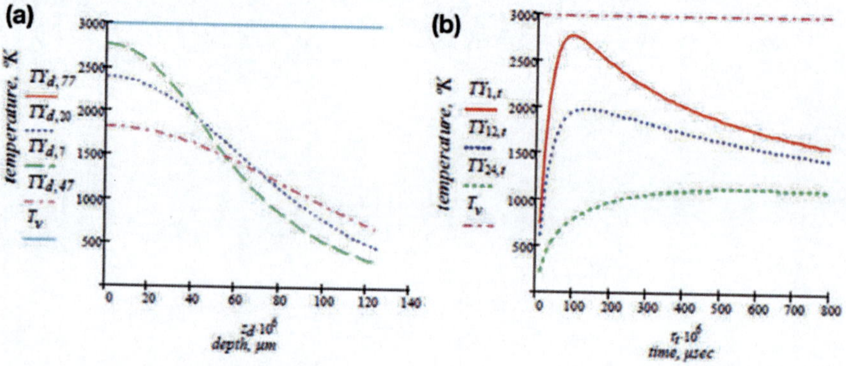

Fig. 5.53 (**a**) Temperature distribution inside the material (z-axis) for few different times; (**b**) transient distribution of the workpiece, at few different depths, with the highest values of thermo-physical parameters calculated using Eq. 5.228

5.9.1.6 Analytical Solution of Governing Heat Transfer Equation, Considering Transfer Evaporative Case

This section presents analysis of the heating process. In the heating analysis, evaporation is considered as the exclusive phenomenon taking place during the ablation process. Based on forgoing discussion, melting without vaporization can be produced only in very narrow range of laser parameters. Usually both phase

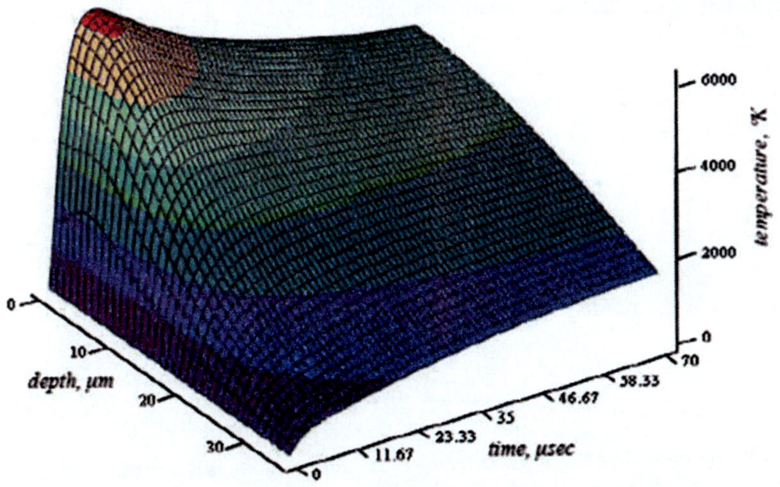

Fig. 5.54 Temperature distribution inside the material (z-axis) with respect to time, with the lowest values of thermo-physical parameters calculated using Eq. 5.228

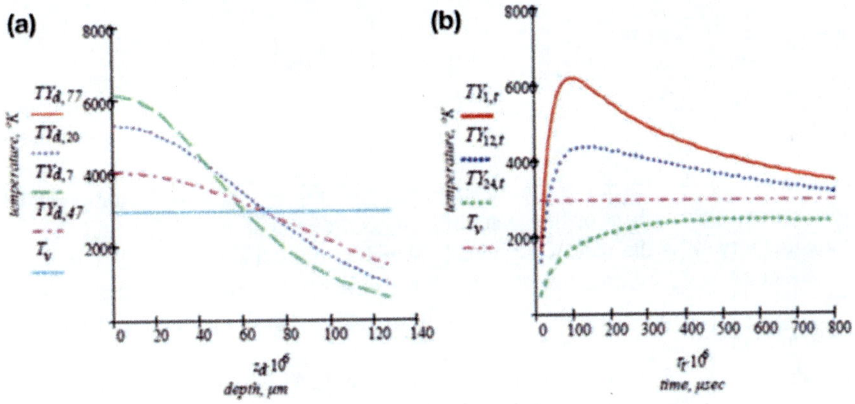

Fig. 5.55 (**a**) Temperature distribution inside the material (z-axis) for few different times; (**b**) transient distribution of the workpiece, at few different depths, with the lowest values of thermo-physical parameters calculated using Eq. 5.228

transitions (melting and vaporization) occur almost simultaneously in standard laser applications. Note that the latent heat of vaporization is larger than that of fusion by a factor of 20–50 times.

Thus, evaporation plays the most important part in the energy balance. To simplify our discussion, we will assume that the vaporization occurs in vacuum and the vapor does not absorb the laser energy. When the vapor does not absorb the laser light, the process in the condensed and gaseous phases can be considered separately. An analytical solution to the problem is obtained with appropriate boundary conditions. In this case, plasma formation, liquid expulsion, and nucleate boiling are omitted.

Solid or liquid evaporates at any temperature greater than 0 °K. The evaporation rates strongly depend on the temperature, T. This dependence can be written in the form:

$$V_n = V_0 \exp\left(-\frac{U}{T}\right) \tag{5.241}$$

where V_n is the normal component of the evaporation front velocity, and:

$$U = \frac{MH_v}{N_a k_B} \tag{5.242}$$

and H_v is the latent heat of vaporization per unit mass, k_B is the Boltzmann constant, M is the atomic mass, V_0 is a constant whose value is in the order of the speed of sound in the condensed phase, and finally N_a is the Avogadro's number. The Fourier differential equation allowing a phase change process can be written as:

$$\rho(T)\frac{\partial}{\partial t}\int_{T_0}^{T} c(T)dT = \frac{\partial}{\partial x}\left[k(T)\frac{\partial T}{\partial x}\right] + \frac{\partial}{\partial y}\left[k(T)\frac{\partial T}{\partial y}\right] + \frac{\partial}{\partial z}\left[k(T)\frac{\partial T}{\partial z}\right] + [\rho(T)c(T)]V\frac{\partial T}{\partial z} + Q \tag{5.243}$$

where $\rho(T)$ is the density of the material of the workpiece, $c(T)$ is the temperature-dependent specific heat of the material, $k(T)$ is the temperature-dependent thermal conductivity, V is the recession velocity which is the velocity of the liquid–vapor interface, $T = T(x, y, z)$ is the resulting three-dimensional time-dependent distribution in the material, t is time, T_0 is the initial temperature, and x, y, and z are the spatial coordinates.

$Q = Q(x, y, z, t)$ is the rate at which heat is supplied to the solid per unit time per unit volume, depends on the laser pulse parameters and optical properties of target material irradiated, and is defined by Eq. 5.179.

Sparks (1975) [54] defined the absorption coefficient of the material as a linear function of the surface temperature in Eq. 5.151. For practical reasons, the absorption factor A we assume as unity. This temperature dependence of A results from the fact that A is proportional to the electron–phonon collision frequency which, in turn, is proportional to the crystal lattice temperature.

The recession velocity of the surface can be formulated from energy balance at the free surface of irradiated workpiece. In this case the energy flux at the free surface can be written as:

$$I_0 = \rho(T)V[c(T)T_s - H_v] \tag{5.244}$$

where:

$$I_0 = I_{\max}(1 - R) \tag{5.245}$$

and T_s is the surface temperature, H_v is the latent heat of vaporization, and R is the surface reflectivity. Rearranging Eq. 5.244 yields:

$$V = \frac{I_0}{\rho(T)[c(T)T_s - H_v]} \tag{5.246}$$

For metals and certain applications, at temperatures above the Debye temperature, it can be assumed that $k(T)$, $\rho(T)$, and $c(T)$ do not change dramatically with temperature.

Therefore, assuming constant specific heat and thermal conductivity for a particular time interval, Eq. 5.243 can be simplified to:

$$\begin{aligned} \rho c \frac{\partial T}{\partial t} = k\left(\frac{\partial^2 T}{\partial z^2}\right) + \rho c V \frac{\partial T}{\partial z} \\ + I_{\max}(x, y, t)_{z=0}(1 - R)\mu e^{-(\mu z + \beta t)} \end{aligned} \tag{5.247}$$

It should be noted that peak power intensity $I_{\max}$ does not vary with time. Since the surface temperature is time-dependent, the recession velocity varies with time. This result is a nonlinear form of Eq. 5.246, which cannot be solved analytically by the Laplace transformation method.

For a known surface temperature, there is a unique value for recession velocity, so an iterative method can be applied here to solve Eq. 5.247 analytically. In our case keeping the recession velocity constant in Eq. 5.246 enables us to determine the surface temperature analytically, and after obtaining the surface temperature, the recession velocity can be calculated using Eq. 5.246. This procedure can be repeated so long as the surface temperature and recession velocity converge to correct results. Rearrangement of Eq. 5.247 gives:

$$\frac{\partial^2 T}{\partial z^2} + \frac{V}{\kappa}\frac{\partial T}{\partial z} + \frac{\mu I_0}{k} e^{-\beta t} e^{-\mu z} = \frac{1}{\kappa}\frac{\partial T}{\partial t} \tag{5.248}$$

where thermal diffusivity $\kappa = \frac{k}{\rho c}$, with boundary conditions:

$$\begin{cases} \left.\frac{\partial T}{\partial t}\right|_{z=0} = \frac{\rho V L}{k} \\ T(z, t)|_{z \to \infty} = 0 \\ T(z, t)|_{t=0} = 0 \end{cases} \tag{5.249}$$

We can apply the Laplace transform, defined as:

$$\overline{T}(z,s) = \int_0^\infty T(z,t)e^{-st}dt \tag{5.250}$$

Laplace transformation of Eq. 5.248 with respect to t and substitution of boundary conditions from Eq. 5.249 give:

$$\frac{\partial^2 \overline{T}(z,s)}{\partial z^2} + \frac{V}{\kappa}\frac{\partial \overline{T}(z,s)}{\partial z} - \frac{s}{\kappa}\overline{T}(z,s) = -\frac{I_0\mu}{k(s+\beta)}e^{(-\mu z)} \tag{5.251}$$

with $\overline{T}(z,s)$ and s the transform variable in Laplace plane, which has a complimentary (homogeneous) and particular (non-homogeneous) solution:

$$\overline{T}(z,s) = \overline{T}(z,s)_c + \overline{T}(z,s)_p \tag{5.252}$$

The homogeneous part of Eq. 5.252, $\overline{T}(z,s)_c$, is:

$$\frac{\partial^2 \overline{T}(z,s)}{\partial z^2} + \frac{V}{\kappa}\frac{\partial \overline{T}(z,s)}{\partial z} - \frac{s}{\kappa}\overline{T}(z,s) = 0 \tag{5.253}$$

The characteristic or auxiliary equation for the homogeneous solution can be written as:

$$r^2 + \frac{V}{\kappa}r - \frac{s}{\kappa} = 0 \tag{5.254}$$

which Eq. 5.254 has two distinctive roots:

$$r_{1,2} = -\frac{V}{2\kappa} \pm \frac{\sqrt{V^2 + 4s\kappa}}{2\kappa} \tag{5.255}$$

The homogeneous solution from Eq. 5.262 then results in:

$$\overline{T}(z,s)_c = c_1 e^{r_1 z} + c_2 e^{r_2 z} \tag{5.256}$$

or:

$$\overline{T}(z,s)_c = e^{-\frac{Vz}{2\kappa}}\left[c_1 e^{-\left(\frac{\sqrt{V^2+4s\kappa}}{2\kappa}\right)z} + c_2 e^{+\left(\frac{\sqrt{V^2+4s\kappa}}{2\kappa}\right)z}\right] \tag{5.257}$$

For the particular solution, one can provide exponential solution as:

$$\overline{T}(z,s)_p = P_0 e^{-\delta z} \tag{5.258}$$

where P_0 is some arbitrary constant coefficient. Substitution of $\overline{T}(z,s)_p$ given by Eq. 5.257 into Eq. 5.251 results in:

$$P_0\delta^2 e^{-\delta z} - \frac{V}{\kappa}P_0\delta e^{-\delta z} - \frac{s}{\kappa}P_0\delta e^{-\delta z} = H_0 e^{-\delta z} \tag{5.259}$$

where:

$$H_0 = I_0\frac{1}{(s+\beta)}\frac{\delta}{k} \tag{5.260}$$

and:

$$P_0 = -\frac{H_0\kappa}{s+V\delta-\kappa\delta^2} \tag{5.261}$$

and substituting Eq. 5.261 into Eq. 5.258, and the end result of that along with Eq. 5.257 into Eq. 5.252, we obtain:

$$\overline{T}(z,s) = e^{-\frac{Vz}{2\kappa}}\left[c_1 e^{-\left(\frac{\sqrt{V^2+4s\kappa}}{2\kappa}\right)z} + c_2 e^{+\left(\frac{\sqrt{V^2+4s\kappa}}{2\kappa}\right)z}\right] - \frac{H_0\kappa}{s+V\delta-\kappa\delta^2}e^{-\delta z} \tag{5.262}$$

Now defining:

$$V_1 = V\delta - \kappa\delta^2 \tag{5.263}$$

and substituting Eq. 5.263 and using boundary Eq. 5.249, then $c_2 = 0$, and all into Eq. 5.262, we get:

$$\overline{T}(z,s) = c_1 e^{\frac{1}{2\kappa}\left[V+\sqrt{V^2+4s\kappa}\right]z} + \frac{I_0\delta\kappa}{k}\frac{e^{-\delta z}}{(s+\beta)(S+V_1)} \tag{5.264}$$

We can calculate c_1 by using boundary condition of Eq. 5.249 in Laplace plane and defining:

$$\varpi_1 = \frac{I_0\delta\kappa}{k} \tag{5.265}$$

$$\begin{aligned}\frac{\overline{T}(z,s)}{\partial z} &= \left\{-\frac{1}{2\kappa}\left[V+\sqrt{V^2+4s\kappa}\right]c_1 e^{\frac{1}{2\kappa}\left[V+\sqrt{V^2+4s\kappa}\right]z} - \frac{\delta\varpi_1\kappa e^{-\delta z}}{(s+\beta)(S+V_1)}\right\}_{z=0} \\ &= \frac{\rho VL}{ks}\end{aligned} \tag{5.266}$$

From Eq. 5.274, c_1 is going to be:

$$c_1 = -\frac{2\delta\varpi_1\kappa}{(s+\beta)(s+V_1)\left(V+\sqrt{V^2+4s\kappa}\right)} - \frac{2\kappa\frac{\rho VL}{k}}{s\left(V+\sqrt{V^2+4s\kappa}\right)} \tag{5.267}$$

Hence:

$$\overline{T}(z,s) = \frac{2\delta\varpi_1\kappa e^{-\frac{1}{2\kappa}\left[V+\sqrt{V^2+4s\kappa}\right]}}{(s+\beta)(s+V_1)\left(V+\sqrt{V^2+4s\kappa}\right)} + \frac{\varpi_1 e^{-\delta z}}{(s+\beta)(s+V_1)} + \overline{T}_{mm}(z,s) \tag{5.268}$$

where:

$$\overline{T}_{mm}(z,s) = -\frac{2\kappa\rho VLe^{-\frac{1}{2\kappa}\left[V+\sqrt{V^2+4s\kappa}\right]}}{ks\left(V+\sqrt{V^2+4s\kappa}\right)} \tag{5.269}$$

For the purpose of easier handling calculation in the future, we define $H_1(s)$, $H_2(s)$, and $H_3(s)$ to be:

$$H_1(s) = \frac{e^{-\frac{1}{2\kappa}\left[V+\sqrt{V^2+4s\kappa}\right]}}{(s+\beta)(s+V_1)\left(V+\sqrt{V^2+4s\kappa}\right)} \tag{5.270}$$

$$H_2(s) = \frac{1}{(s+\beta)(s+V_1)} = \frac{1}{V_1-\beta}\left[\frac{1}{s+\beta} - \frac{1}{s+V_1}\right] \tag{5.271}$$

$$H_3(s) = \frac{e^{-\frac{1}{2\kappa}\left[V+\sqrt{V^2+4s\kappa}\right]}}{s\left(V+\sqrt{V^2+4s\kappa}\right)} \tag{5.272}$$

And in terms of these defined variables, Eq. 5.268 becomes:

$$\overline{T}(z,s) = -2\kappa\mu\varpi_1 H_1(s) + \varpi_1 e^{-\mu z} H_2(s) - 2\kappa\frac{\rho VL}{k} H_3(s) \tag{5.273}$$

Consequently, the solution to Eq. 5.248 is going to be:

$$T(z,t) = \pounds^{-1}\{\overline{T}(z,s)\} \tag{5.274}$$

or:

$$T(z,t) = -2\kappa\mu\varpi_1\pounds^{-1}\{H_1(s)\} + \varpi_1 e^{-\mu z}\pounds^{-1}\{H_2(s)\} - 2\kappa\frac{\rho VL}{k}\pounds^{-1}\{H_3(s)\} \tag{5.275}$$

To obtain the inverse transformation of functions $\pounds^{-1}\{H_1(s)\}$ and $\pounds^{-1}\{H_3(s)\}$, let us introduce:

$$p = V^2 + 4\kappa s \text{or} dp = 4\kappa ds \text{and} s = \frac{1}{4\kappa}(p - V^2)$$

Therefore:

$$\pounds^{-1}\{H_1(s)\} = \frac{1}{2\pi i}\int_{c-i\infty}^{c+i\infty} e^{ts}H_1(s)ds \tag{5.276}$$

or:

$$\pounds^{-1}\{H_1(s)\} = \frac{1}{4\kappa}e^{-\left(\frac{Vz}{2\kappa}+\frac{V^2}{4\kappa}\right)}\frac{1}{2\pi i}\int_{\overline{c}-i\infty}^{\overline{c}+i\infty}\frac{e^{-\left(\frac{Vz}{2\kappa}+\frac{V^2}{4\kappa}\right)}}{\left(\frac{p}{4\kappa}-\frac{V^2}{4\kappa}+\beta\right)\left(\frac{p}{4\kappa}-\frac{V^2}{4\kappa}+V_1\right)\left(V+\sqrt{p}\right)}dp \tag{5.277}$$

where:

$$\overline{c} = 4\kappa c + V_2 \tag{5.278}$$

We can use one more transformation by letting:

$$\vartheta = \frac{1}{4\kappa}p \text{or} dp = 4\kappa d\vartheta \tag{5.279}$$

After tedious and lengthy algebra, we obtain:

$$\pounds^{-1}\{H_1(s)\} = \frac{1}{4\kappa}e^{-\left(\frac{Vz}{2\kappa}+\frac{V^2}{4\kappa}\right)}\frac{1}{2\pi i}\int_{\widehat{c}-i\infty}^{\widehat{c}+i\infty}\frac{4\kappa e^{-\left(\frac{z\sqrt{s}}{\sqrt{\kappa}}\right)}e^{\vartheta t}}{\left(\vartheta-\frac{V^2}{4\kappa}+\beta\right)\left(\vartheta-\frac{V^2}{4\kappa}+V_1\right)\left(V+2\sqrt{\kappa}\sqrt{s}\right)}d\vartheta \tag{5.280}$$

where:

$$\widehat{c} = \frac{\overline{c}}{4\kappa} \tag{5.281}$$

Hence:

$$£^{-1}\{H_1(s)\} = 4\kappa e^{-\frac{V}{2\kappa}\left(z+t\frac{V}{2}\right)} £^{-1}\{H_4(s)\} \tag{5.282}$$

where:

$$H_4(s) = \frac{4\kappa e^{-\left(\frac{z\sqrt{s}}{\sqrt{\kappa}}\right)} e^{\vartheta t}}{\left(\vartheta - \frac{V^2}{4\kappa} + \beta\right)\left(\vartheta - \frac{V^2}{4\kappa} + V_1\right)\left(V + 2\sqrt{\kappa}\sqrt{s}\right)} \tag{5.283}$$

Similarly, $£^{-1}\{H_3(s)\}$ can be obtained, i.e., as:

$$£^{-1}\{H_3(s)\} = 4\kappa e^{-\frac{V}{2\kappa}\left(z+t\frac{V}{2}\right)} £^{-1}\{H_7(s)\} \tag{5.284}$$

where:

$$H_7(s) = \frac{e^{-\left(\frac{z\sqrt{s}}{\sqrt{\kappa}}\right)}}{\left(\sqrt{\kappa}\sqrt{s} - V^2\right)\left(2\sqrt{\kappa s} + V\right)^2} \tag{5.285}$$

or:

$$H_7(s) = \frac{e^{-\left(\frac{z\sqrt{s}}{\sqrt{\kappa}}\right)}}{8\kappa\sqrt{\kappa}\left(\sqrt{s} - \xi\right)\left(\sqrt{s} + \xi\right)^2} \tag{5.286}$$

where:

$$\xi = \frac{V}{2\sqrt{\kappa}} \tag{5.287}$$

Introducing partial fraction and rearrangement yields:

$$H_7(s) = \frac{1}{8V^2\sqrt{\kappa}}\left[\frac{e^{-\left(\frac{z\sqrt{s}}{\sqrt{\kappa}}\right)}}{\left(\sqrt{s} - \xi\right)} - \frac{e^{-\left(\frac{z\sqrt{s}}{\sqrt{\kappa}}\right)}}{\left(\sqrt{s} + \xi\right)} - \frac{e^{-\left(\frac{z\sqrt{s}}{\sqrt{\kappa}}\right)}}{\sqrt{\kappa}\left(\sqrt{s} + \xi\right)^2}\right] \tag{5.288}$$

It is noted from the Laplace inversion that:

$$\pounds^{-1}\left\{\frac{e^{-k\sqrt{s}}}{\sqrt{s}+s}\right\} = \frac{1}{\sqrt{\pi t}}e^{-\frac{k^2}{2t}} - ae^{ak}e^{a^2t}erfc\left(a\sqrt{t}+\frac{k}{2\sqrt{t}}\right) \tag{5.289}$$

where complimentary error function is defined as:

$$erfc(z) = 1 - erf(z) = \frac{2}{\pi}\int_z^\infty e^{-u^2}du \tag{5.290}$$

Therefore:

$$\pounds^{-1}\left\{\frac{e^{-\frac{z}{\sqrt{\kappa}}\sqrt{s}}}{\sqrt{s}-\xi}\right\} = \frac{1}{\sqrt{\pi t}}e^{-\frac{z^2}{2\kappa t}} + \frac{V}{2\sqrt{\kappa}}e^{\frac{Vz}{2\kappa}e^{\frac{V^2}{4\kappa}t}}erfc\left(\frac{z}{2\sqrt{\kappa t}}+\frac{k}{2\sqrt{\kappa}}\sqrt{t}\right) \tag{5.291}$$

$$\pounds^{-1}\left\{\frac{e^{-\frac{z}{\sqrt{\kappa}}\sqrt{s}}}{\sqrt{s}+\xi}\right\} = \frac{1}{\sqrt{\pi t}}e^{-\frac{z^2}{2\kappa t}} - \frac{V}{2\sqrt{\kappa}}e^{\frac{Vz}{2\kappa}e^{\frac{V^2}{4\kappa}t}}erfc\left(\frac{z}{2\sqrt{\kappa t}}+\frac{k}{2\sqrt{\kappa}}\sqrt{t}\right) \tag{5.292}$$

Letting $\sqrt{s}+\xi=\sqrt{p}$ and using the definition of inverse Laplace integral $\pounds^{-1}\left\{\frac{e^{-\frac{z}{\sqrt{\kappa}}\sqrt{s}}}{\sqrt{s}+\xi}\right\}$ as follows:

$$\pounds^{-1}\left\{\frac{e^{-\frac{z}{\sqrt{\kappa}}\sqrt{s}}}{\sqrt{s}+\xi}\right\} = \frac{1}{\sqrt{\pi t}}e^{-\frac{z^2}{2\kappa t}}e^{\frac{V^2}{4\kappa}t}erfc\left(\frac{z\sqrt{t}}{2\sqrt{\kappa}}+\frac{z}{2\sqrt{\kappa t}}\right)\left(1+\frac{Vz}{2\kappa}+\frac{V^{2t}}{2\kappa}\right)$$
$$-\frac{V\sqrt{t}}{2\sqrt{\pi\kappa}}e^{-\frac{z^2}{4\kappa t}} \tag{5.293}$$

Using Eqs. 5.291, 5.292, and 5.293, and after simplification, the Laplace inversion of $H_3(s)$ becomes:

$$\pounds^{-1}\{H_3(s)\} = \frac{1}{16\kappa V}\left[e^{\frac{Vz}{\kappa}}erfc\left(\frac{z-Vt}{2\sqrt{\kappa t}}\right) - \left(1+\frac{Vz}{\kappa}+\frac{V^2t}{\kappa}\right)erfc\left(\frac{z+Vt}{2\sqrt{\kappa t}}\right)\right]$$
$$+\frac{\sqrt{t}}{8\kappa\sqrt{\pi\kappa}}e^{-\frac{(z+Vtz)^2}{4\kappa t}} \tag{5.294}$$

Let $w_2^2=\frac{V^2}{4\kappa}$, $k_1=\frac{z}{\sqrt{\kappa}}$, $w_3^2=\frac{V^2}{4\kappa}-V_1$ and $w_4=\frac{V}{2\sqrt{\kappa}}$; then $H_4(s)$ becomes:

$$H_4(s) = \frac{1}{\left((\sqrt{s})^2 - \mathrm{w}_2^2\right)\left((\sqrt{s})^2 - \mathrm{w}_3^2\right)((\sqrt{s}) - \mathrm{w}_4)} \frac{1}{2\sqrt{\kappa}} e^{-(\mathrm{k}_1\sqrt{s})} \tag{5.295}$$

and after using partial fraction expansion, $H_4(s)$ becomes:

$$H_4(s) = \begin{bmatrix} \frac{D_1}{(\sqrt{s} - \mathrm{w}_2)} + \frac{D_2}{(\sqrt{s} + \mathrm{w}_2)} \frac{D_3}{(\sqrt{s} - \mathrm{w}_3)} \frac{D_4}{(\sqrt{s} - \mathrm{w}_3)} \\ + \frac{D_5}{(\sqrt{s} - \mathrm{w}_4)} \end{bmatrix} \times \frac{1}{2\sqrt{\kappa}} e^{-(\mathrm{k}_1\sqrt{s})} \tag{5.296}$$

where:

$$D_1 = \frac{1}{2\mathrm{w}_2\left(\mathrm{w}_2^2 - \mathrm{w}_3^2\right)(\mathrm{w}_2 + \mathrm{w}_4)} \tag{5.297}$$

$$D_2 = \frac{1}{2\mathrm{w}_2\left(\mathrm{w}_2^2 - \mathrm{w}_3^2\right)(-\mathrm{w}_2 + \mathrm{w}_4)} \tag{5.298}$$

$$D_3 = \frac{1}{2\mathrm{w}_3\left(\mathrm{w}_3^2 - \mathrm{w}_2^2\right)(\mathrm{w}_3 + \mathrm{w}_4)} \tag{5.299}$$

$$D_4 = \frac{1}{2\mathrm{w}_3\left(\mathrm{w}_3^2 - \mathrm{w}_2^2\right)(-\mathrm{w}_2 + \mathrm{w}_4)} \tag{5.300}$$

$$D_5 = \frac{1}{\left(\mathrm{w}_4^2 - \mathrm{w}_2^2\right)\left(\mathrm{w}_4^2 + \mathrm{w}_3^2\right)} \tag{5.301}$$

After substitution of Eqs. 5.297 through 5.301 into Eq. 5.296, then the Laplace inversion $\pounds^{-1}\{H_4(s)\}$ can be written as:

$$
£^{-1}\{H_4(s)\} = \frac{1}{2\sqrt{\kappa}} \times \left\{ \begin{array}{l} D_1\left[\left(\frac{1}{\sqrt{\pi t}}\right)e^{-\frac{k_1^2}{4t}} + \mathrm{w}_2 e^{-\mathrm{w}_2 k_1} e^{\mathrm{w}_2^2 t_1} erfc\left(-\mathrm{w}_2\sqrt{t} + \frac{k_1}{2\sqrt{t}}\right)\right] \\ +D_2\left[\left(\frac{1}{\sqrt{\pi t}}\right)e^{-\frac{k_1^2}{4t}} - \mathrm{w}_2 e^{-\mathrm{w}_2 k_1} e^{\mathrm{w}_2^2 t_1} erfc\left(\mathrm{w}_2\sqrt{t} + \frac{k_1}{2\sqrt{t}}\right)\right] \\ +D_3\left[\left(\frac{1}{\sqrt{\pi t}}\right)e^{-\frac{k_1^2}{4t}} + \mathrm{w}_3 e^{-\mathrm{w}_3 k_1} e^{\mathrm{w}_3^2 t_1} erfc\left(-\mathrm{w}_3\sqrt{t} + \frac{k_1}{2\sqrt{t}}\right)\right] \\ +D_4\left[\left(\frac{1}{\sqrt{\pi t}}\right)e^{-\frac{k_1^2}{4t}} - \mathrm{w}_3 e^{-\mathrm{w}_3 k_1} e^{\mathrm{w}_3^2 t_1} erfc\left(\mathrm{w}_3\sqrt{t} + \frac{k_1}{2\sqrt{t}}\right)\right] \\ +D_5\left[\left(\frac{1}{\sqrt{\pi t}}\right)e^{-\frac{k_1^2}{4t}} - \mathrm{w}_4 e^{-\mathrm{w}_4 k_1} e^{\mathrm{w}_4^2 t_1} erfc\left(\mathrm{w}_4\sqrt{t} + \frac{k_1}{2\sqrt{t}}\right)\right] \end{array} \right\} \tag{5.302}
$$

However, it is known from Eq. 5.274 that $T(z, t)$ can be written as:

$$
\begin{aligned} T(z,t) = £^{-1}\{\overline{T}(z,s)\} &= -2\kappa\mu\varpi_1 e^{-\frac{V}{2\kappa}\left[z+\frac{V}{2\kappa}t\right]} £^{-1}\{H_4(s)\} \\ &\quad +\varpi_1 e^{-\mu z} £^{-1}\{H_2(s)\} \\ &\quad -8\kappa^2\frac{\rho VL}{k} £^{-1}\{H_3(s)\} \end{aligned} \tag{5.303}
$$

Note that:

$$
£^{-1}\{H_2(s)\} = \frac{1}{V_1 - \beta}\left(e^{-\beta t} - e^{-V_1 t}\right) \tag{5.304}
$$

Substituting Eqs. 5.297 through 5.301, into Eq. 5.302, we get:

$$T(z,t) = \frac{\kappa\mu\varpi_1}{\sqrt{\kappa}} e^{-\frac{V}{2t}\left(z+\frac{V}{2}t\right)} \left\{ \begin{array}{l} \frac{1}{\sqrt{\pi t}} e^{-\frac{k_1^2}{4t}} [D_1 + D_2 + D_3 + D_4 + D_5] \\ D_1 \left[\mathrm{w}_2 e^{-\mathrm{w}_2 k_1} e^{\mathrm{w}_2^2 t} erfc\left(-\mathrm{w}_2\sqrt{t} + \frac{k_1}{2\sqrt{t}} \right) \right] \\ +D_2 \left[-\mathrm{w}_2 e^{-\mathrm{w}_2 k_1} e^{\mathrm{w}_2^2 t} erfc\left(\mathrm{w}_2\sqrt{t} + \frac{k_1}{2\sqrt{t}} \right) \right] \\ +D_3 \left[\mathrm{w}_3 e^{-\mathrm{w}_3 k_1} e^{\mathrm{w}_3^2 t} erfc\left(-\mathrm{w}_3\sqrt{t} + \frac{k_1}{2\sqrt{t}} \right) \right] \\ +D_4 \left[-\mathrm{w}_3 e^{-\mathrm{w}_3 k_1} e^{\mathrm{w}_3^2 t} erfc\left(\mathrm{w}_3\sqrt{t} + \frac{k_1}{2\sqrt{t}} \right) \right] \\ +D_5 \left[-\mathrm{w}_4 e^{-\mathrm{w}_4 k_1} e^{\mathrm{w}_4^2 t} erfc\left(\mathrm{w}_4\sqrt{t} + \frac{k_1}{2\sqrt{t}} \right) \right] \\ + \frac{\varpi_1}{V_1 - \beta} e^{-\mu z} \left[e^{-\beta t} - e^{-V_1 t} \right] \end{array} \right\}$$
$$- \frac{\alpha\kappa L}{2k} e^{-\frac{V_2}{\kappa}} erfc\left(\frac{z - Vt}{2\sqrt{\kappa t}} \right)$$
$$- \left[\left(1 + \frac{Vz}{\kappa} + \frac{V^2 t}{\kappa} \right) erfc\left(\frac{z + Vt}{2\sqrt{\kappa t}} \right) \right]$$
$$- \frac{\rho VL}{k} \frac{\sqrt{\kappa t}}{\sqrt{\pi}} e^{-\frac{(z+Vt)^2}{4\kappa t}} \tag{5.305}$$

Knowing that $D_1 + D_2 + D_3 + D_4 + D_5 = 0$, Eq. 5.305 becomes:

$$T(z,t) = \sqrt{\kappa}\mu\varpi_1 e^{-\frac{V}{2\kappa}\left(z+\frac{V}{2}t\right)} \left\{ \begin{array}{l} \frac{e^{\mathrm{w}_2^2 t}}{2\left(\mathrm{w}_2^2 - \mathrm{w}_3^2\right)} \left[e^{-\mathrm{w}_2 k_1} \frac{erfc\left(\mathrm{w}_2\sqrt{t} + \frac{k_1}{2\sqrt{t}} \right)}{\mathrm{w}_2 + \mathrm{w}_4} + e^{-\mathrm{w}_2 k_1} \frac{erfc\left(\mathrm{w}_2\sqrt{t} + \frac{k_1}{2\sqrt{t}} \right)}{\mathrm{w}_4 + \mathrm{w}_2} \right] \\ + \frac{e^{\mathrm{w}_3^2 t}}{2\left(\mathrm{w}_3^2 - \mathrm{w}_2^2\right)} \left[e^{-\mathrm{w}_3 k_1} \frac{erfc\left(\mathrm{w}_3\sqrt{t} + \frac{k_1}{2\sqrt{t}} \right)}{\mathrm{w}_3 + \mathrm{w}_4} + e^{-\mathrm{w}_3 k_1} \frac{erfc\left(\mathrm{w}_3\sqrt{t} + \frac{k_1}{2\sqrt{t}} \right)}{\mathrm{w}_4 - \mathrm{w}_3} \right] \\ + D_5 \mathrm{w}_4 e^{\mathrm{w}_4^2 t} e^{\mathrm{w}_3^2 k_1} erfc\left(\mathrm{w}_4\sqrt{t} + \frac{k_1}{2\sqrt{t}} \right) \end{array} \right\} e^{-\frac{(z+Vt)^2}{4\kappa t}}$$
$$+ \frac{\varpi_1}{V_1 - \beta} e^{-\mu z} \left[e^{-\beta t} - e^{-V_1 t} \right] - \frac{\alpha\kappa L}{2k} e^{\frac{Vz}{\kappa}} erfc\left(\frac{z - Vt}{2\sqrt{\kappa t}} \right)$$
$$- \left[\left(1 + \frac{Vz}{\kappa} + \frac{V^2 t}{\kappa} \right) erfc\left(\frac{z + Vt}{2\sqrt{\kappa t}} \right) \right]$$
$$- \frac{\rho VL}{k} \frac{\sqrt{\kappa t}}{\sqrt{\pi}} \tag{5.306}$$

In order to generate graphical representation of Eq. 5.306, let us define the following variables, which will be later used in Mathcad program:

$$i = 0 \cdots N \cdots\cdots z_i = \frac{h_t i}{N} \tag{5.307}$$

$$j = 1 \cdots L \cdots\cdots t_j = \frac{\tau_t j}{L} \tag{5.308}$$

$$\tau_j = \kappa\mu^2 t \tag{5.309}$$

$$V^0 = \frac{V}{\kappa\mu} \text{or} V^0 = 1 + \frac{V_1}{\kappa\mu^2} \tag{5.310}$$

$$\beta^0 = \frac{1}{\kappa\mu^2}\beta \tag{5.311}$$

$$\mathrm{w}_2 = \frac{V^0}{4\kappa} - \beta^0 \tag{5.312}$$

$$\mathrm{w}_3 = \left(\frac{V^0 - 2}{2}\right)\kappa\mu^2 \tag{5.313}$$

$$\mathrm{w}_4 = \frac{V^0}{2}\sqrt{\kappa\mu} \tag{5.314}$$

$$D_5 = \frac{1}{\beta^0 \left(V^0 - 1\right)\left(\kappa\mu^2\right)^2} \text{or} D_5 = \frac{1}{\beta V_1} \tag{5.315}$$

One can use the Mathcad program to perform calculations, so our function $T(z, t)$ can be formulated as T_{ij}, and substituting above parameters, we can obtain:

$$
\begin{aligned}
T_{i,j}(z,t) &= \frac{-I_{\max}}{k\mu}\exp\left[\frac{-V^{0^2}}{2}\left(z_d-\frac{V^0}{2}\tau_t\right)\right] \\
&\times\Bigg\{\frac{\exp\left[\frac{V^{0^2}}{4}(V^0-1)\right]t_j}{[V^0-(1+\beta^0]\}}\Bigg[\frac{\exp\left[\left[\frac{V^{0^2}}{4}-(V^0-1)\right]t_j\right]}{[V^0-(1+\beta^0)]}\Bigg[\exp\left[-\left(\sqrt{\frac{V^0}{4}}-\beta\right)\right]z_i\frac{1-erf\left[-\sqrt{\left(\frac{V^{0^2}}{4}-\beta^0\right)}\right]t_j}{\sqrt{V^{0^2}-\beta^0}+V^0}+\frac{z_i}{2\sqrt{t_j}} \\
&\qquad+\exp\left[\left(\sqrt{\frac{V^{0^2}}{4}}-\beta^0\right)z_i\right]\frac{1-erf\left[\sqrt{\left(\frac{V^{0^2}}{4}-\beta^0\right)t_j+\frac{z_i}{2\sqrt{t_j}}}\right]}{\left(V^0-\sqrt{V^{0^2}-\beta^0}\right)}\Bigg] \\
&\quad+\frac{\exp\left[\left[\frac{V^{0^2}}{4}-(V^0-1)\right]t_j\right]}{[\beta^0-(V^0-1)]}\Bigg[\exp\left[\sqrt{\frac{V^{0^2}}{4}-(V^0-1)}\right]\frac{1-erf\left[-\sqrt{\left[\frac{V^{0^2}}{4}-(V^0-1)\right]t_j+\frac{z_i}{2\sqrt{t_j}}}\right]}{\sqrt{V^{0^2}-4(V^0-1)}+V^0} \\
&\qquad+\exp\left[\sqrt{\frac{V^{0^2}}{4}-(V^0-1)}\right]\frac{1-erf\left[\sqrt{\left[\frac{V^{0^2}}{4}-(V^0-1)\right]t_j+\frac{z_i}{2\sqrt{t_j}}}\right]}{-\sqrt{V^{0^2}-4(V^0-1)}+V^0} \\
&\qquad+\frac{-1}{2\beta^0(V^0-1)}\exp\left[\frac{V^{0^7}}{4}t_j\right]\exp\left[\frac{V^0}{2}(z_i)\right]\left(1-erf\left(\frac{V^0}{2}t_j+\frac{z_i}{2\sqrt{t_j}}\right)\right)\Bigg] \\
&\quad+\frac{\kappa\rho L}{2k}\Bigg[e^{V^0 z_i}\left(1-erf\left(\frac{z_i-V^0t_j}{2\sqrt{t_j}}\right)\right)-\left(1+V^0z_1+V^{0^2}t_j\right)\left(1-erf\left(\frac{z_i+V^0t_j}{2\sqrt{t_j}}\right)\right) \\
&\qquad+\frac{2V^0t_j}{\sqrt{\pi}}\exp\left[-\left(\frac{V^{0^2}}{4\tau_j}-\frac{V^0z_i}{2}+\frac{V^{0^2}t_j}{4}\right)\right]\Bigg] \\
&\quad+\left[\frac{I_{\max}}{k\mu}\frac{1}{(V^0-1)-\beta^0}e^{-z_i}\left[\exp\left(e^{-\beta^0t_j}-e^{-(V^0-1)t_j}\right)\right]\right]\Bigg]
\end{aligned}
\tag{5.316}
$$

$$E_{1_{i,j}} = \sqrt{\left(\frac{V^{0^2}}{4}-\beta^0\right)t_j+\frac{z_i}{2\sqrt{t_j}}} \tag{5.317}$$

$$E_{2_{i,j}} = \sqrt{\left[\frac{V^{0^2}}{4}-(V^0-1)\right]t_j+\frac{z_i}{2\sqrt{t_j}}} \tag{5.318}$$

$$E_{3_{i,j}} = \sqrt{V^{0^2}-4(V^0-1)} \tag{5.319}$$

$$E_{4_{i,j}} = e^{-\left[\sqrt{\frac{V^{0^2}}{4}-(V^0-1)}\right]z_1} \tag{5.320}$$

$$E_{5_{i,j}} = e^{-\left[\sqrt{\left(\frac{V^{0^2}}{4} - \beta^0\right)}\right] z_1} \tag{5.321}$$

$$E_{6_{i,j}} = \frac{V^0}{2} t_j + \frac{z_i}{2\sqrt{t_j}} \tag{5.322}$$

$$E_{7_{i,j}} = \frac{z_i + V^0 t_j}{2\sqrt{t_j}} \tag{5.323}$$

$$E_{8_{i,j}} = \frac{V^{0^2}}{4} - \left(V^0 - 1\right) \tag{5.324}$$

$$\begin{aligned}
T(z,t) = {} & \frac{-I_{\max}}{k\mu} \exp\left[\frac{-V^{0^2}}{2}\left(z_d - \frac{V^0}{2}\tau_t\right)\right] \times \\
& \left[\frac{\exp\left(E_{8_{d,t}}\right)\tau_t}{\left[V^0 - \left(1 + \beta^0\right)\right]} \left[E_{5_{d,t}} \frac{1 - erf\left(-\operatorname{Re}\left(E_{1_{d,t}}\right)\right)}{\sqrt{V^{0^2} - \beta^0} + V^0} + E_{5_{d,j}} \frac{1 - erf\left(-\operatorname{Re}\left(E_{1_{d,t}}\right)\right)}{V^0 - \sqrt{V^{0^2} - \beta^0}} \right] \right. \\
& + \frac{e^{E_{8_{d,s}}\tau_t}}{\left[\beta^0 - \left(V^0 - 1\right)\right]} \left(E_{4_{d,t}} \frac{1 - erf\left(-E_{2_{d,t}}\right)}{E_{3_{d,t}} + V^0} + E_{4_{d,t}} \frac{1 - erf\left(-E_{2_{d,t}}\right)}{-E_{3_{d,t}} + V^0} \right) \\
& - \frac{1}{2\beta^0\left(V^0 - 1\right)} \exp\left[\frac{V^{0^2}}{4}\tau_t\right] \exp\left[\frac{V^0}{2} z_d\right] \left(1 - erf\left(E_{6_{d,t}}\right)\right) \\
& - \frac{\kappa\rho L}{2k} \left[\begin{array}{l} e^{-V^0 z_d}\left(1 - erf\left(\frac{z_d - V^0\tau_t}{2\sqrt{\tau_t}}\right)\right) \\ -\left(1 - V^0 z_d + V^{0^2}\tau_t\right)\left(1 - erf\left(E_{7_{d,t}}\right)\right) \end{array} \right] \\
& + \frac{2V^0\tau_t}{\sqrt{\pi}} \exp\left[-\left(\frac{V^{0^2}}{4\tau_t} - \frac{V^0 z_d}{2} + \frac{V^{0^2}\tau_t}{4}\right)\right] \\
& \left. + \left[\frac{I_0}{k\mu} \frac{1}{\left(V^0 - 1\right) - \beta^0} \exp\left[-z_d\left(e^{-\beta^0\tau_t} - e^{-\left(V^0 - 1\right)\tau_t}\right)\right]\right] \right]
\end{aligned} \tag{5.325}$$

Temperature distribution with respect to depth of the workpiece and time is graphically shown in Fig. 5.56, where recession velocity of vaporization surface was estimated to be about 33 m/s.

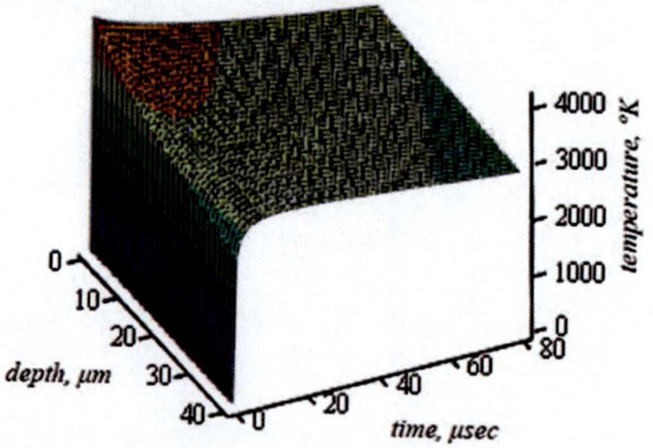

Fig. 5.56 Temperature distribution of the workpiece with consideration of transfer evaporative case calculated using Eq. 5.326

A closed form solution illustrated in Fig. 5.56 is limited to the surface ablation only and does not consider the plasma formation and liquid expulsion from the irradiated workpiece.

The temperature gradient attains relatively lower values in the surface vicinity of the substrate material than that corresponding to some depth below the surface. In this case, internal energy gain in this region due to absorption of laser beam energy becomes more important as compared to diffusional heat transfer from the surface vicinity to the bulk of the substrate material due to temperature gradient. As the heating period progresses, the point of minimum temperature gradient moves toward the solid bulk of the substrate material.

In the region beyond the point of minimum temperature gradient, the diffusional energy transport plays an important role inside the substrate material.

The large change in the power intensity occurs with increasing β^0; however, the magnitude of temperatures in the surface region does not alter considerably. This is because of the rate of evaporation, which increases with increasing power intensity. Consequently, the convective boundary condition at the surface suppresses the temperature rise in the surface vicinity at high power intensities.

5.9.1.7 Analytical Solution of Heat Transfer Equation with Kinetic Theory Approach

This section examines the pulsed laser heating process by considering both Fourier conduction and electron–phonon kinetic theory approaches, which is described in Sect. 7.9 of this book. A one-dimensional kinetic theory approach is presented. More specifically, this section is focused on development of equations governing an interaction of a Gaussian laser beam with materials, where an analytical solution of these equations based on both the Fourier heat conduction theory and kinetic theory approach using a Laplace integral transformed method is utilized. A comparison of the analytical solution of the Fourier theory and closed form solution of the kinetic theory approach is introduced in Sect. 5.11. The temperature distribution for the heated material predicted from the kinetic theory is compared with the Fourier theory findings.

Most theoretical work relating to laser beam interaction with materials is based on the solution of the classical heat conduction equation derived from the Fourier theory. It has been shown that the Fourier theory of heat conduction is not fully applicable to short pulsed laser heating due to the assumptions made in the theory.

These assumptions include:

1. Matter is assumed as continuous and homogeneous.
2. The heat flux across any plane is a function of only the temperature gradient at that plane.

The first assumption is not valid for distances less than inter-atomic spacing, and the second is only true if all the energy crossing the isothermal plane is accounted for.

Conduction in metals occurs due to subsequent collisions between excited electrons and lattice site atoms. The electron motion in the substrate is random, which means that electrons move from the surface to the bulk as well as from the bulk to the surface. Moreover, the amount of energy transferred to lattice site atoms depends on the electron energy distribution in a particular region. Harrington (1967) showed, based on kinetic theory considerations, that electrons within five times of the electron mean free path contribute to 98.5% of the total energy transported provided that $\partial T/\partial z$ is constant over this distance. Consequently, attempting to generate the Fourier equation, the application of $q = -k(\partial T/\partial z)$ is limited to planes in excess of ten times the electron mean free path apart.

In addition, the absorption depth of the metals is of the order of ten electron mean free paths; therefore, the gradient $\partial T/\partial z$ is not uniform over the spatial increment ($\Delta z \geq 10\lambda$) and λ being the electron mean free path. In this case, the higher-order gradients ($\partial^3 T/\partial z^3$), which are neglected in the Fourier heating model, become important, and the validity of the Fourier heating model comes into question. Therefore, it becomes necessary to examine the laser-induced conduction heating on a microscopic scale.

The applicability of the Fourier equation in laser heating is limited to cases in which low-power laser intensities are employed (Yilbas et al. 2000) [56].

This is due to the following facts:

1. In the analysis of the Fourier heat conduction model, the heat flux through a given plane is considered as a function of the spatial temperature gradient at that plane. This depends upon the assumption that the temperature gradient remains almost constant between two successive and closely spaced planes. The distance between these planes is finite; therefore, error occurs when high-order terms, which are neglected, become important at high-power laser intensities.

2. The heat flux through a given plane depends on the electron energy distribution through the material; therefore, the material cannot be considered as a homogeneous continuum.

Consequently, a new model may be required to be developed for heating mechanism, which is appropriate to high-power laser heating process. A model considering a kinetic theory approach describing the transport of energy by electrons within the electron mean free path may be suitable in this case. The basis of this model was introduced by Yilbas (2001) [57] for one-dimensional heating. It was shown that the predictions made from the new model agree well with the experimental findings. He adopted the electron motion in metals to formulate the laser pulse heating process. The heating process was established based on a kinetic energy transfer mechanism, which occurred during the collisions between excited electrons and lattice site atoms. In this case, the excess electron energy was transferred to lattice site atoms resulting in increased amplitude of lattice site vibrations during the collision process.

In this section, comparisons of the electron kinetic theory approach are made with the Fourier theory models for a pulse laser heating process. The temperature field due to each model is predicted for step intensity as well as exponentially decaying intensity pulses. The study is extended to include the analytical solution to the electron kinetic theory approach for Gaussian intensity pulses. Electron kinetic theory approach is based on electron and phonon movements.

5.9.2 Melting Process

Consider first the case of a semi-infinite slab as before, melted, with instantaneous melt removal, as indicated in Fig. 5.57.

We solve this by considering ourselves moving along with interface. But to do this, we have to reconsider our heat flow equation, which is:

$$\frac{\partial^2 T(z,t)}{\partial z^2} - \frac{2}{\kappa}\frac{\partial T(z,t)}{\partial t} = 0 \tag{5.326}$$

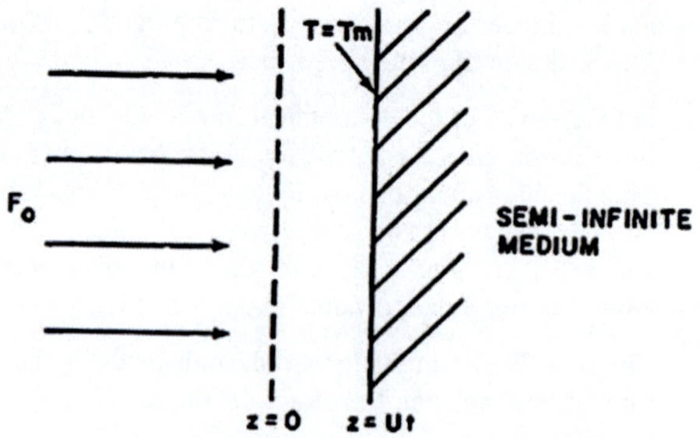

Fig. 5.57 Irradiation of a semi-infinite slab with instantaneous melt removal

Recall that this was derived by noting the rate at which heat accumulated in an elemental volume:

$$\rho c\frac{\partial T(z,t)}{\partial t}+\frac{\partial}{\partial z}\left(-K\frac{\partial T(z,t)}{\partial z}\right)=0 \tag{5.327}$$

If the medium is moving, then an additional amount of heat ρcT is flowing in at rate V; therefore:

$$-K\frac{\partial T}{\partial z}\text{becomes}-K\frac{\partial T}{\partial z}+V\rho cT$$

As result of this, Eq. 5.328(a) turns out to be:

$$\rho c\frac{\partial T(z,t)}{\partial t}+\frac{\partial}{\partial z}\left(-K\frac{\partial T(z,t)}{\partial z}+V\rho cT(z,t)\right)=0 \tag{5.328(a)}$$

or for simplicity, we write it as:

$$\rho c\frac{\partial T}{\partial t}+\frac{\partial}{\partial z}\left(-K\frac{\partial T}{\partial z}+V\rho cT\right)=0 \tag{5.328(b)}$$

This equation is valid, providing that we are not generating any heat in the solid. Assuming that $\kappa = K/\rho c$ as before, then Eq. 5.328(b) yields to:

$$\frac{\partial T}{\partial t}-\kappa\frac{\partial^2 T}{\partial z^2}+V\frac{\partial T}{\partial z}=0 \tag{5.329}$$

Therefore, we can find at least a steady-state solution to the problem by basing our coordinate system at the interface and letting the material flow in at some rate to be determined. Call the rate $-U$, where U is a positive number. The solution will be:

$$T=T_m e^{-Uz\prime/\kappa} \tag{5.330}$$

where κ is diffusivity of the solid, T_m is the melting temperature, and z' is the distance from melting front. That is, this is solution to:

$$\frac{\partial T(z',t)}{\partial t} - \kappa \frac{\partial^2 T(z',t)}{\partial z'^2} - U \frac{\partial T(z',t)}{\partial z'} = 0 \tag{5.331}$$

as we can verify, first, $\partial T/\partial t = 0$ because this is the steady-state profile, or:

$$\kappa \frac{\partial^2 T(z',t)}{\partial z'^2} + U \frac{\partial T(z',t)}{\partial z'} = 0 \tag{5.332}$$

of course $T = T_m$ at $z' = 0$ and $T(z',t) \to 0$ as $z' \to 0$. It is trivial to show, by differentiation, that Eq. 5.329 is the solution to Eq. 5.331.

To calculate U, we use the energy balance. F_0 must raise the material to T_m, its melting point, and note that $T_m\,^\circ$C above ambient or environment, and then melt it. Thus in time Δt, the energy put into thickness $\Delta z'$ (where $U = \Delta z'/\Delta t$) must be given by:

$$\frac{F_0 \Delta t}{\Delta z'} = L\rho + cT_m\rho \tag{5.333(a)}$$

or:

$$F_0 = \rho[L + cT_m]U \tag{5.333(b)}$$

Thus, we can write, for the steady-state temperature profile:

$$T(z',t) = T_m \exp\left\{-\frac{z'}{\kappa}\left[\frac{F_0}{\rho(L + cT_m)}\right]\right\} \tag{5.334}$$

Definition of Sublimation

Sublimation is a physical process in which a solid directly converts into a gaseous (vapor) state without going through a liquid state. The latent heat of sublimation at a particular temperature is the amount of heat required to convert a unit mass of solid into gas. For example, when ice sublimates into vapor, the amount of heat required at 0 °C is estimated to be equal to 2838 kJ/kg, which is the latent heat of sublimation of ice at 0 °C. In the crystal growth of ice and snow in atmosphere, this process plays a dominant role. See Fig. (d).

(continued)

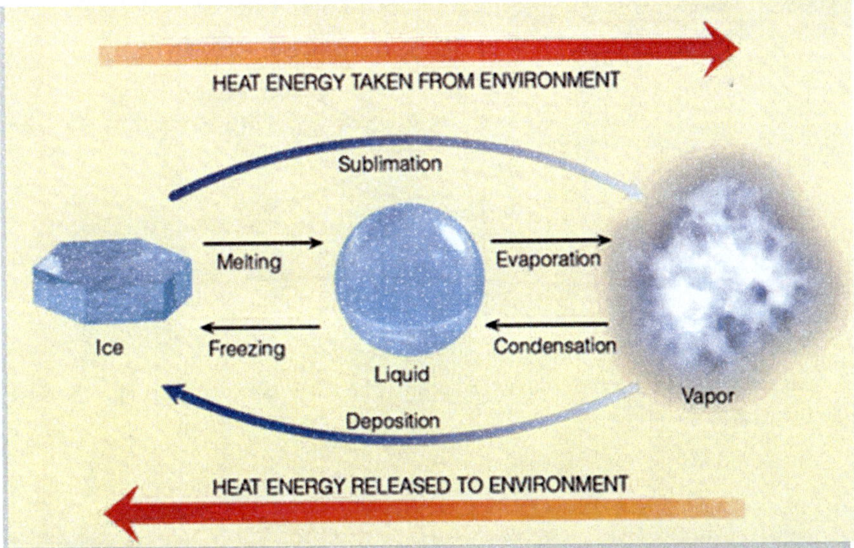

Fig. (d) Depiction of latent heat

This process is opposite to the process of deposition. Also bear in mind that when a substance changes from one state to another, by definition, latent heat is added or released in the process.

where $F_0/[\rho(L + cT_m)]$ is the velocity of the melting front. This solution also would be appropriate to sublimation, where L is then the latent heat of sublimation and T_m the sublimation temperature. Note that this is the semi-infinite slab approximation and cannot be used to estimate the time to penetrate a slab of given thickness.

Let us consider aluminum, with $F_0 = 200$ W/cm^2. Taking a more accurate value of the melting point than in earlier examples, $T_m = 640^{\circ}$C (the actual melting point of 660 °C minus room temperature of 20 °C). If we put in the other values of the parameters, T falls off as shown in Fig. 5.58.

As another illustration, consider Plexiglas. Plexiglas is rapidly eroded by 10.6 μm laser radiation, by a process that is essentially sublimation. Since it couples extremely well ($A \approx 1.0$) and has a very low thermal diffusivity ($\kappa \approx 10^{-4}$cm^2/s), it can be used to make "burn patterns" of the beam. That is, the depth of the erosion at a given point is linearly proportional to the energy density incident at that point. We can understand this by applying the heat flow concept. Let us apply Eq. 5.338 and interpret L as the heat of erosion. Since $c \approx$ J/g $-$ $^{\circ}$C and $T_m = 200^{\circ}$C and $L \approx 1000$ J/g, for a crude calculation, we can ignore cT_m. The density is about 1.1 g/cm^3. We then have:

$$T(z', t) \approx 200\ e^{-9F_0 z'} \tag{5.335}$$

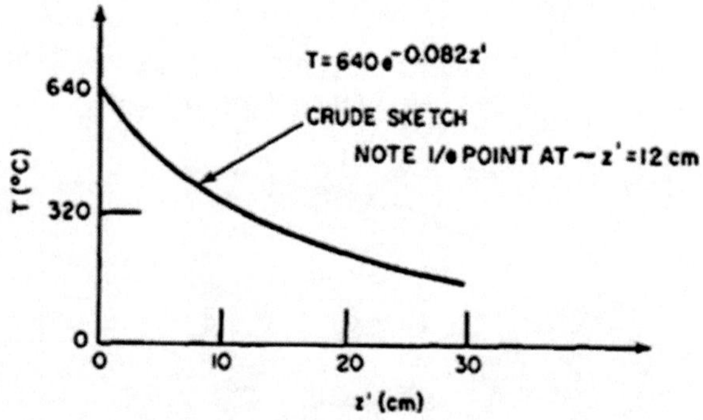

Fig. 5.58 Temperature profile in aluminum with instantaneous melt removal

Since $A \approx 1.0$, for a typical power density like 5 kW/cm^2, we have $F_0 = 5 \times 10^3$, so that:

$$T \approx 200\ e^{-4.5\times10^4 z'} \tag{5.336}$$

Hence the temperature profile is confined to an extremely narrow region near the eroding surface. The rate of erosion is, by Eq. 5.333(b) with cT_m, ignored as:

$$U \approx 4.5\ \text{cm/s} \tag{5.337}$$

Now we can see why Plexiglas is useful for monitoring the beam profile and why it can pick up a fairly fine structure in the beam. After irradiation for a time t, the penetration depth is $Ut = (F_0 t)/(\rho L)$, and this should be large compared to a thermal diffusion length $D = 2\sqrt{\kappa t}$ if the pattern is to reveal a fine structure. Otherwise thermal diffusion would "wash out" the pattern by distributing the energy in a radial direction. Thus:

$$2\sqrt{\kappa t} \ll \frac{F_0 t}{\rho L} \tag{5.338}$$

For the numbers we used above at $t = 1/4$ s, we have a depth of 1 cm. Thus:

$$2\sqrt{10^{-4} \times \frac{1}{4}} = 0.1\ \text{mm} \ll 1\ \text{cm} \tag{5.339}$$

and we see the criterion is well satisfied.

Let us look at a more complete problem, namely, melting by laser radiation of a slab of material. One basic problem is what happens to the melted material. (We will ignore the vaporization question for now.) There are two eases which are fairly amenable to the numerical solution. They are the "fully retained liquid" case, in which all the liquid is presumed to stay in place, and the "full ablation" case, in which the melt is presumed to disappear magically as soon as it forms. The latter case might correspond to the presence of a heavy wind stream which blows away the melt.

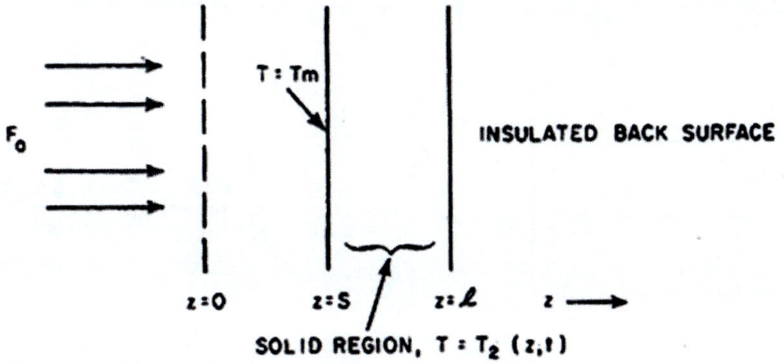

Fig. 5.59 Fully ablated case

Looking first at the full ablation case, we have the situation shown in Fig. 5.59. T_2 is the temperature in the solid, and the front surface at $z = 0$ first warms up to the melting point T_m (above ambient temperature), and then begins to move to the right. S denotes its position as a function of time. When $S = \ell$, the process is over, and we call this time t_f. We denote by t_m the time at which the front surface begins to melt. The field equation is:

$$\frac{\partial^2 T_2}{\partial z^2} - \frac{1}{\kappa_2}\frac{\partial T_2}{\partial t} = 0 \text{for} 0 \le t \le t_f \tag{5.340}$$

The boundary conditions are:

$$K_2 \frac{\partial T_2}{\partial z}\bigg|_{z=0} = -F_0 \quad \text{for} \quad 0 \le t \le t_m \tag{5.341}$$

$$K_2 \frac{\partial T_2}{\partial z}\bigg|_{z=S} = -F_0 + \rho L \frac{dS}{dt} \quad \text{for} \quad t_m \le t \le t_f \tag{5.342}$$

$$K_2 \frac{\partial T_2}{\partial z}\bigg|_{z=\ell} = 0 \quad \text{for} \quad 0 \le t \le t_f \tag{5.343}$$

The boundary conditions are:

$$\begin{aligned} T_2(z,0) &= T_{20} \\ S|_{t \le t_m} &= 0 \end{aligned} \tag{5.344}$$

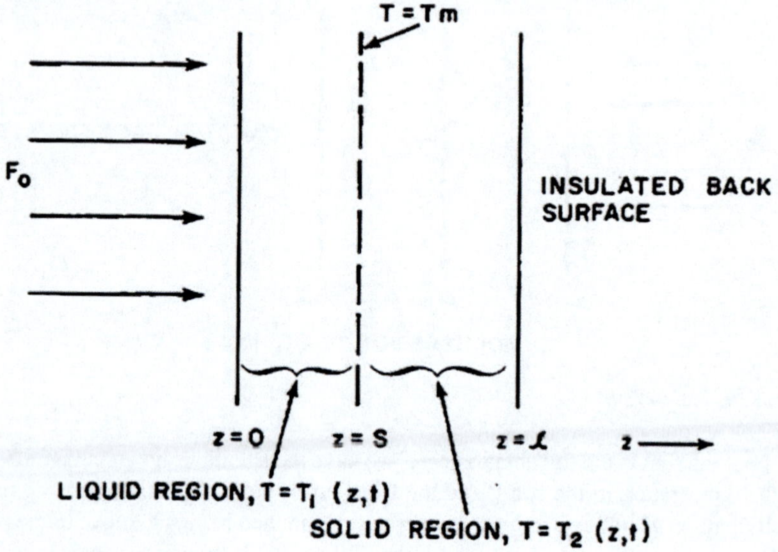

Fig. 5.60 Fully retained liquid case

The above boundary conditions are nonlinear, and a solution in an analytical form is very difficult. This is due to the presence of the moving boundary and appears in the second boundary condition, which states that the boundary moves at a rate *dS*/*dt* determined by a balance between the heat of melting *L*, the heat input F_0, and the heat flow by thermal conduction.

One relationship must hold for this problem; it follows from energy balance. The total energy put in per unit area is $F_0 t_f$, and, since the material is simply heated to T_m and melted, this energy goes solely to those processes. Thus:

$$F_0 t_f = \rho(L + cT_m) \tag{5.345}$$

This is convenient, for one can check numerical solutions. More important, it gives a first-order estimate of the time needed to melt through materials by laser radiation.

Turning now to the fully retained liquid case, we have the following set of equations. The definitions are the same as above, except that subscript 1 now refers to the molten state, whereas subscript 2 is still the solid state (see Fig. 5.60). The field equations are:

$$\frac{\partial^2 T_2}{\partial z^2} - \frac{1}{\kappa_2}\frac{\partial T_2}{\partial t} = 0 \quad \text{for} \quad 0 \le t \le t_f \qquad \text{(solid)} \tag{5.346(a)}$$

$$\frac{\partial^2 T_1}{\partial z^2} - \frac{1}{\kappa_1}\frac{\partial T_1}{\partial t} = 0 \quad \text{for} \quad t_m \le t \le t_f \qquad \text{(liquid)} \tag{5.346(b)}$$

The boundary conditions are:

$$K_2 \frac{\partial T_2}{\partial z}\bigg|_{z=0} = -F_0 \quad \text{f} \quad 0 \le t \le t_m \tag{5.347(a)}$$

$$K_1 \frac{\partial T_1}{\partial z}\bigg|_{z=0} = -F_0 \qquad 0 \le t \le t_f \tag{5.347(b)}$$

$$K_1 \frac{\partial T_1}{\partial z}\bigg|_{z=S} - K_2 \frac{\partial T_2}{\partial z}\bigg|_{z=S} = -\rho L \frac{dS}{dt} \qquad t_m \le t \le t_f \tag{5.347(c)}$$

$$K_2 \frac{\partial T_2}{\partial z}\bigg|_{z=\ell} = 0 \qquad 0 \le t \le t_f \tag{5.347(d)}$$

$$T_2|_{z=S} = T_1|_{z=S} = T_m \qquad t_m \le t \le t_f \tag{5.347(e)}$$

The initial conditions are:

$$T_2(z, 0) = T_{20} \tag{5.348(a)}$$

$$S|_{t \le t_m} = 0 \tag{5.348(b)}$$

Rather than discussing this problem in detail, we pass on to the more practical, although more complex, case of vaporization. For the fully retained liquid case, suffice it to say that the retained liquid has a shielding effect, and this causes the time to reach melting at the back surface to be longer than in the ablated model. Some typical values of melt through time for a 0.2-cm-thick material with $h\,\lambda\,2\ \text{kW/cm}^2$ are given below.

	Ablated (s)	Retained (s)
Aliminum	0.32	0.37
Stainless steel	4.0	4.5

5.9.3 *Melting and Vaporization*

We present here without derivation some results for the case of a slab of material, insulated on the surfaces, subjected to uniform and continuous irradiation [58]. These are one-dimensional calculations. It is assumed that the melt is fully retained until it reaches the vaporization temperature, where it disappears. Then we have the case illustrated in Fig. 5.61.

Here S_2 is the position of the liquid–vapor interface and S_1 the position of the solid–liquid interface. This problem has been solved numerically at NRL [58], and we shall show sonic results. The assumptions are that in each phase, the thermal properties are independent of temperature.

In the curves, the following definitions are used:

$$\alpha\widehat{Q} = \frac{F_0 \ell}{\rho L \kappa_{\text{solid}}} \tag{5.349}$$

$$\tau = \frac{\kappa_{\text{solid}}}{\ell^2} \tag{5.350}$$

$$\theta = (T - T_m) c_{\text{solid}} / L \tag{5.351}$$

where:

$$L = \text{latent heat of fusion}$$

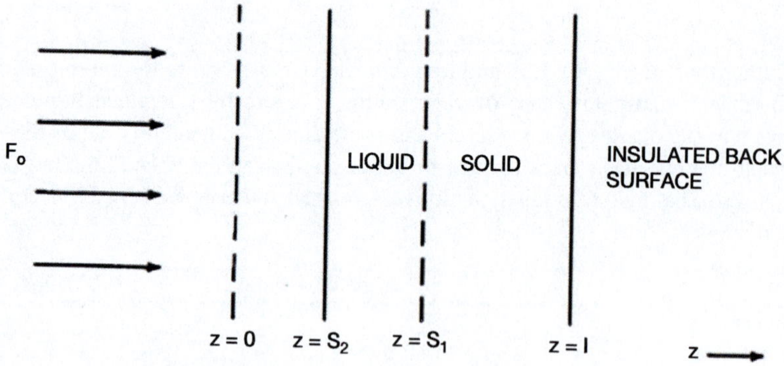

Fig. 5.61 Melting and vaporization, with fully retained liquid

Subscription 0 is = ambient or environment temperature.

$$\omega = L_v / L$$

with L_v as the latent heat of vaporization.

In these equations T is understood to be in degrees Celsius and represents the actual temperature. Although the thermal conductivities of liquid and solid are allowed to differ, the specific heats are assumed to be the same.

Note that when $F_0 \to 0$ and $F_0 \to \infty$, we get certain easy limits. For $F_0 \to 0$ no vaporization can take place, the melting is small, and we approach to fully ablated limit. On the other hand, as $F_0 \to \infty$, all the liquid should be vaporized by f, the time the back surface melts.

So, in this limit:

$$F_0 t_f = \rho \ell \left[L + (T_m - T_0) c_{\text{solid}} + (T_v - T_m) c_{\text{liquid}} + L_v \right] \quad \text{for} \quad F_0 \to \infty \tag{5.352}$$

whereas:

$$F_0 t_f = \rho \ell [L + (T_m - T_0) c_{\text{solid}}] \quad \text{for} \quad F_0 \to 0 \tag{5.353}$$

The limiting values are represented by asymptotes of the curves Figs. 5.62, 5.63, 5.64, 5.65, 5.66, and 5.67 indication 5.64 indicating by dashed lines on the plots of $\alpha\widehat{Q}$ vs. τ. Note on these plots that the dashed lines are at 45° or have a slope of -1. These are log-log plots, so the asymptotes can be described by:

$$\log\left(\alpha\widehat{Q}\right) = \log C - \log \tau_f \tag{5.354}$$

where $C \to C_\infty$ as $\alpha\widehat{Q} \to \infty$ and $C \to C_0$ as $\alpha\widehat{Q} \to 0$. If we take the antilog of Eq. 5.361, we get the following:

$$\alpha\widehat{Q} = \frac{C}{\tau_f} = C(\tau_f)^{-1} \tag{5.355}$$

Substituting in the definition of $\alpha\widehat{Q}$ Eq. 5.349 and τ Eq. 5.350 gives:

$$\frac{F_0 \ell}{\rho L \kappa_{\text{solid}}} = C \frac{\ell^2}{\kappa_{\text{solid}} \tau_f} \tag{5.356(a)}$$

or:

$$F_0 \tau_f = \rho \ell L C \tag{5.356(b)}$$

By comparison with the $F_0 \to 0$ and $F_0 \to \infty$ limits above, we can see that:

$$C_0 = \frac{1}{L}\left[L + (T_m - T_0)c_{\text{solid}}\right] \tag{5.357}$$

and:

$$C_\infty = \frac{1}{L}\left[L + (T_m - T_0)c_{\text{solid}} + (T_v - T_m)c_{\text{liquid}} + L_v\right] \tag{5.358}$$

The numerical values of the thermal parameters which were used in generating these solutions are included as a separate table (Table 5.4) in addition to the graphs (Figs. 5.62, 5.63, 5.64, 5.65, 5.66, and 5.67).

Provided one knows F_0, these solutions are reasonable estimates for the time to penetrate a metal specimen with a laser beam. In application, however, one must consider the actual size of the beam. These solutions will be useful for effects in the center of the beam if the diffusion length is small compared to the beam radius or if for times up to and including the melt-through time t_f the beam radius $R > 2\sqrt{\kappa t_f}$ is as follows:

$$R > 2\sqrt{\kappa t_f} \tag{5.359}$$

In terms of the parameter τ_f, this becomes, upon squaring both sides:

$$\tau_f =< \frac{1}{4}\left(\frac{R}{\ell}\right)^2 \tag{5.360}$$

5.9.4 Electron–Phonon Analytical Solution

The kinetic theory approach essentially deals with the kinetic energy transfer mechanism that occurs when electrons and lattice atoms with different energies interact. In order to simplify the phenomenon, some useful assumptions are made. These include omission of thermionic emission, attainment of steady space charge, and mean free path of molecules being independent of temperature. A net flow of electrons occurs in the substrate by the presence of an electron source at infinity, and the electron gives a fraction of its excess energy to the lattice site atoms during an electron–phonon collision, and this fraction is assumed to be constant throughout the successive collisions.

Table 5.4 Numerical solutions

Metal or alloy	T_m (°C)	T_v (°C)	L (cal/g)	L_v (cal/g)	ρ (g/cm^3)	K_s (cal/s-cm-°C)	K_ℓ (cal/s-cm-°C)	c_s (cal/g °C)	c_ℓ (cal/g-°C)	κ_s (cm^2/s)	K_ℓ (cm^2/s)
304 Stainless steel	1454	3000	65	1112	8.0	0.062	0.062	0.148	0.148	0.0523	0.0523
Ti (6Al, 4V)	1649	3285	94	2124	4.47	0.0455	0.0455	0.199	0.199	0.051	0.051
2024 Alum.	638	2480	95	2565	2.77	0.414	0.207	0.249	0.249	0.6	0.3

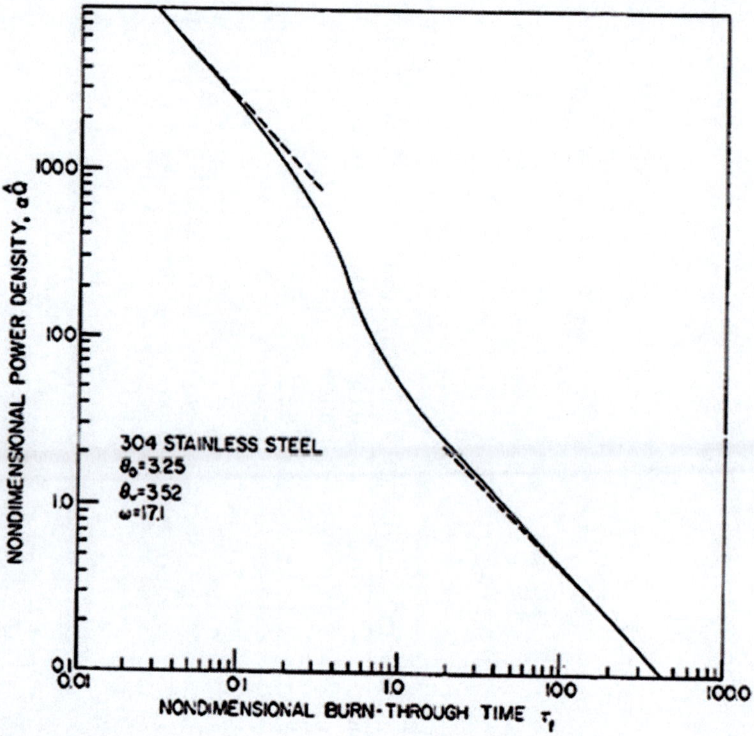

Fig. 5.62 Power density vs. burn-through time for ctainless steel [58]

When the electron absorbs the incident laser energy, some excess energy of the electron is transferred to the lattice site atoms during electron–phonon collision process. This energy manifests itself as an increase in the amplitude of the atomic vibration (phonon). As a result, neighboring atoms in the lattice are forced away to new equilibrium positions and absorb some of this extra energy in the process. A stage may be reached where eventually the lattice site atoms in the localized region around the original collision site are all in equilibrium and have increased their vibrational energies. It is this energy mechanism, which defines the conduction process in the solid substrate when subjected to a laser heating pulse. The amount of energy which electrons from Section I transfer to lattice site atoms in the same section (Fig. 5.68) can be calculated as follows.

The number of electrons leaving Section I is $N_z A_z \overline{V}_z dt$ where $A_z = dxdy$, N_z is the number density of electrons which transfer energy to dz from $d\zeta$, and $\overline{V}_z$ is the average electron velocity entering the control volume in the z-axis across area A_z at time dt.

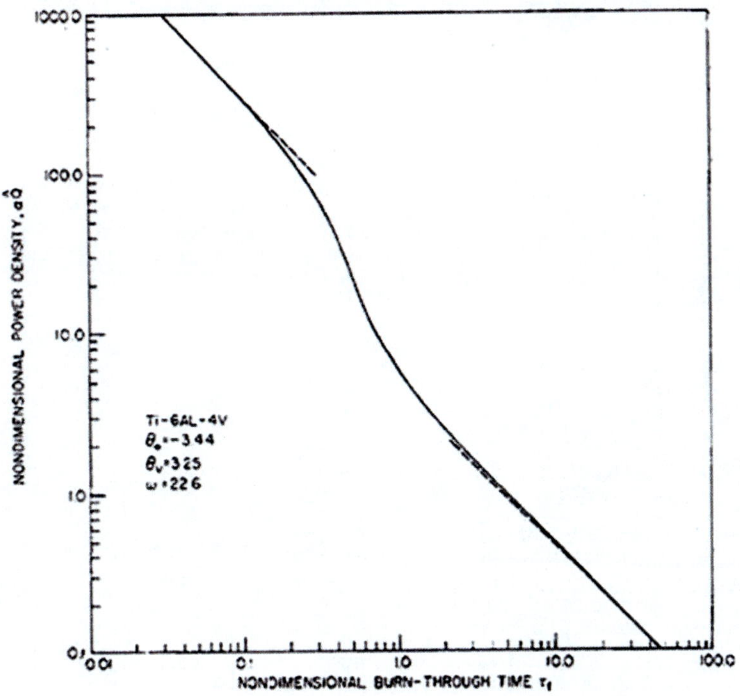

Fig. 5.63 Power density vs. burn-through time for titanium alloy [58]

The energy of N atoms in one direction due to lattice vibration is $E = Nk_BT$ which can be described as the phonon energy. The number of collisions, which takes place between electrons and phonons through the material, can be assessed by the total collision probability of electrons. Probability of electron traveling a distance z, where $z \ll 2\lambda$ (λ being the mean free path), without making a collision is $\exp\left(-\frac{z}{\lambda}\right)$; the probability of an electron having just collided in dz is dz/λ.

In general, thermal conductivity is usually defined with respect to steady heat transfer through a homogeneous medium in a random process. Thus, the thermal conductivity in one direction can be defined as:

$$k = \frac{N\overline{V}_z k_B \lambda}{6} \tag{5.361}$$

Note that N is the number of free electrons per unit volume. It has been shown that for temperatures higher than the Debye temperature, thermal conductivity can be assumed constant. The number of electrons which have just collided in Section I is $N_{\zeta z} A_z \overline{V}_z \frac{d\zeta}{\lambda}$. Therefore, the number of electrons which have just collided in $d\zeta$ during dt is:

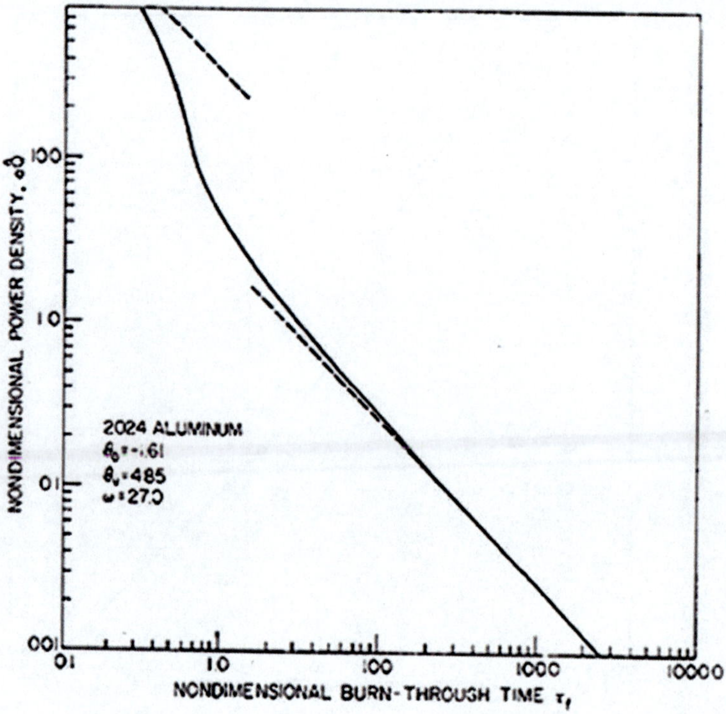

Fig. 5.64 Power density vs. burn-through time for aluminum alloy [58]

$$N_{\zeta z} A_z \overline{V}_z \int_{-\infty}^{+\infty} \exp\left(-\frac{|z-\zeta|}{\lambda}\right) \frac{d\zeta}{\lambda} \frac{dz}{\lambda} \tag{5.362}$$

and which then travel to Section *II* before colliding in *dz*, where $\int_{-\infty}^{+\infty} \exp\left(-\frac{|z-\zeta|}{\lambda}\right) \frac{d\zeta}{\lambda} \frac{dz}{\lambda}$, is the total electrons-lattice site atoms collision probability as described by Yilbas (2001) [57].

The negative sign of the integral is due to a mirror image introduced at a surface. This mirror image represents reflected electrons from the free surface. The net transfer of energy during the electron–phonon collision through the entire body can be written as:

$$\Delta E_{z,t} = \int_{-\infty}^{+\infty} \exp\left(-\frac{|z-\zeta|}{\lambda}\right) \frac{dz}{\lambda} \frac{d\zeta}{\lambda} (f) \cdot \left[E_{\zeta,t} - E_{z,t}\right] \tag{5.363}$$

where $E_{\zeta,\, t}$ and $E_{z,\, t}$ are energy of electrons and photons at a considered region, respectively, in the *z*-axis and parameter *f* is a fraction of electron energy. It is suggested that the rate of transfer of energy between the electrons and the molecules will be determined only by the difference in temperature of the electrons and lattice

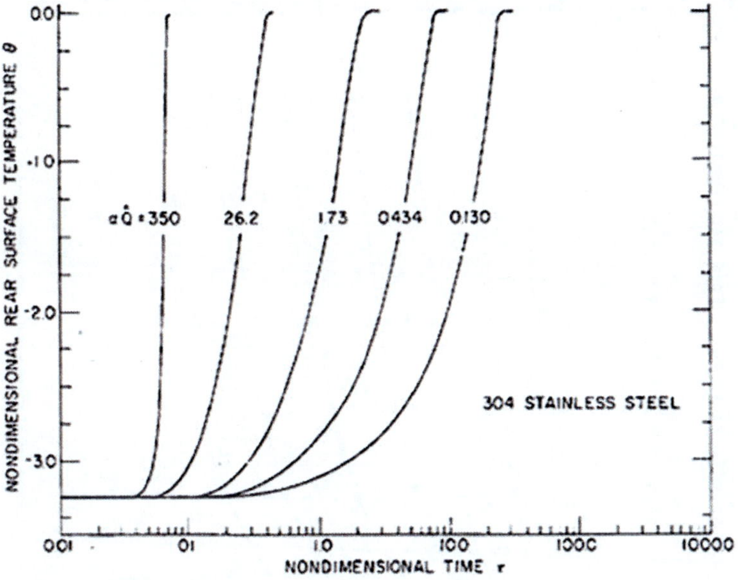

Fig. 5.65 Rear surface temperature rise for stainless steel [58]

vibrations. If the temperature of the lattice site atoms in dz is $\Theta(\zeta, t)$, then the energy transfer to the lattice site atoms in dz from collisions with electrons in which the electrons give up a fraction "f" of their excess energy is:

$$N_{\zeta z} A_z \overline{V}_z \frac{d\zeta}{\lambda} \frac{dz}{\lambda} \exp\left(-\frac{|z-\zeta|}{\lambda}\right) f \cdot k_B [\Theta(\zeta, t) - \Theta(z, t)] \tag{5.364}$$

Summing the contributions from all such sections to obtain the energy in Section II gives:

$$\Delta E_{z,t} = A_2 dz dt \int_{-\infty}^{+\infty} \frac{N_{\zeta z} \overline{V}_z}{\lambda} \frac{f \cdot k_B}{\lambda} \exp\left(-\frac{|z-\zeta|}{\lambda}\right) [\Theta(\zeta, t) - \Theta(z, t)] d\zeta \tag{5.365}$$

During electron–phonon collision, some fraction f of the electron excess energy is transferred to the phonon. For any inelastic collision, the conservation of energy in any section may be written as:

$$\begin{aligned} \text{Electron energy entering the section} = &\text{Electron energy entering the section} \\ &+ \text{Energy transfer to photons in the section} \end{aligned} \tag{5.366}$$

This gives:

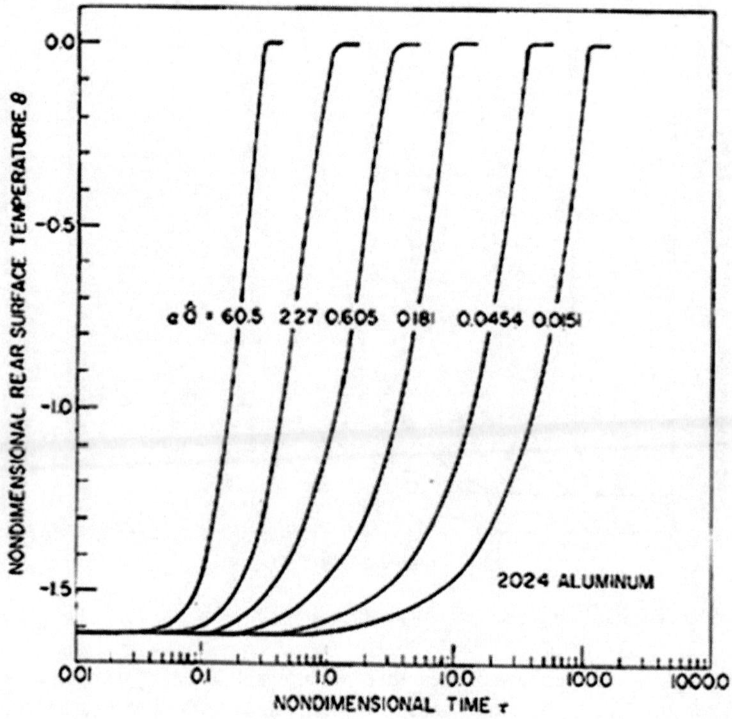

Fig. 5.66 Rear surface temperature rise for aluminum alloy [58]

$$f = \frac{\text{Electron energy in} - \text{Electron energy out}}{\text{Excess electron energy in}} \tag{5.367(a)}$$

$$f = \frac{(E_{el})_{\text{in}} - (E_{el})_{\text{out}}}{(E_{el})_{\text{in}} - E_{\text{photon}}} \tag{5.367(b)}$$

providing that $0 \leq f \leq I$ from the energy conservation, where $(E_{el})_{\text{excess}} = (E_{el})_{\text{in}} - E_{\text{photon}}$ and E_{photon} = mean energy of photon. The effective *f* value over a region sufficiently large to allow many collisions approaches unity, and this corresponds to attainment of thermal equilibrium. In the case of a single collision, *f* depends only upon the masses of the colliding particles, according to:

$$f = \frac{2M \cdot m}{(M + m)^2} \tag{5.368}$$

where *M* and *m* are the masses of an atom and an electron, respectively. Substituting appropriate values shows that the *f* value is of the order of 10^{-4}. The most energetic

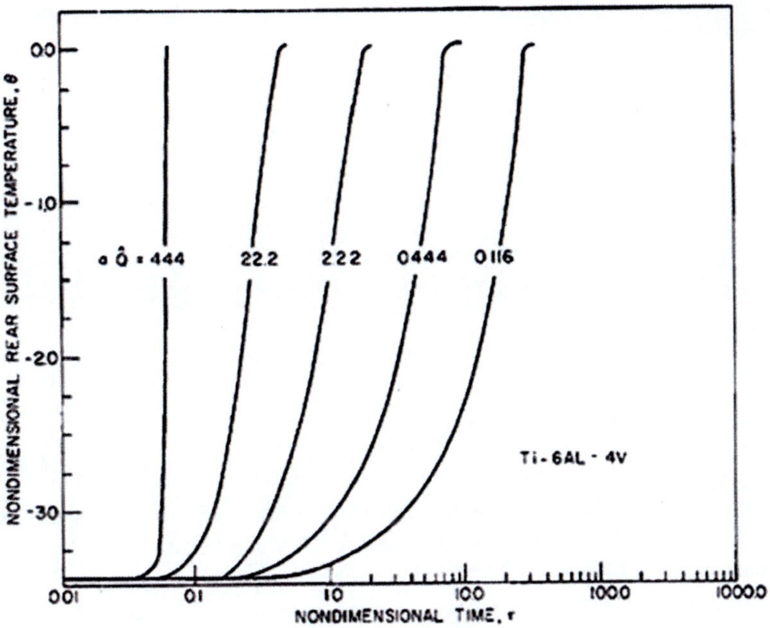

Fig. 5.67 Rear surface temperature rise for titanium alloy [58]

phonon is only 0.01 eV, assuming $c_s = 10^5$cm/s (velocity of sound in the solid), but electrons near the Fermi level have energies of several eV; hence, when such electrons are scattered, only a small fraction of their energy can be given during an electron–phonon collision. In the present analysis, f is equal to 10^{-4} and assumed as a constant over successive collisions. The change of irradiance of the laser beam passing through a homogeneous medium (metal), as a function of distance, is given by $dI_z/dz = -\mu I_z(z)$, where μ is the absorption coefficient. The negative sign indicates the reduction in beam irradiance due to absorption as μ is a positive quantity. The absorption of the incident laser beam takes place in the z-axis. Integrating defined dI_z/dz, the intensity of the incident beam at any plane z inside the substrate is given by $I_z = I_0 \exp(-\mu z)$, where I_0 is the peak intensity of incident irradiance. The limit for a small section Δz wide at z, the energy absorbed.

$$\frac{dI_z}{dz} = \frac{d}{dz}(\mu I_0 \exp(-\mu z)) \tag{5.369(a)}$$

or:

$$I_z = -I_0 \frac{d}{dz} |f(z)| \tag{5.369(b)}$$

or:

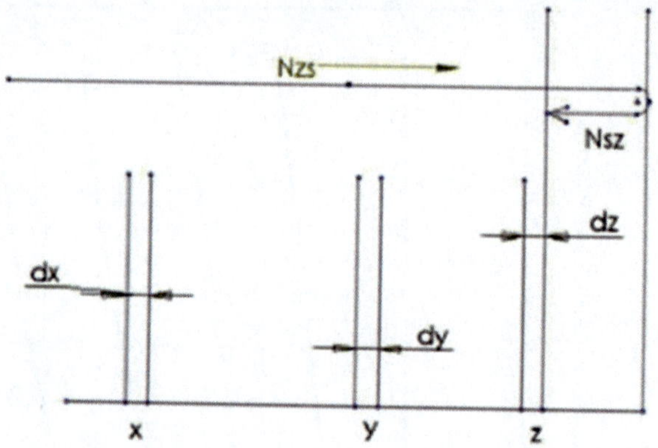

Fig. 5.68 Electron movement at the metal surface vicinity ($z = 0$ is the surface)

$$I_z = -I_0 f'(z) \tag{5.369(c)}$$

where $f'(z)$ is the absorption function. Since the mirror image situation is considered (Fig. 5.68) for electron movement at the surface, the laser beam will be absorbed in a manner described by: $-f'(z) = \frac{d}{dz}|\exp(-\mu|z|)|$ for all z. Using Eq. 5.376 for the intensity of the laser beam, the rate of applied external energy at dz during the time interval dt can be given as:

$$\Delta E_{z,t}|_{abz} = I_0 \mu e^{-\mu z} A dt dz \tag{5.370}$$

The total energy increases in the material at dz during dt is:

$$NA(E_{z,t+dt} - E_{z,t})dz = \Delta E_{z,t} + E_{z,t}|_{abs} \tag{5.371}$$

The total amount of energy, which is absorbed in an element $\partial\zeta$, area A, in time dt is $-I_0 A dt d\zeta f'(\zeta)$, since all the beam energy is absorbed in the z-direction. One must allow for the possibility that electron densities may vary throughout the material and, in particular, the number traveling from $d\zeta$ to dz may not be the same as that from dz to $d\zeta$. Therefore, the proportion of energy which is absorbed by the electrons which travel from $d\zeta$ to dz in dt is: $-I_0 A dt d\zeta f'(\zeta) \frac{N_{\zeta z}}{N_{\zeta z} + N_z \zeta}$

The average energy absorbed by one electron in $d\zeta$ in a time dt is: $-I_0 \frac{f'(\xi) d\xi}{(N_{\zeta z} + N_{z\zeta})\overline{V}_z}$, , and the total amount absorbed by this electron from dz to $d\zeta$ is: $\int_z^\zeta I_0 \frac{f'(\xi) d\zeta}{(N_{\zeta z} + N_{z\zeta})\overline{V}_z}$

The final temperature of the electrons in dz after the collision process can be readily found from the conservation of energy, i.e.:

Total electron energy after collision = total electron energy in during dt
− change of lattice site energy

The assumption that all directions of travel are equally probable gives $N_{\zeta z} = N_{z\zeta} = \frac{N}{6}$where N is the number of free electrons per unit volume; therefore, we can write:

$$\begin{aligned}\frac{\partial}{\partial t}(\rho c\Phi(z,t)) = &\int_{-\infty}^{\infty}\frac{fk}{\lambda^3}\exp\left(-\frac{|z-\zeta|}{\lambda}\right)\Theta(\zeta,t)d\zeta\\ &-\int_{-\infty}^{\infty}\frac{fk}{\lambda^3}\exp\left(-\frac{|z-\zeta|}{\lambda}\right)\Theta(z,t)d\zeta\\ &+\int_{-\infty}^{\infty}\frac{I_0 f}{\lambda^2}\exp\left(-\frac{|z-\zeta|}{\lambda}\right)\int_z^{\zeta}f'(z)d\xi d\zeta\end{aligned} \tag{5.372}$$

and:

$$\begin{aligned}&\int_{-\infty}^{\infty}\frac{fk}{\lambda^3}\exp\left(-\frac{|z-\zeta|}{\lambda}\right)[\Theta(\zeta,t)d\zeta - f\Phi(z,t)d\zeta]\\ &=\int_{-\infty}^{\infty}\frac{fk}{\lambda^3}\exp\left(-\frac{|z-\zeta|}{\lambda}\right)(1-f)\Theta(\zeta,t)d\zeta\\ &+\int_{-\infty}^{\infty}(1-f)\exp\left(-\frac{|z-\zeta|}{\lambda}\right)\int_z^{\zeta}f'(z)d\xi d\zeta\end{aligned} \tag{5.373}$$

Equations 5.372 and 5.373 are of interest to laser machining. The method of solution to be used in the following analysis is the transformation of the simultaneous differential integral Eqs. 5.372 and 5.373 using the Fourier integral transformation, with respect to A. The resulting ordinary differential equations may then be handled much more conveniently. The Fourier transformation of a function $f(z)$ is defined in (Yilbas 2001) [57] by:

$$F[f(z)] = \int_{-\infty}^{\infty}\exp(-i\omega z)f(z)dz = F(\omega) \tag{5.374}$$

and the Fourier inversion is given as:

$$f(z) = \frac{1}{2\pi}\int_{-\infty}^{\infty}F(\omega)[\exp(-\omega z)d(\omega)] = F(\omega) \tag{5.375}$$

The Fourier transformation of convolution integral $\int_{-\infty}^{\infty}f(\xi)g(z-\zeta)d\zeta$ is the product of the transforms $F(\omega)\overline{g}(\omega)$, and the transform of function $\exp\left(\frac{|z|}{\lambda}\right)$ is $\frac{2\lambda}{1+\omega^2\lambda^2}$. Applying Fourier transformation to Eqs. 5.370 and 5.371 yields:

$$\left(f+\omega^2\lambda^2\right)\frac{\partial}{\partial t}(\rho c\Phi)=-\omega^2 kf\overline{\Theta}+I_0\delta f\frac{2\delta}{\delta^2+\omega^2} \tag{5.376}$$

The multiplication in the transform domain by $(i\omega)^2$ corresponds to a second-order differential in the real plane. Hence, the inversion of Eq. 5.376 gives:

$$\left(f-\frac{\lambda^2\mu^2}{\partial z^2}\right)\rho c\frac{\partial\Phi}{\partial t}=kf\frac{\partial^2\Phi}{\partial z^2}+I_0\mu f\exp\left(-\mu|z|\right) \tag{5.377}$$

If the term $\left(\frac{\lambda^2\mu^2}{f\partial z^2}\right)\rho c\frac{\partial\Phi}{\partial t}$ is neglected for all f values, Eq. 5.375 becomes:

$$\rho c\frac{\partial\Phi}{\partial t}=k\frac{\partial\Phi}{\partial z^2}+I_0\mu\exp\left(-\mu|z|\right) \tag{5.378}$$

which is the same as a Fourier heat conduction equation, shown in Eq. 5.182. It is apparent that the electron kinetic theory equations for the heat conduction process are much more general than the Fourier equation. The new model of the conduction process is valid in regions close to the surface where an absorption process takes place, and therefore, the temperature profiles, which are obtained in these regions, can be expected to be valid.

One further advantage of this new approach is that the problem is completely specified, together with spatial boundary conditions, by the final equations, and that these equations can be solved using the method of Fourier transformation.

The Eq. 5.383 in the transform plane for the Gaussian intensity pulse is:

$$\Phi(z,t)=\frac{AI_0\mu}{\rho c}\left[\frac{1}{\beta\left(1-\mu^2\lambda^2\right)+\kappa\mu^2}\right]\left\{\begin{array}{l} i\mu\sqrt{\frac{\kappa-\beta\lambda}{\beta}}\exp\left[-\frac{\beta\kappa t}{\kappa-\beta\lambda^2}\right]\exp\left[i|z|\sqrt{\frac{\beta}{\kappa-\beta\lambda^2}}\right] \\ \qquad erfc\left[\frac{|z|}{2\sqrt{\kappa t}}+i\sqrt{\frac{\beta\kappa t}{\kappa-\beta\lambda^2}}\right] \\ -\exp\left[i|z|\sqrt{\frac{\beta}{\kappa-\beta\lambda^2}}\right]erfc\left[\frac{|z|}{2\sqrt{\kappa t}}+i\sqrt{\frac{\beta\kappa t}{\kappa-\beta\lambda^2}}\right] \\ +\exp\left(\kappa\mu^2 t+\mu|z|\right)erfc\left[\mu\sqrt{\kappa t}+\frac{|z|}{2\sqrt{\kappa t}}\right] \\ -\exp\left(\kappa\mu^2 t-\mu|z|\right)erfc\left[\mu\sqrt{\kappa t}-\frac{|z|}{2\sqrt{\kappa t}}\right] \\ -2\exp\left[-\frac{\beta\kappa t}{\kappa-\beta\lambda^2}+\mu|z|\right] \end{array}\right\} \tag{5.379}$$

and gives the closed form result of the electron kinetic theory approach, which graphically is shown in Fig. 5.69.

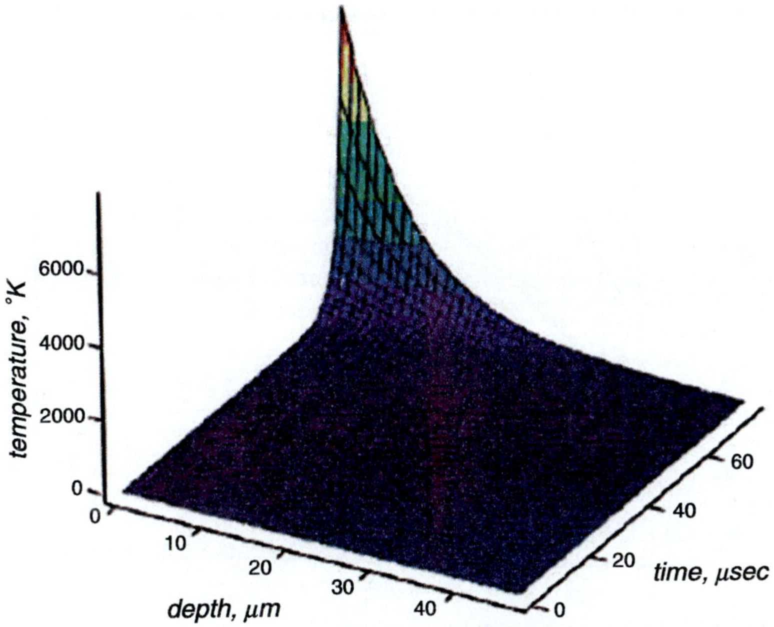

Fig. 5.69 Temperature, closed form solution of the electron kinetic theory approach, with respect to depth of the sample and time

5.10 Comparison of Fourier and Kinetic Theory

Comparison of analytical results of the Fourier relation given by Eq. 5.226, and kinetic theory Eq. 5.320, shows that the two are identical when the following two conditions are met: $\mu^2\lambda^2 \ll 1$ and $\kappa \gg \beta\lambda^2$. For most materials, μ and λ are of the same order of magnitude. Since μ is of the order of 10^{-5} and λ is of the order of 10^{-8}, then $\beta \gg 10^{11}$. This scenario corresponds to laser micromachining for laser pulses with picosecond rise times. Generally, $\beta \gg \frac{\kappa}{\lambda^2}$, and Eq. 5.386 reduces to exactly the analytical solution obtained from the Fourier theory for the exponential pulse, provided that $I \gg \mu^2\lambda^2$. When "β " approaches 0, the pulse solution reduces to that for a constant intensity analytical solution of Fourier heating Eq. 5.235. The comparison of the temperature gradient $\partial\Phi/\partial z$ predicted from the Fourier and electron kinetic theories with the distance in the z-axis for two pulse lengths is shown in Fig. 5.70. The temperature gradients predicted from both theories are similar for the long pulse length $t_{\text{pulse}} = 6 \times 10^{-10}$. In general, the temperature gradient decreases sharply in the surface vicinity to reach its minimum.

As the distance from the point of minimum increases further inside the substrate, $\partial\Phi/\partial z$ increases gradually. The sharp decrease of $\partial\Phi/\partial z$ in the surface vicinity is due to the rapid increase of the temperature in this region. In this case, the energy absorbed by the electrons in the surface vicinity is converted into the internal energy

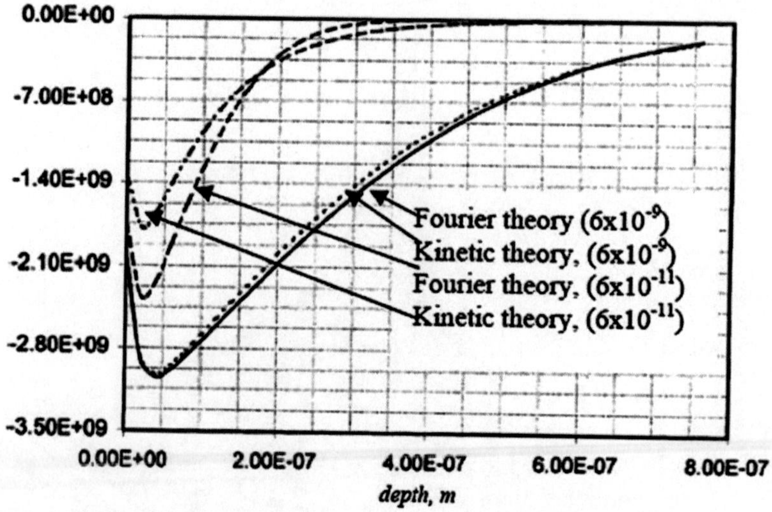

Fig. 5.70 ***dT/dz*** predicted from the Fourier and kinetic theory, along the –axis for two pulse length: 6×10^{-9} s and 6×10^{-11} s (Yilbas 2001) [57]

gain of the substrate through collision process. This gives rise to a sharp increase of the lattice site temperature. The energy balance attains among the absorbed energy internal energy gain and the conduction process at the point of minimum $\partial\Phi/\partial z$. As the distance increases beyond the point of minimum, the gradual increase in $\partial\Phi/\partial z$ reveals that the conduction effect due to phonon relaxation dominates. However, as the pulse length reduces, less $t_{pulse} < 6 \times 10^{-10}$s s, the temperature gradients predicted from both theories differ considerably. The temperature gradient predicted from the Fourier theory reduces significantly as compared to its counterpart predicted from electron kinetic theory. The difference in $\partial\Phi/\partial z$ predicted from both theories is due to the temperature response of the material for a short laser heating pulse as indicated earlier.

In summary, one can deduct a following conclusion: Temperature profiles predicted from Fourier heating model with kinetic theory approach for a one-dimensional model were compared in this study. In general, temperature profiles for one-dimensional Fourier heating case and kinetic theory approach are very similar. Although the Fourier heating model fails to predict correct temperatures for the short pulse heating case, the equilibrium time exists for a given material, and the balance occurs between the internal energy gain due to laser irradiation and conduction losses. The analytical solution of the electron kinetic theory approach obtained for the exponentially decaying pulse reduces to a step input intensity solution when β approaches 0. Moreover, the closed form solution of the electron kinetic theory for the step point intensity approaches the Fourier solution when $\mu^2 \frac{\lambda^2}{f} \leq 1$. The electron kinetic theory results deviate considerably from the Fourier theory results when $\beta \gg 10^{11}$.

The difference in temperature profiles occurs because of the fact that the electrons in the surface vicinity absorb the incident laser energy and the excited electrons do not make sufficient collisions with the lattice site atoms to transfer their excess energy in the surface region. Thus, lattice site temperature in this region becomes lower than the electron temperature as evident from Fig. 5.70, in which the electron temperature distribution inside the substrate is shown. Therefore, the Fourier theory fails to predict the temperature rise in the surface vicinity accurately for heating time of $t_{\text{pulse}} < 6 \times 10^{-10}$s.

5.11 Finite Difference Methods

In this section, the solution of governing equation is accomplished using an explicit finite difference approximation and the corresponding boundary conditions. The representative results for computational investigations are presented and discussed. To solve general governing equation, Eq. 5.124, finite difference method (FDM) was used, and the approximations to model the governing equation were based on a forward-difference in time and central-difference in space. For more detailed information, please refer to Nowakowski [40].

5.12 Effects of Pulsed Wave Laser Radiation

In this section and related sub-sections, we are going to deal with power levels of pulsed laser, material vaporization effects, and finally the effects from absorption of radiation in the plume which we did touch before.

5.12.1 Power Levels of Pulsed Wave Laser

Highly intense pulses of short duration can be produced in a variety of ways. Typically, it is done by creating a large population inversion by the injection of electrical energy from the discharge of large, highly charged capacitors. In these systems, all the energy is produced in a burst, the duration of which can be made quite short. We shall not discuss the various techniques by which these pulses are created, but in Table 5.5, we simply note some commonly obtained values (Ready) [2].

From the above table, it is apparent that with beam areas of the order of 1 cm^2, extremely high power densities can be obtained, and, although the pulse lengths are short, the total energy in each pulse is considerable. The available power densities range as high as 10^{12} W/cm^2.

Table 5.5 Some commonly obtained values for different pulsed laser

Laser type	Pulse length	Power	Energy per pulse
Ruby (normal mode)	0.1 ms	1–100 kW	1–50 J
Ruby (Q-switched)	10^{-8} s	1–10 GW	1–10 J
Ruby (mode-locked)	10^{-11} s	0.1–1 TW	0.1–1 J
CO_2 TEA	10×10^{-6} s	100 MW	100 J
CO_2 e-beam	2×10^{-5} s	50 MW	1000 J
CO_2 shock tube gdl	3×10^{-4} s	0.3 MW	100 J

Practically speaking, one is usually interested only in power densities below the breakdown threshold of air because at higher power densities, the energy never reaches the target. These breakdown levels are functions of wavelength, spot size, and pulse length and depend as well on the contaminants in the air. Typical values are 10^9 W/cm^2 in "clean" air at standard temperature and pressure (STP) for CO_2 laser pulses with a duration of about 10^{-6} s and longer. At shorter pulse lengths, the threshold is somewhat higher, becoming 10^{10} W/cm^2 at 10^{-8} s and 10^{11} W/cm^2 at 10^{-10} s. In the infrared region, the breakdown threshold scales with the square of the frequency.

5.12.2 *Material Vaporization Effects*

We shall first discuss the effect of high-power-density laser pulses on materials from the point of view of target vaporization and shall assume that the vaporizing surface is not shielded from the radiation by the vapor. In this case we can show that, in addition to thermal input to the target, there is a strong pressure built up on the target surface due to recoil from the blowoff of the vapor. The integral of this pressure over the time of the laser pulse imparts a net impulse to the target. There arises then the possibility of inducing stresses large enough to create gross mechanical changes, such as spall and deformation, by pulsed laser irradiation.

To calculate the pressure applied to a surface by a laser pulse, we start with a consideration of the vaporization process. We use a one-dimensional calculation because in most cases of interest, the beam radius R is larger than the thermal diffusion length during the pulse time t_p, or:

$$R \gg 2\sqrt{\kappa t_p} \tag{5.380}$$

We shall avoid consideration of thin targets, so that ℓ is also large compared to the diffusion length. In this case we can calculate the time t_b required for the front surface to reach the vaporization temperature T_v from the semi-infinite slab result of Eq. 5.160, which is:

$$T_v = \frac{2F_0}{K}\sqrt{\frac{\kappa t_b}{\pi}} \tag{5.381(a)}$$

or:

$$t_b = \frac{\pi K^2 T_v^2}{4F_0^2 \kappa} \tag{5.381(b)}$$

or, since $\kappa = \frac{K}{\rho c}$, then Eq. 5.388(b) yields:

$$t_b = \frac{\pi}{4}\frac{K\rho c T_v^2}{F_0^2 \kappa} \tag{5.381(c)}$$

In applying Eq. 5.160 in this way, we ignore the molten layer and assume that the values of K, ρ, and c appropriate to the solid can be used. This is not as gross an approximation as it may seem, because at these power densities, the molten layer is very thin.

Once the material on the surface reaches the boiling point, the surface begins to erode at a rate U_s given by energy consideration, as we saw in Sects. 5.7 and 5.8 and their subsequent sub-sections:

$$U_s = \frac{F_0}{\rho_s\left[c_{\text{solid}}T_m + L_m + c_{\text{liquid}}(T_v - T_m) + L_v\right]} \tag{5.382}$$

To simplify the calculation, we take $c_{\text{solid}} = c_{\text{liquid}} = c$ and ignore L_m by comparison to L_v, and then Eq. 5.389 reduces to:

$$U_s = \frac{F_0}{\rho_s(L_v + cT_v)} \tag{5.383}$$

Here we have used ρ_s, for the density of the solid. So, after the time t_b given by Eq. 5.390, the surface begins to evaporate, and it recedes at the rate U_s. By conservation of momentum, it must be true that:

$$U_v\rho_v = U_s\rho_s \tag{5.384}$$

where ρ_v and U_v designate the density and velocity, respectively, of the evaporation products. Thus, we have:

$$U_v\rho_v = \frac{F_0}{L_v + cT_v} \tag{5.385}$$

by combining Eqs. 5.383 and 5.385.

To see how the pressure exerted on the surface is related to density and velocity, note that the pressure on the surface is just the pressure of the evaporation products. To calculate this pressure, consider particles which move a distance Δz in time Δt under the pressure P and thereby acquire a velocity V. The pressure (force per unit area) must equal the rate of change of momentum (per unit area) so that:

$$P = \frac{(\rho \Delta z)V}{\Delta t} \tag{5.386}$$

That is, $\rho\Delta z$ is the mass per unit area which is brought to velocity V in time Δt by the force per unit area P. $V = \Delta z/\Delta t$, and so $P = \rho V^2$. Thus, in our specific case of density ρ_v and velocity U_v, there is an associated pressure, given by:

$$P = \rho_v U_v^2 \tag{5.387}$$

We could compute the pressure from this expression if we knew ρ_v and U_v. However, we only know the product $\rho_v U_v$ from Eq. 5.392. We need another relationship, which we simply take from the ideal gas law:

$$P = \rho_v \frac{R}{A} T_v \tag{5.388}$$

where R is the gas constant and A the molecular weight. Denote R/A by C', and use Equation 3-389; then:

$$\rho_v U_v^2 = \rho_v C' T_v \tag{5.389}$$

or:

$$U_v = \sqrt{C' T_v} \tag{5.390}$$

Upon combining Eqs. 5.384 through 5.388, we get the desired relationship:

$$P = \frac{F_0 \sqrt{C' T_v}}{L_v + c T_v} \tag{5.391}$$

Since the specific heat of metal is typically $3R/A$, we can approximate C' by $(1/3)c$ to yield:

$$P = \frac{1}{\sqrt{3}} \frac{F_0 \sqrt{c T_v}}{L_v + c T_v} \tag{5.392}$$

Finally, we compute the specific impulse delivered during the pulse, which is the force per unit area multiplied by the time over which it acts, and we get:

$$I_m = P\left(t_p - t_b\right) \tag{5.393}$$

or:

$$I_m = \frac{1}{\sqrt{3}} \frac{F_0\sqrt{cT_v}}{L_v + cT_v}\left(t_p - \frac{\pi}{4}\frac{K\rho cT_v^2}{F_0^2}\right) \tag{5.394}$$

A word about units is in order. It has become conventional to quote impulse in units of dyne-s and specific impulse in dyne-s/cm^2. If we use J/cm^2 for energy density, J/g.°C for specific heat, and J/g for heat of vaporization, we have:

$$\begin{aligned} I_m &= \frac{J/cm^2}{\sqrt{J/g}} \\ &= cm^{-2}\sqrt{J-g} \\ &= cm^{-2}\sqrt{10^7 erg \cdot g} \end{aligned} \tag{5.395}$$

or:

$$I_m = \sqrt{10^7}(dyne - s)/cm^2 \tag{5.396}$$

This unit, dyne-s/cm^2, is called a tap. Thus, in taps:

$$I_m = 1.83 \times 10^3 E_0 \frac{\sqrt{cT_0}}{[L_v + cT_v]}\left(1 - \frac{\pi}{4}\frac{K\rho cT_v^2}{E_0^2}t_p\right) \tag{5.397}$$

Note that Eq. 5.397 predicts a threshold value of D_F for impulse production at a given pulse length t_p. This is due to the criterion we introduced for vaporization; vaporization must commence before the end of the pulse, or there will be no impulse. The threshold is given by:

$$\frac{\pi}{4}\frac{K\rho cT_v^2}{E_0^2}t_p = 1 \tag{5.398}$$

or:

$$E_0|_{th} = \frac{\sqrt{\pi}}{2}T_v\sqrt{K\rho ct_p} \tag{5.399}$$

This vaporization model also predicates that, at very large E_0, the impulse per unit area is directly proportional to the energy density with a constant *coupling coefficient*, given by:

$$\left(\frac{I_m}{E_0}\right)_{\max} = 1.83 \times 10^3 \frac{\sqrt{cT_v}}{(L_v + cT_v)} \tag{5.400}$$

This is the limit at which vaporization begins essentially instantaneously with respect to the pulse length and vapor produces are produced for the entire pulse.

Some numerical values are illustrated below and in Fig. 5.71. E_0 is in J/cm^2 and t_p in μs, so that I_m is in taps:

For titanium:

$$I_m = 8.04 E_0 \left[1 - (6.23)\frac{t_p}{E_0^2}\right] \tag{5.401}$$

For aluminum:

$$I_m = 6.49 E_0 \left[1 - (33.9)\frac{t_p}{E_0^2}\right] \tag{5.402}$$

The above model illustrates the principals involved in generating impulse by laser vaporization. In fact, in predicting threshold values, it gives results which are within a factor of two of experimental measurements. It has been refined (Anisimov) [59] by a calculation which accounts for the fact that T_v is probably not a handbook value that comes from measurements at atmospheric pressure, but rather a different value appropriate to the dynamic and high-pressure situation created by the laser-induced vaporization. In this refinement, T_v is determined from the kinetic model of vaporization, which predicts that $U_s = c_a \exp\left[-L_v/(R'T_v)\right]$ where c_a is the speed of sound in the solid and R' the gas constant per gram. When this is done, the thresholds agree very well with theory. However, as E_0 (and hence F_0, since t_p is constant) is

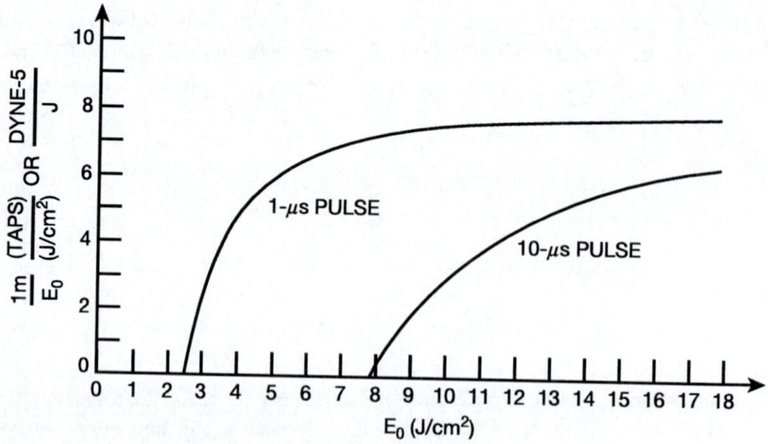

Fig. 5.71 Coupling coefficient vs. energy density for titanium

increased, experiments show that delivered impulse does not increase indefinitely but falls off. This is due to the onset of absorption of laser energy by the vapor products and/or the heated air near the target. We turn now to a consideration of this problem.

5.12.3 *Effects from Absorption of Radiation in the Plume*

The plume of vaporized material blown off the target becomes at some power density hot enough that it or the air begins to absorb the laser radiation. The onset of this process is not thoroughly understood, and the ignition of these so-called absorption waves is then subject of a great deal of research. Proper treatment of the problem depends on, among other things, computing the onset of ionization and the rate of absorption of light by the electrons and also accounting for both cascade processes and relaxation processes in a full dynamic sense. We shall not treat this problem here. Rather we shall look at some crude models which show, in a semi-quantitative way, various features of the absorption process.

First note that the decoupling of the absorption from the material surface due to shielding by the plume depends on the wavelength of the radiation. Recall from Eq. 5.100(a) that at the plasma frequency of the ν_p, reflectivity of a "free-electron" metal drops sharply from a value near unity to essentially 0. If we assume that the coupling to the plume is due to the light interacting with electrons, Eq. 5.100(a) is valid. Using the mass of the free electron gives:

$$\nu_p = 8.97 \times 10^3 N^{1/2} \tag{5.403}$$

For the plasma frequency ν_p in hertz, when N is in electrons per cubic centimeter, this can be rewritten in terms of the corresponding wavelength λ_p to yield:

$$N = \left(1.12 \times 10^{13}\right)/\lambda_p^2 \tag{5.404}$$

where λ_p is in centimeters. At a given wavelength, the plume is transparent until the electron density reaches the value given by Eq. 5.404, where there will be a transition to a condition in which the plume absorbs and reflects the radiation and thereby shields the material. For 10.6 μm CO_2 radiation, shielding begins at 10^{19} electrons per cm^3, for 1.06 μm (Nd laser) at 10^{21} electron per cm^3, and for 0.6943 μm (ruby laser) at 2.3×10^{21} electrons per cm^3.

When the electron density reaches a high enough value, the beam decouples from the surface, and presumably the pressure due to material blowoff will drop. To get some idea of the order of magnitude of the energy density for a given pulse length where this process begins, let us simply assume that the cutoff begins when the front surface reaches the temperature at which the material is fully ionized. 's should predict an upper limit, for full ionization is obviously not required. For example, solids have $N \approx 10^{23}\text{cm}^{-3}$, whereas we only require, at 10.6 μm, $N > 10^{19}\ \text{cm}^{-3}$. For simplicity, assume that melting and vaporization processes can be ignored, and again use for the front surface temperature raises the simple expression:

$$T = \frac{2F_0}{K}\sqrt{\frac{\kappa t}{\pi}} \tag{5.405}$$

A typical ionization temperature for a metal would be about 75,000 °C. Using simply the values of K and κ for the solid, we get for titanium:

$$F_0\sqrt{t} \approx 5.7 \times 10^4\ \text{Ws}^{1/2}/\text{cm} \tag{5.406}$$

Using $E_0 = F_0 t$, this can be rewritten as:

$$E_0 \approx 57\sqrt{t_{\mu s}} \tag{5.407}$$

where $t_{\mu s}$ is understood to be time in microseconds.

Fig. 5.72 shows some data that is taken by Dr. Rudder of the Air Force Weapons Laboratory, at two pulse lengths, 1.2 and 11 μs, with 1.06 pm radiation and titanium targets (Canavan et al.) [61].

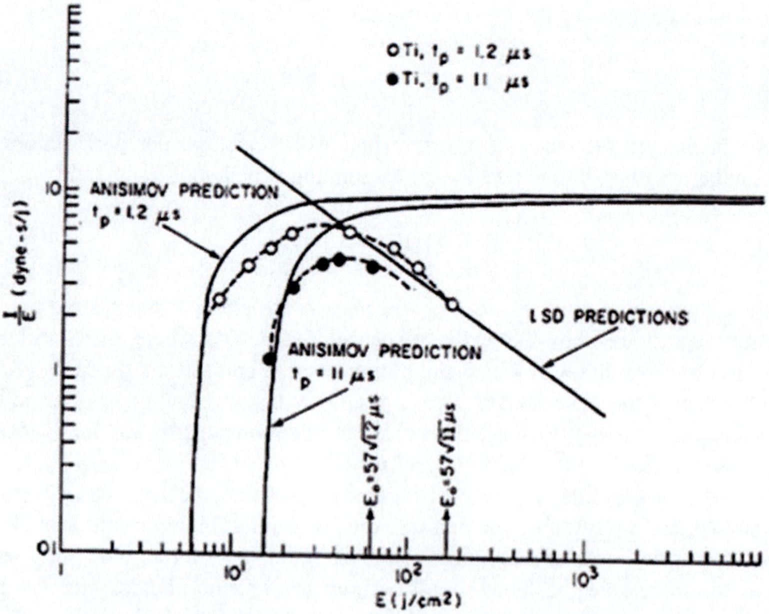

Fig. 5.72 Specific impulse delivered to solid targets by 1.06 μm laser radiation [61]

The lines marked "Anisimov predictions" are calculated from the vaporization model of the previous section with the refined method for determining T_v (this was first done in the Soviet Union by (Anisimov) [55]. The experimental data agree very well at values of E near threshold. Note that Eq. 5.414 for estimating the onset of shielding is roughly consistent with these data, although the experimental onset of shielding is, as one might expect, fairly gradual. The line on the graph marked LSD. Predictions refer to a theoretical estimate based on the idea that the laser light, when it couples into the blowoff, can create an explosion like shock wave in the air which travels up the beam, absorbing the radiation energy in the process. This laser-supported detonation (or LSD) wave is one form of laser-supported absorption wave. We will discuss these waves next.

Once the coupling of the radiation with the ejected vapor (and perhaps the air) reaches a sufficient level, the absorption region begins to behave in a fashion characterized by hydrodynamic dissuasion of the energy coupled into it. For now, let us ignore the ignition problem. The absorption region typically propagates up the laser beam in a way that is determined by the medium in which it propagates (usually air) and also by the balance between the power being fed in by the laser and the relaxation processes which dissipate the power. Three types of laser-supported absorption waves are usually identified. Typical power level at which they appear and their typical velocities of propagation are indicated below for a 10.6 μm radiation and targets in air at standard temperature and pressure (Nielson et al.) [61].

Type of wave	Power level of laser flux (W/cm^2)	Velocity of propagatio (cm/s^2)
Laser-supported detonation (LSD) wage	10^7	10^5
Laser-supported combusstion wave (LSCW)	10^4	10^3
Plasmatron	10^4	0

The LSD wave propagates as a shock wave, i.e., at supersonic velocity, whereas the LSC wave moves more slowly and relaxes by thermal conduction. The plasmatron is at rest, with the energy input being balanced by re-radiation and convective losses into the atmosphere. Although we discuss these effects here in the section on pulsed lasers, they are just valid for continuous radiation. Since pulsed lasers are the most convenient devices for reaching these power levels, especially for LSD waves, absorption waves are usually considered under pulsed effects.

Hydrodynamic theory can be applied to model these waves. The problem was first solved in the Soviet Union by Raizer [62]. Detonation waves can be discussed most. Readily because the hydrodynamic equations reduce to fairly simple expressions, we shall consider them in some detail. A few remarks about combustion waves will come later.

We can derive conditions for the steady-state behavior of a detonation wave by considering conservation of mass, momentum, and energy at the detonation front.

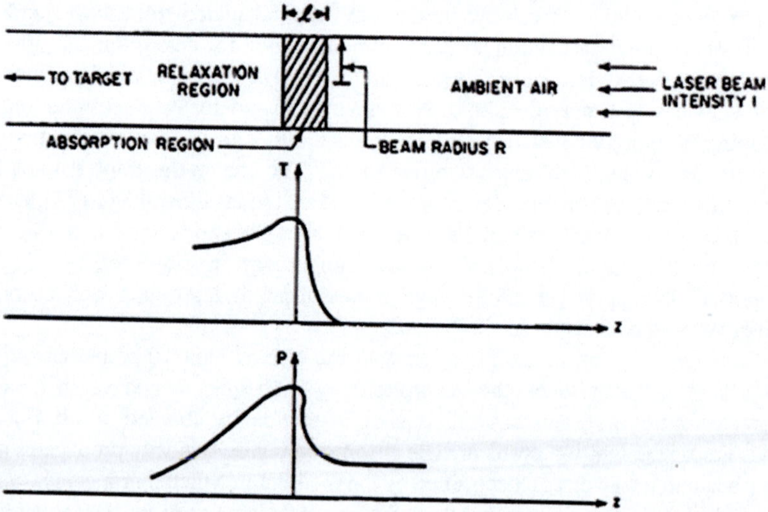

Fig. 5.73 Temperature and density profiles typical of a laser-supported detonation wave [60]

For this purpose, we do not concern ourselves with how the process starts but presume that a detonation wave has been formed and is propagating at some steady rate as sketched in Fig. 5.73. The absorption region is propagating to the right at a steady velocity u. We assume that it is very thin and can be treated as a detonation front. Thus u is the detonation velocity. The temperature, density, etc. of air go through very rapid changes in a very short distance. Note that this wave propagates, in this treatment in air, and thus our results will be independent of target material.

In this discussion "behind the front" refers to the high-temperature-and-pressure region immediately to the left of the absorption region in Fig. 5.73. "Ahead of the front" is to the right in the sketch and refers to ambient air conditions. Note that we have given the beam a finite radius R and thus will have to consider lateral expansion. First let us do the one-dimensional problem and assume that the detonation front propagates simply as a plane wave.

Behind the front, let ρ, P, and e be the density, pressure, and internal energy *per unit mass*, respectively, and ρ_0, P_0, and e_0 be the same variables ahead of the front. Define the velocities with respect to a coordinate system moving with the front at the detonation rate u. Then the ambient gas moves into the front with the speed of u, and we define v as the speed with which the high-pressure gases leave the front. We can now write down the conservation equations for mass, momentum, and energy across the detonation front. These equations are based on *flow*, that is, they are in the terms of "per unit mass area, per unit time." The equation for mass is:

$$\text{Mass} \quad \rho_0 u = \rho v \tag{5.408}$$

The conservation-of-momentum condition results from the impulse to the change in momentum. Now impulse is force multiplied by time, but in the "per unit area, per unit time" sense, this becomes simply pressure. Since mass, in this flow concept, is ρv, momentum is $(\rho v)v = \rho v^2$. Hence we have:

$$P - P_0 = -[(\rho v)v - (\rho_0 u)u] \tag{5.409}$$

and, if we ignore P_0, which is much smaller than $\underline{P}$, we have for momentum:

$$\text{Momentum} \quad P + \rho v^2 = \rho_0 u^2 \tag{5.410}$$

The conservation-of-energy condition follows from similar considerations. The difference in energy flow on each side of the front must be balanced by the work done on the gas ($P_0 u - Pv$) and the energy absorbed from the laser beam, which, using our earlier notation, is F. $F = AI$, where A is the absorptance of the gas in the absorption region. Thus, we have:

$$\rho u\left(e + \frac{1}{2}v^2\right) - \rho_0 u\left(e_0 + \frac{1}{2}u^2\right) = P_0 u - Pv + F \tag{5.412}$$

If we use $P_0 \approx 0$ and $e_0 \approx 0$ and substitute from Eq. 5.408, we get for energy:

$$e + \frac{1}{2}v^2 - \frac{1}{2}u^2 = -\frac{P}{\rho} + \frac{F}{\rho_0 u} \tag{5.412}$$

or:

$$\text{Energy} \quad e + \frac{P}{\rho} + \frac{1}{2}v^2 = \frac{1}{2}u^2 + \frac{F}{\rho_0 u} \tag{5.413}$$

Our goal is to use these conservation laws to predict the pressure P behind the front and ultimately the pressure transmitted to the target. For now, assume that F is known, and, of course, the ambient air density ρ_0 is known. Thus, we have three equations and five unknown, P, ρ, v, e, and u. To proceed we need to invoke some equation of state for the gases, and we shall simply assume that the ideal gas law holds. Thus, we have:

$$P = \rho R'T \tag{5.414}$$

where R' is the gas constant per unit mass, or $R' = R/\mathfrak{M}$, where $\mathfrak{M}$ is the molecular weight. Since the wave is presumed to be in air in this treatment, $\mathfrak{M}$ would be the average molecular weight of air. Taking $\mathfrak{M}$ for air to be 29.4/mol gives $R' = 2.84 \times 10^6$erg/(g. °C) and is consistent with the ideal gas law and with $\rho_0 = 1.29 \times 10 - 3$ g/cm^3 for air at 0 °C and 1 atmosphere = 10^6 dyne/cm^2. Now

Eq. 5.421 essentially introduces another unknown, the temperature T, so we need to add the expression for the energy of an ideal gas, which is:

$$e = \frac{R'T}{\gamma - 1} = \frac{P}{(\gamma - 1)\rho} \tag{5.415}$$

where γ is the ratio of the specific heat, $\gamma = c_p/c_v$. For our purpose it is sufficient to take $\gamma = 1.4$.

Now we have four equations that are Eqs. 5.415, 5.417, and 5.420, and 5.422 in five unknowns that are P, ρ, v, e, and u. To get the final condition, we use the criterion for detonation, which is the velocity of the high-pressure gases behind the front, relative to the front, is equal to or greater than the local speed of sound. Intuitively this seems reasonable, for propagation of shock waves is, by definition, in excess of the speed of sound. The criterion can be properly derived from a consideration of the thermodynamics of the situation, but we shall not do so here; however you can refer to (Ya. B. Zel'dovich and Yu P. Raizer) [63]. Since we shall be interested in the minimum value of F (or I) required to sustain in detonation wave, we take v equal to the speed of sound. For an ideal gas, the sound speed is $(\gamma P/\rho)^{1/2}$, so we have our last condition as:

$$v^2 = \frac{\gamma P}{\rho} \tag{5.416}$$

Before discussing the algebra, let us collect the equations as it is written in Eqs. 5.408, 5.410, 5.413, 5.424, and 5.416:

$$\text{Mass} \quad \rho_0 u = \rho v \tag{5.408}$$

$$\text{Momentum} \quad P + \rho v^2 = \rho_0 u^2 \tag{5.410}$$

$$\text{Energy} \quad e + \frac{P}{\rho} + \frac{1}{2}v^2 = \frac{1}{2}u^2 + \frac{F}{\rho_0 u} \tag{5.413}$$

$$\text{Energy of ideal gas} \quad e = \frac{P}{(\gamma - 1)\rho} \tag{5.424}$$

$$\text{Velocity behind shock} \quad v^2 = \frac{\gamma P}{\rho} \tag{5.416}$$

Combine Eqs. 5.408 and 5.410 to yield expressions for u^2 and v^2:

$$u^2 = \frac{P\rho}{\rho_0(\rho - \rho_0)} \tag{5.417}$$

$$v^2 = \frac{P\rho}{\rho(\rho - \rho_0)} \tag{5.418}$$

Now use Eqs. 5.414 and 5.416 to eliminate v^2 and ρ so that we can get:

$$\frac{\rho}{\rho_0} = \frac{1+\gamma}{\gamma} \tag{5.419}$$

which is one of the equations we need, namely ρ in terms of the known quantities ρ_0 and γ. Now we can use Eq. 5.410 to get P in terms of u by eliminating v^2 with Eq. 5.416 to get:

$$P = \frac{\rho_0 u^2}{1+\gamma} \tag{5.420}$$

We need one more relation to complete the solution, namely, u in terms of F. This will, by Eq. 5.420, give us P in terms of F. To get this, we use Eq. 5.413 and replace e via Eq. 5.415. Thus Eq. 5.413 becomes:

$$\frac{P}{(1+\gamma)\rho} + \frac{P}{\rho} + \frac{1}{2}\frac{\gamma P}{\rho} = \frac{1}{2}u^2 + \frac{F}{\rho_0 u} \tag{5.421}$$

If we use Eq. 5.420 to eliminate P, we get:

$$\frac{u^2}{1+\gamma}\left[\frac{1}{(\gamma-1)}\frac{\rho_0}{\rho} + \frac{\rho_0}{\rho} + \frac{1}{2}\frac{\gamma\rho_0}{\rho}\right] = \frac{1}{2}u^2 + \frac{F}{\rho_0 u} \tag{5.422}$$

or:

$$\left(\frac{\rho_0}{\rho}\right)\left(\frac{u^2}{1+\gamma}\right)\left[\frac{\frac{1}{2}\gamma(1+\gamma)}{\gamma-1}\right] = \frac{1}{2}u^2 + \frac{F}{\rho_0 u} \tag{5.423}$$

or:

$$\frac{1}{2}\frac{\rho_0}{\rho}u^2\left(\frac{\gamma}{\gamma-1}\right) = \frac{1}{2}u^2 + \frac{F}{\rho_0 u} \tag{5.424}$$

If we use Eq. 5.419 for ρ/ρ_0, we get:

$$\frac{1}{2}u^2\left(\frac{\gamma^2}{\gamma^2-1}\right) = \frac{1}{2}u^2 + \frac{F}{\rho_0 u} \tag{5.425}$$

Finally, we arrive at:

$$u^3\left(\frac{1}{\gamma^2-1}\right)=\frac{2F}{\rho_0} \tag{5.426}$$

or:

$$u=\left\{\frac{2(\gamma^2-1)F}{\rho_0}\right\}^{1/3} \tag{5.427}$$

The equations which represent the solution for the detonation wave, then, are:

$$\frac{\rho}{\rho_0}=\frac{1+\gamma}{\gamma} \tag{5.428}$$

$$P=\frac{\rho_0 u^2}{1+\gamma} \tag{5.429}$$

$$u=\left[\frac{2(\gamma^2-1)F}{\rho_0}\right]^{1/3} \tag{5.430}$$

These three equations, together with the ideal gas law, represent the formal solution to the propagation of the laser-supported detonation wave. Given the temperature behind the front, and since Eq. 5.428 defines ρ, we could calculate P and hence u and finally the F required to support it. However, this does not really solve the problem. What we wish to discover is: given the laser intensity I, will an LSD wave be supported? To answer this question, we need to consider the distance it takes for the laser radiation to be absorbed. We also need a more realistic situation than the simple plane wave.

First we note that the beam has a finite radius R and that lateral expansion can take place. The order of magnitude of the radial expansion velocity will be the speed of sound c_a. To maintain the detonation, we must replace the energy lost to expansion by energy put into the absorption region. To simplify, let us assume all of the laser beam energy is absorbed in the distance ℓ. (Actually, the beam intensity only falls by $1/e$ in the distance ℓ.) We define Δt as the time for the shock front to move a distance ℓ, or $\Delta t=\ell/u$. In this time the radial expansion is the amount $c_a\Delta t$. Now $\pi R^2 I\Delta t$ is the energy deposited by the beam in the cylindrical volume shown in Fig. 5.74 ($c_a\Delta t \ll R$). But the energy in this volume after expansion is approximately equal to its volume multiplied by its internal energy p: unit volume. Thus:

$$\pi R^2 I\Delta t \approx \rho_0 e\left[\pi R^2\ell+2\pi R c_a\Delta t\ell\right] \tag{5.431}$$

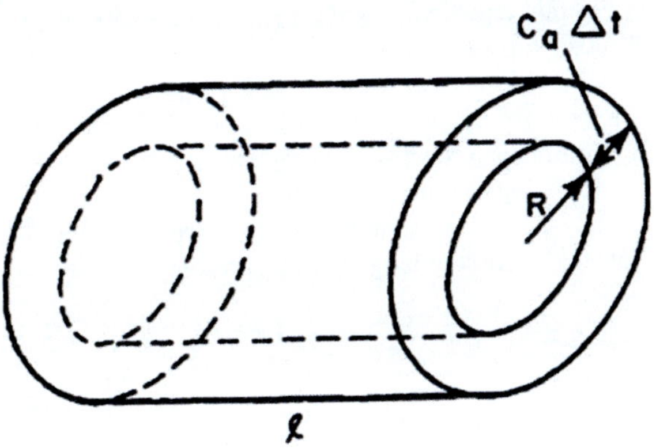

Fig. 5.74 Cylindrical volume which absorbs beam energy via expansion

Since $\Delta t = \ell/u$:

$$I\pi R^2 \approx \rho_0 eu\left[\pi R^2 + 2\pi R c_a \frac{\ell}{u}\right] \tag{5.432}$$

Note that $\gamma \approx 1.4$. Eq. 5.435 predicts that $\rho \approx 2\rho_0$, and hence, using $\rho v = \rho c_a \equiv \rho_0 u$, $u \approx 2c_a$. So, we have, after some algebra:

$$\rho_0 eu \approx \frac{I}{1 + \ell/R} \tag{5.433}$$

But this equation simply represents the rate at which we must put energy into the absorption volume in order to maintain the conditions we assume to carry out our detonation wave calculation, namely, a plane wave propagating by absorption of laser energy in a distance ℓ. Thus, the energy flow per unit area from the laser beam is:

$$F = \frac{I}{1 + (\ell/R)} \tag{5.434}$$

Finally, we can complete the problem if we know the absorption length ℓ. To compute ℓ we need to invoke some model of the ionized air. For this purpose, it is sufficient to assume that free electrons absorb the light and that the electrons come from singly ionized atoms. We shall not derive the expressions which we need but simply quote them. There are two relationships. The first of these is that which is known as Saha equation [63] (Eq. 5.424) which relates the fraction of atoms ionized α to the absolute temperature $\mathcal{T}$ and the ionization potential $\mathcal{J}$ of a single atom:

$$\frac{\alpha^2}{1-\alpha} = 2\frac{g_1}{g_0}\frac{m}{\rho}\left(\frac{2\pi m_0 kT}{h^2}\right)^{3/2} e^{-\mathcal{J}/kt} \tag{5.435}$$

In this equation $_m$ is the mass of the atom, m_0 is that of the electron, k is the Boltzmann constant, and h is Plank's constant. The statistical weights of ground state of the atom and its first ionized state are g_0 and g_1, respectively. Typically, $g_1 = g_0 = 1$. In terms of known constants, then, the Saha equation gives us the degree of ionization as a function of temperature.

Knowing the degree of ionization, we can get the absorption length. Again, we simply quote the relationship [60], which assumes that the light is absorbed by inverse bremsstrahlung. The expression is:

$$\frac{1}{\ell} = \frac{4}{3}\left(\frac{2\pi}{3m_0kT}\right)^{1/2} \frac{e^6h^2}{m_0c(h\nu)^3}\left(\frac{\rho^2\alpha^2}{m^2}\right)\left(1 - e^{-h\nu/hT}\right) \tag{5.436}$$

which, at the temperature of interest and for 10.6 μm radiation, becomes (for $h\nu/hT \ll 1$):

$$\frac{1}{\ell} = \frac{4}{3}\left(\frac{2\pi}{3m_0}\right)^{1/2} \frac{e^6h^2}{m^2m_0c(h\nu)^2} \cdot \frac{\rho^2\alpha^2}{(kT)^{3/2}} \tag{5.438}$$

In Eq. 5.437, ν is of course the frequency of the laser radiation, c is the speed of light, and e is the electronic charge. By combining Eqs. 5.436 and 5.437, we can calculate ℓ in terms of temperature T, density ρ, and known parameters. A typical value of $\mathcal{I}$ for air is of the order of 13 eV. For O_2, $\mathcal{I}$ is 12.1 eV; for N_2 it is 15.6 eV; thus we have ℓ as a function of ρ and T as:

$$\ell = f(\rho, T) \tag{5.438}$$

or, since $\rho = \rho_0(1 + \gamma)/\gamma \approx 2\rho_0$, we can get a relationship between and ℓ and hence between, via Eq. 5.428 and Eq. 5.432. Typically, results are shown in Fig. 5.75.

$_I$ and T

Here we have assumed a beam radius of 10 cm. The important point is that there is a *minimum* in the I vs. T relation. We identify this as the minimum flux I_m required to maintain LSD wave. Associated with it is the temperature T_m of the high-pressure region at the detonation front.

We can then use our detonation wave relationships, Eqs. 5.428 and 5.430 as well Eq. 5.419, to get the pressure behind the front, or equivalently, and get pressure from T_m via $\rho \approx 2\rho_0$ and ideal gas law.

We shall turn to a calculation of the pressure on the target in a moment. First note that the radial expansion concept imposes a natural criterion for the difference between a combustion wave and a detonation wave. The time for radial expansion is R/c_a, whereas the time for passage of the absorption region is V/u. If the detonation condition is to be maintained, radial expansion times must e larger than propagation times in order for the high-pressure region to move as a shock front and not dissipate itself radially. Hence $R/c_a > \ell/u$. We have already noted that $u \approx 2c_a$ crudely, $u \approx c_a$,

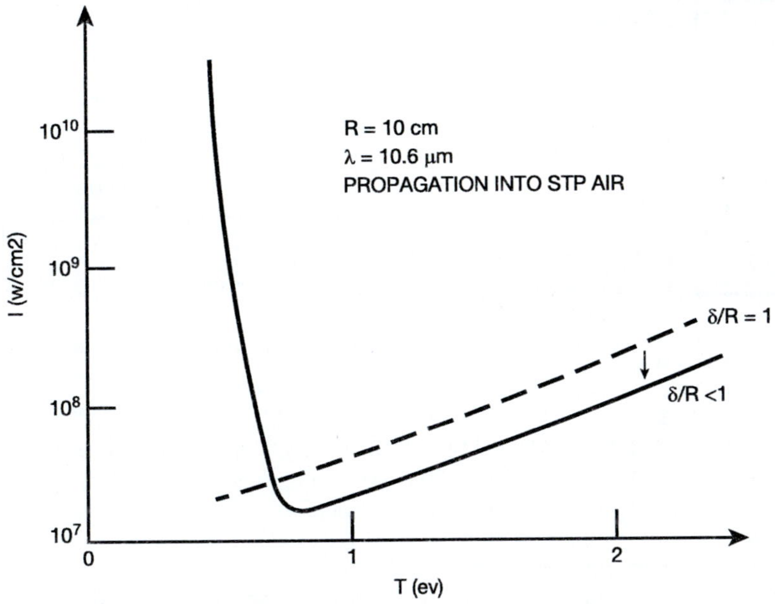

Fig. 5.75 Relationship between $_I$ and T for a laser-supported detonation wave [60]. This plot is based on a more realistic expression for the equation of state of the gases than the ideal gas law, but the ideal gas law gives a similar result

so that $R > \ell$, or $\ell/R < 1$, is the condition for detonation waves. If ℓ becomes larger than R, the absorption region is large, the relaxation in the radial direction is important, and the process called a combustion wave takes place. This can be treated in a similar fashion to the detonation wave, but the hydrodynamic equations do not take the simple form of Eqs. 5.408, 5.410, and 5.412. We shall not treat combustion waves in this book. The solution in Fig. 5.75 is for a detonation wave and hence is valid for $\ell/R < 1$. The limit $\ell/R = 1$ is shown in the figure by a dashed line.

Finally, compute the impulse delivered to target by a laser beam of intensity I just sufficient to maintain a detonation wave [61]. The beam has a pulse duration t_p. We wish to calculate that on the target due to the "explosion products" behind the absorption region. These, of course, expand in all directions and create a pressure on the target. To demonstrate the effect, we shall use a very simple model, namely, a model of cylindrical expansion. We consider that the absorption region has propagated a distance Z by the end of the laser pulse, and at that time, we have created a cylinder of high-pressure gas which has a radius equal to the beam radius R, a length Z, and a pressure P_d given by Eqs. 5.429 and 5.430, with $P = P_d$. This cylinder is then allowed to expand radially at a speed estimated to be the speed of sound c_a. Then we get the impulse delivered to the target by integrating the force on the target due to the pressure in the expanding cylinder during the time the cylinder expands from R to the target radius R_T. For R_T is very large, the integration is stopped when the cylinder pressure drops to atmospheric pressure. The model is sketched in Fig. 5.76 at the time $t = t_p$.

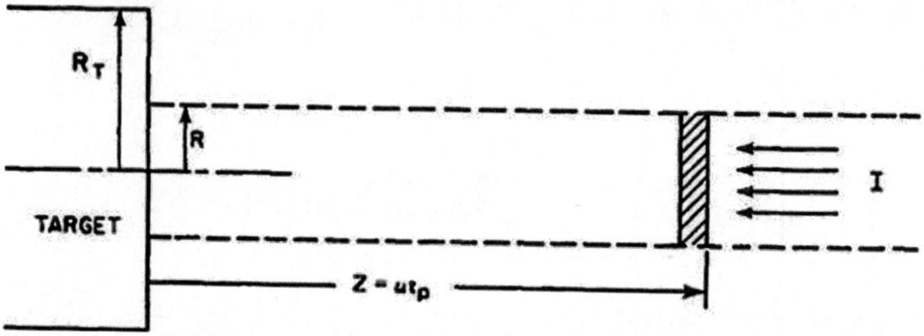

Fig. 5.76 Radial expansion model for impulse delivered to a target from a laser-supported detonation (LSD) wave

The model might be expected to be valid if $Z \gg R$ and if $t_p <<$ the time required for radial expansion, either to R_T or to atmospheric pressure, to take place. We also are assuming that impulse due to target vaporization is negligible, i.e., that the detonation wave is formed very early in the laser pulse.

We take the radial expansion to be at constant temperature. Then P times the volume O of the cylinder is a constant. Since our model presumes only cylindrical expansion, we have the condition that:

$$Pr^2 = \text{Constant} \tag{5.439}$$

where r is the radius of the cylinder and $R \leq r \leq R_T$. We shall need this relationship in the derivation f the impulse. Let the impulse be I'_m, and let F be the force on the target due to the pressure. This gives the following:

$$I'_m = \int_{t(r=R)}^{t(r=R_T)} Fdt \tag{5.440}$$

where the upper limit is understood to be valid only where P is greater than the atmospheric pressure at $r = R_T$. Since the radial expansion rate is c_a, from the ideal gas relation for the speed of sound, we have:

$$c_a = \sqrt{\frac{\gamma P}{\rho}} \tag{5.441}$$

we have:

$$dt = \sqrt{\frac{\rho}{\gamma P}} dr \tag{5.442}$$

Thus:

$$I'_m = \int_R^{R_T} F\sqrt{\frac{\rho}{\gamma P}}dr \tag{5.443}$$

Now at any time:

$$F = \left(\pi r^2\right)P = \pi r^2 P \tag{5.444}$$

but, since r^2P is constant, then we can evaluate F from the initial pressure P_d at $r = R$, or:

$$F = \pi R^2 P_d \tag{5.445}$$

Now the impulse becomes, by taking F out of integral sign of Eq. 5.443:

$$I'_m = \pi R^2 P_d \int_R^{R_T} \sqrt{\frac{\rho}{\gamma P}}dr \tag{5.446}$$

again invoking $r^2P =$ constant gives:

$$r^2 P = P_d R^2 \tag{5.447}$$

or:

$$P = P_d \frac{R^2}{r^2} \tag{5.448}$$

So, the impulse is then:

$$I'_m = \pi R\sqrt{P_d} \int_R^{R_T} \sqrt{\frac{\rho}{\gamma}}rdr \tag{5.449}$$

Recall from Eq. 5.428 that $\rho = \rho_0(1 + \gamma)/\gamma$. This gives:

$$I'_m = \pi R\sqrt{\frac{P_d\rho_0(1+\gamma)}{\gamma^2}}\left(\frac{R_T^2}{2} - \frac{R^2}{2}\right) \tag{5.450}$$

Since $R < R_T$, we shall ignore the $R^2/2$ term. Thus, specific impulse I_m, which is I'_m divided by the area of the beam πR^2, is then:

$$I_m = \frac{I'_m}{\pi R^2} = \frac{R_T^2}{2R}\sqrt{P_d\rho_0}\frac{\sqrt{1+\gamma}}{\gamma} \tag{5.451}$$

Recalling the expression in Eqs. 5.429 and 5.430 for P_d, and ignoring ℓ/R with respect to unity, we can write:

$$I_m = \frac{R_T^2}{2R}\sqrt{\left(\frac{1+\gamma}{\gamma^2}\right)\rho_0\left(\frac{\rho_0}{1+\gamma}\right)\left[\frac{2(\gamma^2-1)}{\rho_0}I\right]^{2/3}} \tag{5.452}$$

This simplifies to:

$$I_m = \frac{R_T^2}{2R}\left\{\frac{1}{\gamma}\left[2(\gamma^2-1)\right]^{1/3}\right\}\rho_0^{2/3}I^{1/3} \tag{5.453}$$

The expression in braces is nearly equal to unity for the typical value of γ (say, $\gamma = 1.4$), so, Eq. 5.453 reduces to:

$$I_m = \frac{R_T^2}{2R}\rho_0^{2/3}I^{1/3} \tag{5.454}$$

Now the energy per unit area in the beam is $E_0 = It_p$, and we write the coupling coefficient I_m/E_0 as:

$$\frac{I_m}{E_0} = \frac{R_T^2}{2R}\rho_0^{2/3}/\left(t_p^{1/3}E_0^{2/3}\right) \tag{5.455}$$

This is the equation of the straight line marked LSD Predictions shown in Fig. 5.72, where the calculations were done for the parameters appropriate to the 1.2 μs pulse length.

Several important consequences of the LSD wave are seen in Eq. 5.444. One is that the coupling coefficient is reduced as E_0 becomes larger, which tells us that we cannot create an arbitrarily large impulse at a target by simply increasing the energy in the laser beam. In fact, when Eq. 5.455 is considered to be correct at high E_0 and the results of the vaporization model (see Eq. 5.397 and Fig. 5.72) are used at lower values of E_0, there is, for a given pulse length, an optimum value of E_0 for transferring the largest amount of impulse to a target. For the 1.2 μs pulse illustrated in Fig. 5.72, the optimum value of E_0 is about 22 J/cm^2, and this is in reasonable accord with the data. Of course, the specific impulse I_m per se goes as $I_m/E_0^{1/3}$, so larger E_0 will create larger I_m. However, this slow increase of 1m with E_0 is a very inefficient way to impart stress to a material. A better scheme, perhaps, would be to use multiple pulses at the optimum E_0 value.

Another consequence of the LSD wave is a lack of dependence of impulse on the parameters of the target material. The same impulse is produced independently of the target. This is in accord with the experiment. When I is well into the range where LSDs are formed, the measured impulse is the same for all target materials. Some data taken by the Naval Research Laboratory (NRL) [64] are shown in Fig. 5.77. In

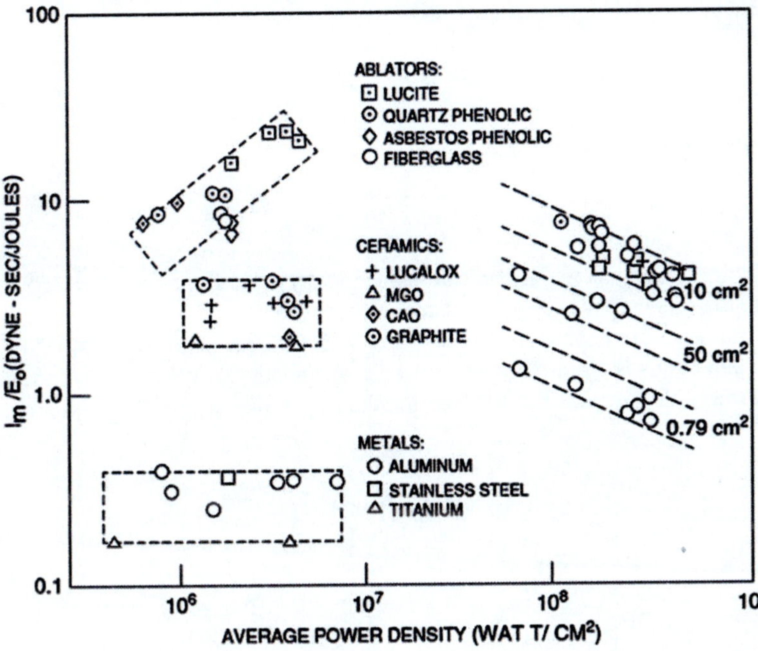

Fig. 5.77 Coupling coefficient as a function of power density [64]

this graph, we see, in accord with the vaporization model, a strong dependence of I_m/E_0 on material type at lower power densities, whereas at the high power densities typical of LSD formation, the values of I_m/E_0 are the same for all materials. In this range, however, a target area dependence appears. The target area dependence shown is for aluminum. Here again the general behavior predicted by Eq. 5.444 can be seen.

We have not yet considered the radius at which the expanding cylinder reaches atmospheric pressure. Call this radius R_0. As explained above, Pr^2 is constant, so:

$$P_0 R_0^2 = P_d R^2 \tag{5.456}$$

where P_0 is atmospheric pressure, 106 dyne/cm^2. Thus:

$$R_0 = R P_0^{-1/2} P_d^{1/2} \tag{5.457}$$

From Eqs. 5.418 and 5.419, for P_d, this becomes:

$$R_0 = R P_0^{1/2} \sqrt{\frac{\rho_0}{1+\gamma}\left[\frac{2(\gamma^2-1)t}{\rho_0}\right]^{2/3}} \tag{5.458}$$

Again, the factors involving γ are nearly in unity, so:

$$R_0 = RP_0^{1/2}\rho_0^{1/6}I^{1/3} \tag{5.459}$$

Upon substituting $I = E_0/t_p$, we get:

$$R_0 = RP_0^{1/2}\rho_0^{1/6}E_0^{1/3}t_p^{-1/3} \tag{5.460}$$

So, if R_0 is less than R_T, one can replace R_T by R_0 in Eq. 5.455. For target sizes larger than R_0 will all receive the same impulse.

A numerical illustration is useful. Let $E_0 = 1000$ J/cm^2 and $t_p = 100$ μs. Suppose the beam radius is 1 cm. The power density I is about 10^7 W/cm^2, so we expect an LSD wave. If we take $R_T = 5$ cm for the target radius, Eq. 5.453 yields (with $\rho_0 = 1.29 \times 10^{-3}$ g/cm^3):

$$\frac{I_m}{E_0} = \frac{25}{2}\left(1.29 \times 10^{-3}\right)^{2/3} / \left[\left(10^{-4}\right)^{1/3}\left(10^{10}\right)^{2/3}\right] \text{ dyne-s/erg}$$

$$\frac{I_m}{E_0} \approx 7 \times 10^{-7} \text{ dyne-s/erg}$$

or:

$$\frac{I_m}{E_0} \approx 7 \text{ dyne-s/J}$$

In this example, expansion to atmospheric pressure would take place at a radius given by Eq. 5.461:

$$R_0 = \left(10^6\right)^{-1/2}\left(1.29 \times 10^{-3}\right)^{1/6}\left(10^{10}\right)^{1/3}\left(10^{-4}\right)^{1/3} \tag{5.461}$$

where we have used erg/cm^2 for E_0. Then:

$$R_0 \approx 15 \text{ cm}$$

Thus, targets with radii of 15 cm and larger would exhibit a maximum coupling coefficient of $(15/5)^2$ time 7, or 65 dyne-s/J. If we compute I_m, we have 63,000 dyne-s/cm^2, that is, 63,000 taps, as the maximum specific impulse from this laser pulse. Since this impulse is delivered in times of the order of magnitude of the laser pulse time, this corresponds to a pressure of roughly $(6.3 \times 10^4)/10^{-4} \approx 6 \times 10^8$ dyne/cm^2, or about 600 atmospheres.

5.13 Effects of Continuous Wave Laser Radiation

The choice of laser system depends on the application at hand. First of all, the structuring of metals, semiconductors, and dielectrics is very diverse. Since ultra-short pulse lasers operate at very high intensities, they can employ nonlinear absorption and are therefore an obvious choice for laser processing of transparent materials, but it all depends on the specifications of the application.

Laser can be categorized most easily according to their medium, which are divided into three basic categories as defined by the state of the lasing material:

(a) Gas
(b) Liquid
(c) Solid

Furthermore, all laser types operate in one or two temporal modes:

1. Continuous wave (CW)
2. Pulse mode

In CW mode, the laser beam is emitted without interruption. In pulsed mode as we described in previous sections, the laser beam is emitted periodically.

If structures on the tens of micrometers scale are the aim of the application, a nanosecond laser will often provide sufficient accuracy and will typically have a greater throughput, thereby making it the preferred solution. If, however, features on the nanometer scale are required, a pico or femtosecond laser system would be a better choice.

For nanosecond laser systems, the resolution of the generated features is limited by the heat diffusion length, $\ell \approx \sqrt{\kappa\tau}$, where κ is the thermal diffusivity of the material and τ is the laser pulse duration [65]. Metals typically have a high thermal diffusivity, and precise micro-structuring must therefore be undertaken with shorter pulses [66, 67]. It should also be noted that nanosecond laser processing is associated with a heat-affected zone, i.e., an area around the laser-generated hole where the temperature has been high enough for the material to undergo melting (but not vaporization). Though the material re-solidifies, the strength of this area may have been greatly reduced.

The output of a laser can be a continuous wave (CW) or pulsed. In pulsed operation, much higher peak powers can be achieved since the energy stored in the laser gain medium is released in a short burst. Laser ablation is the process of removing material from a surface by laser beam irradiation. Short pulse laser ablation is advantageous since the material can be heated up to the temperature of vaporization in a very short time. This means that the energy does not have time to spread into the deeper parts of the material, and thereby the energy is localized where it is needed.

For a sufficiently low-power flux, the thin surface layer will be heated to the fluid state but will stay beyond the evaporation temperature, while the solid–fluid interface will slowly progress into the bulk material by heat conduction. In iron the typical progress rate is about 10^{-2} cm ms^{-1}, and the power flux for this situation may be around 10^6 W cm^{-2}.

At a somewhat higher-power flux, between 10^6 and 10^8 W/cm^2, the thin absorbing surface layer is heated up to its evaporation temperature before the solid–fluid interface has progressed appreciably into the material by heat conduction. Thus, with continuing laser power, a less than μ-thick layer of material will continually evaporate, with the material–air interface progressing into the material. Typically, a gas jet develops. The gas jet ejects also part of the molten material; thus the progress rate of the hole is faster than with evaporation of all the material.

At a still higher flux of 10^9 to 10^{10} Wcm^{-2}, after initial evaporation of the surface layer, the gas jet is thermally ionized and absorbs most of the incident radiation, which is blocked away from the material. The surface layer explodes with an ultrasonic jet, and its temperature may rise beyond 10^5 °K and its pressure beyond 10^3 at material properties or select from a set of predefined materials.

As we said at the beginning of this chapter, light contains energy. When that light is absorbed by a material, its energy goes toward heating the material. However, the faster the material gains heat, the faster it sheds heat. If the rate of heat loss gets high enough to equal the rate of heat gain, the death ray can no longer raise the temperature of the surface. Otherwise, the surface continues to gain heat, and its temperature rises. The temperature to which a beam of light can bring a material depends on the intensity of the light and the rates and mechanisms by which the material responds to the heat.

There are generally considered to be three mechanisms by which heat moves around: heat conduction, heat convection, and radiation. Heat conduction is what you get when heat moves through a material, transferred from atom to atom by the atoms bumping into each other and exchanging energy. Convection is the movement of heat by moving the material it is stored in. Radiation transfers heat by the emission or absorption of electromagnetic waves. When a surface is exposed to a death ray, it is being heated by radiation.

For continuous wave (CW) laser beams of relatively low intensity, the only effect is heating from absorption of electromagnetic radiation. The temperature of the absorbing material is raised until a balance exists between the absorbed power and thermal losses from conduction, convection, and radiation.

It has been known since prehistory that sunlight can make absorbing surface hot, and in Greek antiquity, it was known that concentrated sunlight, having traversed spherical water-filled flasks or convex pieces of glass, could kindle fires [69]· Archimedes proposed to concentrate the sun's rays by means of reflecting the harbor of Syracuse. He hoped to ignite the ropes, sails, and spars of the vessels of enemy's fleet.

The power flux density of the sun in zenith on the earth's surface is about 0.13 W/cm^2 leading to the solar furnaces with temperatures near 3000 °C. If the irradiated area is large enough, lateral heat conduction may be ignored. The radiative loss of a blackbody surface is given by the Stefan-Boltzmann law. If the irradiated area is large enough, lateral heat conduction may be ignored. The power flux density, σT^4, corresponds to 1 kW/cm^2 at $T = 3644$ K. Convective cooling by airflow over a 3000 K surface at Mach number unity is only a few hundred W/cm^2. At pressures prevailing at booster burnout altitudes, 80–160 km, convective cooling is completely negligible. From the foregoing, it is clear that the temperature of most materials may be raised above the melting temperature T_m and the vaporization temperature T_v for CW laser flux densities in the range of 1–100 kW/cm^2. In the early days of laser history, in 1961, when the pulsed ruby laser was the most powerful available, it was established that a focused ruby laser pulse of about 1 joule energy could punch a hole in a razor blade [70]. Two very simple cases serve to establish the order of magnitude of fluxes and fluencies on target required for lethality [68]. Most analysis that we have showed so far are applicable for the CW laser type as well.

5.14 Similarity and Self-Similarity Mathematical Solutions

Although we have suggested few close solution to problem of solving heat conduction differential equation for simple cases of boundary and initial conductions associated with the problem by using either Laplace or Fourier transformation, yet for more complex cases, we can apply methods of similarity or self-similarity approach [71–72].

In most complex senior, where the heat conduction equation needs to be solved, due to phase changes of material during interaction of high-power laser energy, the problem and its boundary conditions no longer fall in the schema of Eulerian condition, rather the solution should be found under Lagrangian conditions due to the fact that we need to solve the problem, where boundary and initial conditions may move along the frame of reference. In that case, utilization of self-similarity method and technique like the asymptotic method is required and makes life much easier than it is with the problem of interest.

In laser dynamics, the diffusion of the absorbed energy by electronic heat conduction and by thermal radiation and subsequent expansion of the hot plasma into the vacuum or a surrounding low-density gas and plasma cooling by expansion are of particular interest. The dynamics studied in the laboratory are nonstationary and at best exhibit rotational symmetry around the normal to the target onto which the laser beam is focused. This is a typical case for numerical computation. Only much simpler idealized, so to say, few purely one-dimensional cases are accessible to an analytical treatment. Such solutions are of great relevance for the laser plasma in plane and spherical geometry.

5.15 Summary

In summary, the main aspects of high-power laser–matter interaction in the intensity range 10^{10}–1022 W/cm^2 are subject of interest here, and we offer a guide to this topic for scientists and students who have just discovered the field as a new and attractive area of research and for scientists who have worked in another field and want to join now the subject of laser plasmas.

Being aware of the wide differences in the degree of mathematical preparation the individual candidate has acquired, we tried to present the subject in an almost self-contained manner. To be more specific, a bachelor's degree in physics enables the reader in any case to follow without difficulty.

Generally fluid or gas dynamics and its relativistic version are not a part of this education; it is developed in the context where it is needed. Basic knowledge in theoretical mechanics, electrodynamics, and quantum physics is the only prerequisite we expect from the reader. Throughout the book, the main emphasis is on the various basic phenomena and their underlying physics. Not more mathematics than necessary is introduced. The preference is given to ideas. A good model is the best guide to the adequate mathematics.

More details can be found in the book by Mulser and Bauer [70] as well.

References

1. Schriempf, J.T.: Response of materials to laser radiation a short course. NRL Report 7728, July 10, 1974. Department of Navy, Office of Naval Research Arlington, VA 22217 (1974)
2. Ready, J.F.: Effects of High-Power Laser Radiation. Academic, New York (1971)
3. Carslaw, H.S., Jaeger, J.C.: Conduction of Heat in Solids, 2nd edn. Clarendon Press, Oxford (1959)
4. Nowakowski, KA.: Laser beam interaction with materials for microscale applications. PhD dissertation, Worcester Polytechnic Institute, 22 November 2005 (2005)
5. Ross, D.: Lasers: Light Amplifiers and Oscillators, p. 72. Academic, New York (1969)
6. Wester, R.: In: Poprawe, R. (ed.) Tailored Light 2: Laser Application Technology. Springer Publication, New York (2011)
7. Born, M., Wolf, E.: Principles of Optics. Cambridge University Press, Cambridge (1999)
8. Zettili, N.: Quantum Mechanics, Concept and Applications, 2nd edn. Wiley, New York (2009)
9. John, F.: Ready, Effects of High-Power Laser Radiation. Academic, New York (1971)
10. Wieting, T.J., Schriempf, J.T.: Free electron theory and laser interaction with metals. Report of NRL Progress, June 1972, pp. 1–13 (1972)
11. Bonch-Bruevich, A.M., Imas YaA (1967) Zh Tekh Fiz 37:1917 (English transl.: Sov PhysTech Phys 12:1407 (1968))
12. Zhang, S.Y., Ren, Y.H., Lupke, G.: Appl. Opt. **42**(4), 715 (2003)
13. Romero, L.A., Dickey, F.M.: J. Opt. Soc. Am. A. **13**(4), 751 (1996)
14. Momma, C., Nolte, S., Kamlage, G., von Alvensleben, F., Tunnermann, A.: Appl. Phys. A Mater. Sci. Process. **67**(5), 517 (1998)
15. Sanner, N., Huot, N., Audouard, E., Larat, C., Huignard, J.P., Loiseaux, B.: Opt. Lett. **30**(12), 1479 (2005)
16. Nemoto, K., Nayuki, T., Fujii, T., Goto, N., Kanai, Y.: Appl. Opt. **36**(30), 7689 (1997)

17. McLeod, E., Hopkins, A.B., Arnold, C.B.: Opt. Lett. **31**(21), 3155 (2006)
18. Heinemann, S.: Opt. Commun. **119**(5–6), 613 (1995)
19. Dickey, F.M., Holswade, S.C. (eds.): Laser Beam Shaping: Theory and Techniques. Marcel Dekker, New York (2000)
20. Bauerle, D.: Laser Processing and Chemistry. Springer, Berlin (2000)
21. Heller, J., Bartha, J.W., Poon, C.C., Tam, A.C.: Appl. Phys. Lett. **75**(1), 43 (1999)
22. Toulemonde, M., Unamuno, S., Heddache, R., Lampert, M.O., Hageali, M., Siffert, P.: Appl. Phys. A Mater. Sci. Process. **36**(1), 31 (1985)
23. Toulemonde, M., Unamuno, S., Heddache, R., Lampert, M.O., Hageali, M., Siffert, P.: Appl. Phys. A Mater. Sci. Process. **36**(1), 31 (1985)
24. Arnold, C.B., Aziz, M.J., Schwarz, M., Herlach, D.M.: Phys. Rev. B. **59**(1), 334 (1999)
25. Weeber, J.C., Krenn, J.R., Dereux, A., Lamprecht, B., Lacroute, Y., Goudonnet, J.P.: Phys. Rev. B. **64**(4), 045411 (2001)
26. Baeuerle, D.: Laser Processing and Chemistry, 3rd edn. Springer, New York (2000). isbn:3-540-66891-8
27. Lide, D.R.: CRC Handbook of Chemistry and Physics, 82nd edn. CRC, Boca Raton (2001)
28. Slusher, R.E., Eggleton, B.J.: Nonlinear Photonic Crystals, 1st edn. Springer, Berlin (2004)
29. Ghofraniha, N., Conti, C., Ruocco, G., Trillo, S.: Phys. Rev. Lett. **99**(4), 043903 (2007)
30. Staudt, W., Borneis, S., Pippert, K.D.: Phys. Status Solidi. A Appl. Res. **166**(2), 743 (1998)
31. Mori, N., Ando, T.: Phys. Rev. B. **40**(9), 6175 (1989)
32. Columbia: Laser micromachining, 10 November 2005. www.mrl.columbia.edu (2005)
33. Lasag (1997) Operator's Manual for KLS 126 Laser Source. LASAG Corporation, Thun
34. Semak, V., Matsunawa, A.: The role of recoil pressure in energy balance during laser materials processing. J. Phys. D Appl. Phys. **30**, 2541–2552 (1997)
35. Bronski, M.T.: Development of a process characterization of Nd:YAG crystals. MS thesis, Center for Holographic Studies and Laser micro-mechaTronics, Mechanical Engineering Department, Worcester Polytechnic Institute, Worcester, MA (2003)
36. Bonch-Bruevich, A.M.: Imas YaA (1967) Zh Tekh Fiz 37:1917 (English transl.: Sov Phys Tech Phys 12:1407 (1968))
37. Nowak, T.: Theoretical and experimental investigation of laser drilling in a partially transparent medium. MS thesis, Center for Holographic Studies and LasermicromechaTronics, Mechanical Engineering Department, Worcester Polytechnic Institute, Worcester, MA (1990)
38. Yilbas, B.S.: Study of liquid and vapor ejection processes during laser drilling of metals. J. Laser Appl. **7**, 147–152 (1995)
39. Tokarev, V.N., Lunney, J.G., Marinea, W., Sentis, M.: Analytical thermal model of ultraviolet laser ablation with single-photon absorption in the plume. J. Appl. Phys. **78**(2), 1241–1246 (1995)
40. Yilbas, B.S., Yilbas, Z., Akcakoyun, N.: Investigation into absorption of the incident laser beam during Nd:YAG laser processing of metals. Opt. Laser Technol. **28**(7), 503–511 (1996)
41. Yilbas, B.S., Yilbas, Z., Sami, M.: Thermal processes taking place in the bone during CO_2 laser irradiation. Opt. Laser Technol. **28**(7), 513–519 (1996)
42. Tabor, D.: Gases, Liquids and Solids and Other States of Matter, 3rd edn, p. 272. Cambridge University Press, Cambridge (1991)
43. Bejan, A.: Heat Transfer. Wiley, New York (1993)
44. Mazumder, J., Steen, W.M.: Heat transfer model for CW laser material processing. J. Appl. Phys. **51**, 941–947 (1980)
45. Gordon, R., Cobonpue, J.: Heat transfer between a flat plate and jets of air impinging on it. Heat Transfer Pt. II:454–460. ASME, New York (1961)
46. Niedrig, R., Bostanjoglo, O.: Imaging and modeling of pulse laser induced evaporation of metal films. J. Appl. Phys. **81**(1), 480–485 (1996)
47. Semak, V., Damkroger, B., Kemka, S.: Temporal evaluation of the temperature filed in the beam interaction zone during laser material processing. J. Phys. D Appl. Phys. **32**, 1819–1825 (1999)

48. Zohuri, B.: Dimensional analysis and self-similarity methods for engineers and scientists. Springer Publishing Company, New York (2015)
49. Parker, W.J., Jenkins, R.J., Butler, C.P., Abbott, G.L.: J. Appl. Phys. **32**, 1679 (1961)
50. Heckman, R.C.: Thermal diffusivity finite pulse time corrections. Sandia Laboratories Research Report SC-RR-710280, May 1971 (1971)
51. Schriempf, J.T.: Rev. Sci. Instrum. **43**, 781 (1972)
52. Sparks, M.: Theory of laser heating of solids: metals. J. Appl. Phys. **47**(3), 837–849 (1975)
53. Yilbas, B.S., Sami, M.: Liquid ejection and possible nucleate boiling mechanism in relation to the Laser drilling process. J. Phys. D Appl. Phys. **30**, 1996–2005 (1997)
54. Sparks, M.: Theory of laser heating of solids: metals. J. Appl. Phys. **47**(3), 837–849 (1975)
55. Anisimov, S.I., Khokhlov, V.A.: Instabilities in Laser-Matter Interaction. CRC, Boca Raton (1995)
56. Yilbas, B.S., Gbadebo, S.A., Sami, M.: Laser heating: an electro-kinetic theory approach and induced thermal stresses. Opt. Laser Eng. **33**, 65–79 (2000)
57. Yilbas, B.S., Arif, A.F.M.: Material response to thermal loading due to short pulse laser heating. Int. J. Heat Mass Transfer. **44**, 3787–3798 (2001)
58. Nash, G.E.: Thermal response calculation. Naval Research Laboratory, Washington, DC, unpublished data.
59. Anisimov, S.I.: High Temp 6:110. Translated from Teplofizika Vysokikh Temperature 6:116 (1968). Original article submitted December 6, 1966 (1968)
60. Nielsen, P.E., Canavan, G.H.: Laser absorption waves. Air Force Weapons Laboratory Laser Division Digest LRD-71-2, p 110, December 1971 (1971)
61. Canavan, G.H., Nielsen, P.E., Harris, R.D.: Radiation of momentum transfer to solid targets by plasma ignition. Air Force Weapon Laboratory Laser Division Digest LRD-72-1, p 125, June 1972 (1972)
62. Raizer, Y.P.: Sov. Phys. JETP. **21**, 1009 (1965)
63. Zel'dovich, Y.B., Raizer, Y.P.: Physics of Shock Waves and High-Temperature Hydrodynamic Phenomena, p. 346. Academic, New York (1966)
64. Metz, S.M., Hettche, L.R., Stegman, R.L.: Intensity dependence of target response to high intensity pulsed 10.6 μ laser radiation. In DOD Laser effects/hardening conference, Monterey, CA, 23–26 October 1973 (1973)
65. Bauerle, D.: Laser Processing and Chemistry, 3rd edn. Springer, Berlin (2000)
66. Dausinger, F., Lichtner, F., Lubatschowski, H.: Femtosecond Technology for Technical and Medical Applications. Springer, Berlin (2004)
67. Balling, P.: In: Kane, D.M. (ed.) Laser Cleaning II, p. 257. World Scientific Publishing Company, Singapore (2006)
68. The American Physical Society: Report to the American Physical Society of the Study Group on Science and Technology of Directed Energy Weapons. The American Physical Society (1987)
69. Zohuri, B.: Directed Energy Weapons, Physics of High Energy Lasers (HEL), 1st edn. Springer Publishing Company (2016)
70. Mulser, P., Bauer, D.: High Power Laser Matter Interaction. Springer Tracts in Modern Physics Volume 238, 1st edn. Published by Springer Publishing Company (2010)
71. Zohuri, B.: Dimensional Analysis Beyond the Pi Theorem. 1st edn. Published by Springer Publishing Company, November 3, 2016
72. Zohuri, B.: Dimensional Analysis and Self-Similarity Methods for Engineers and Scientists. Springer; 2015 edition April 16, 2015

Chapter 6
Laser Surface Processing

High-power lasers (HPLs) have been used to process materials to improve their wear resistance since the late 1960s. Laser surface processing encompasses established techniques, such as laser transformation hardening, which has achieved production-level capabilities, and techniques that are still in the development stage, such as laser cladding, which is emerging as a viable alternative to other hard-facing processes, such as plasma deposition. The laser beam is a chemically clean light source that delivers a precisely controlled quantity of energy to localized regions. By this means, only those areas exposed to wear are processed for improvement in wear resistance. The laser beam is easily maneuvered by optical elements and can be adapted to automation. High power densities and low interaction times result in rapid heating and cooling, which produces a shallow heat-affected zone, low distortion of the workpiece, and minimal deterioration of bulk properties. Rapid solidification effects produce refined and novel microstructures that result in improved properties.

6.1 Introduction

Laser characteristics can be described by saying that most laser surface processing utilizes continuous wave (CW) lasers. In limited instances, a pulsed laser is used, for example, to harden the tip of a tool bit. The most common lasing medium for materials processing is CO_2 gas. The 10.6 m wavelength CO_2 laser couples poorly with metals, but generally has a high average power output between 1 and 25 kW.

Note: This chapter write-up was provided to this author by Dr. Khershed P. Cooper of Naval Research Laboratory (NRL), while this author was employed at Lockheed Missiles and Space in Sunnyvale, California, during the time period of 1979–1984 as chief scientist and was involved with laser hardening and radiation hardening during the era of SDI. These materials later on were published by Dr. Cooper in ASM International ASM Handbook, Volume 18, *Friction, Lubrication, and Wear Technology* during the 1992 time frame. Now it is republished here [1].

B. Zohuri, *Thermal Effects of High Power Laser Energy on Materials*,
https://doi.org/10.1007/978-3-030-63064-5_6

With the help of radiation-absorbing agents, metals with high melting points can be processed. The excitation-mode structure of the CO2 laser is usually "doughnut-shaped," which, after being modified by optics, produces a near-Gaussian energy distribution on the sample surface.

In its simplest form, laser–material interaction involves the heating and melting of a moving substrate with a stationary beam. This "spot" energy source is typically 1 to 2 mm (0.04 to 0.08 in.) in diameter and produces a melt pass with a width from 2 to 3 mm (0.08 to 0.12 in.). By overlapping several of these passes, usually by 25 to 50%, larger areas can be processed. In addition to being slow, this technique can often cause tempering and cracking within the overlap regions, thereby lowering the quality of the product surface.

An alternative method is the use of beam-shaping optics to produce broad energy distributions on the sample surface. Scanning the beam over the substrate, either by oscillating a flat mirror or by rotating a polygonal mirror, produces a "line" energy source. An integrated beam, which is formed by an integrated mirror, produces a "rectangular" energy source. When interacting with a moving workpiece, either energy source produces a strip of modified surface. The use of beam-shaping optics represents an efficient method of achieving the hardening of selected areas on the workpiece, which is required by most wear applications.

In addition to shaping the beam, optical elements can be designed and arranged to direct it onto the workpiece in many ways. For example, a reflecting mirror can be used to project the laser beam onto the inner wall of a rotating tube. Toric mirrors also can be designed to form a continuous band of radiation to treat inner or outer surfaces of cylinders and cones. Prisms can also be used to split the laser beam and direct the two parts at different angles to a shaped surface, such as a gear groove.

Toric Mirrors or Lens A toric lens is a lens with different optical power and focal length in two orientations perpendicular to each other. One of the lens surfaces is shaped like a "cap" from a torus (see Photo 6.1), and the other one is usually spherical. Such a lens behaves like a combination of a spherical lens and a cylindrical lens. Toric lenses are used primarily in eyeglasses, contact lenses, and intraocular lenses to correct astigmatism.

By definition, torus is defined as follows: A torus is the spatial body resulting when a circle with radius r rotates around an axis lying within the same plane as the circle, at a distance R from the circle's center (see Photo 6.2.). If $R > r$, a *ring torus* is produced. If $R = r$, a *horn torus* is produced, where the opening is contracted into a single point. $R < r$ results in a *spindle torus*, where only two "dips" remain from the opening; these dips become less deep as R approaches 0. When $R = 0$, the torus degenerates into a sphere with radius r.

(continued)

Photo 6.1 Toric lens surface as "cap" (top–right) from a torus (here with $R = 1.2\ r$)

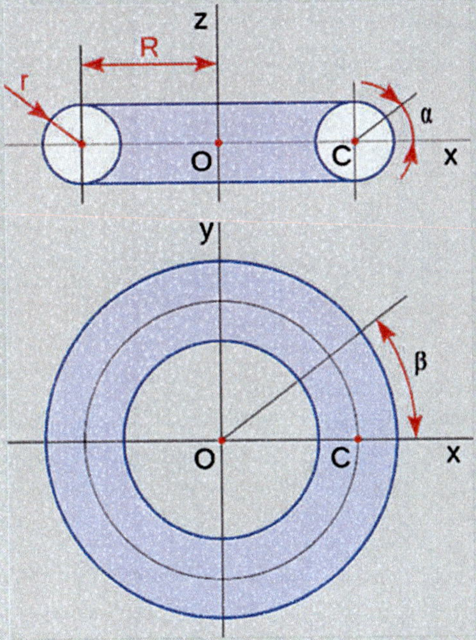

Photo 6.2 A torus results when a circle with radius r rotates around an axis lying in the same plane as the circle (here the z axis) at a distance R from the center of the circle

(continued)

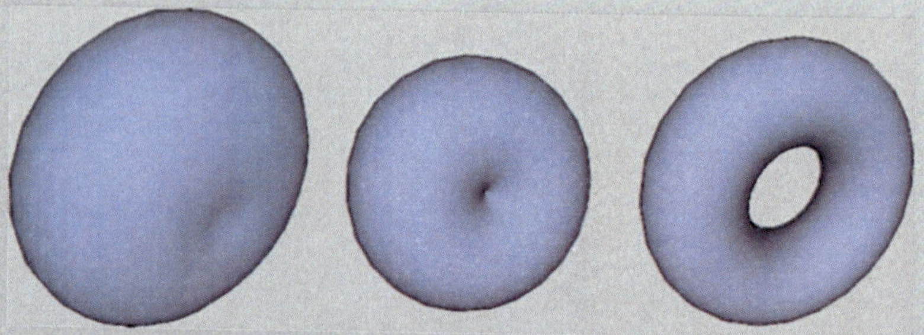

Photo 6.3 When the major radius R approaches 0 (here from right to left), the torus becomes a sphere.

Note that when the major radius R approaches 0 from right to left as illustrated in Photo 6.3 here, the torus becomes a sphere.

Radius of curvature and optical power is being explained here. The greatest radius of curvature of the toric lens surface, $R + r$, corresponds to the smallest refractive power, S, given by:

$$S = \frac{n - 1}{R + r} \tag{6.1}$$

where n is the index of refraction of the lens material.

The smallest radius of curvature, r, corresponds to the greatest refractive power, s, given by:

$$s = \frac{n - 1}{r} \tag{6.2}$$

Since $R + r > r$, $S < s$, the lens behaves approximately like a combination of a spherical lens with optical power s and a cylindrical lens with power $s - S$. In ophthalmology and optometry, $s - S$ is called the *cylinder power* of the lens.

This is used for correcting astigmatism. In this context, the term *cylinder* is based on a mathematical approximation, which is only valid for small corrective powers.

The description of light ray and its refractive power is given below.

Light rays within the (x, y)-plane of the torus (as defined in the figure above) are refracted according to the greatest radius of curvature, $R + r$, which means that it has the smallest refractive power, S.

Light rays within a plane through the axis of revolution (the z axis) of the torus are refracted according to the smallest radius of curvature, r, which means that it has the greatest refractive power, s.

(continued)

As a consequence, there are two different refractive powers at orientations perpendicular to each other. At intermediate orientations, the refractive power changes gradually from the greatest to the smallest value, or reverse. This will compensate for the astigmatic aberration of the eye.

With modern computer-controlled design and grinding and polishing techniques, good vision corrections can be achieved for even wider angles of view by allowing certain deviations from the toric shape. This is called an *atoric lens* (literally, non-toric lens) [2, 3].They are related to toric lenses in the same way that aspheric lenses are related to spherical lenses.

Laser surface modification techniques that are used to improve the wear resistance of metals include transformation hardening, melting, alloying, cladding, and particle injection. The processing conditions, examples of microstructures, wear characteristics, and applications associated with each technique are discussed in detail in this article.

Parameters for laser surface processing depend on the type of modification (e.g., transformation hardening or cladding) and thermophysical properties of the material (such as thermal diffusivity or heat capacity). The important laser processing parameters are power density and interaction time, which are functions of the average laser power, the sample translation rate, and the beam diameter. Rapid melting or glazing requires either high power densities and short interaction times or high power levels and high translation rates, whereas transformation hardening requires either low power densities and long interaction times or low power levels and low translation rates.

Other processing variables, such as the composition of the atmosphere or the rate of material addition, depend on the particulars of the process. For example, consideration must be given to the need to protect against oxidation and the desire to have certain characteristics of the surface layer, for example, thickness, composition, and microstructure. Depending on the processing conditions, surface layer thicknesses can range from 0.01 to 5 mm (0.0004 to 0.2 in.). The longer the interaction time, the deeper the treated layer will be.

Processing parameters for cladding and surface melting typically modify the surface to depths from 0.5 to 5 mm (0.02 to 0.2 in.) with microstructures that are similar to welded structures. The parameters for rapid surface melting or glazing modify surfaces from 0.02 to 0.6 mm (0.0008 to 0.02 in.) deep, but the resulting rapid quench rates from 105 to 107 K/s allow for the synthesis of novel microstructures. Material addition practices will depend on the specific laser surface processing technique under consideration. Steen [4] has summarized the process variables and their methods of measuring (Table 6.1).

Table 6.1 Process variables and methods of measuring

Process variable	Method of measuring
Laser power	Flowing cone calorimeter
Beam diameter	Rotating rod
Beam mode structure	Acrylic prints, fluorescent screen
Sample speed	Electronic timer
Material addition flow rate	Load cell on feed hopper, acoustic emission sensor on feed line
Particle velocity	Rotameters, fast-action photography, photographic tracer technique

6.2 Laser Transformation Hardening

Processing The wear resistance of highly stressed ferrous alloy components, such as gears and bearings, can be improved by transformation hardening. Because ferrous materials are very good heat conductors, the high heat fluxes generated by lasers are most suitable to heat the surface layer to austenization levels without affecting the bulk temperature of the sample. The ensuring self-quenching is rapid enough to eliminate the need for external quenching to produce the hard martensite in the heated surface. Because ferrous alloys have high reflectivity, absorptive coatings such as manganese phosphate and graphite are applied to the workpiece for efficient laser heating. Most of these coatings burn off and normally do not affect the microstructure. Processing conditions for laser transformation hardening are typically power densities that range from 5 to 100 MW/m2 (3.2 to 64.5 kW/in.2) and interaction times ranging from 0.01 to 10 s. Usually, inert gas shielding is used.

A rectangular beam is often used for laser transformation hardening. According to Sandven [5], the case depth depends on the hardenability of the material and rarely exceeds 2.5 mm (0.1 in.). For example, in medium-carbon steel, a case depth of 1.25 mm (0.05 in.) is common, whereas in mild steel, a case depth of only 0.25 mm (0.01 in.) is possible for the same processing conditions. Because of the rapid heating and cooling rates, steels that are not hardenable with other heat treatment methods can be hardened with laser processing.

The hardened case depth and the transition zone also depend on the coarseness of the initial microstructure. With a coarse starting structure, the case depth is shallower, and the transition from full martensite hardness to base metal hardness is more gradual than with a fine starting structure. According to Seaman [6], coarser structures provide longer diffusion paths to achieve uniform austenization. To facilitate complete transformation during laser transformation hardening, materials with fine starting microstructures, such as closely spaced pearlite, are recommended. In many cases, the part is conventionally quenched and tempered prior to laser processing.

Because laser processing can occur too rapidly to allow for complete diffusion of carbon in the austenite, the interaction times are kept longer. The laser power density is kept low to prevent surface melting, which can form harmful eutectic carbides and can blunt the sharp edges and corners of a complex workpiece. Melting can also change the surface topography, requiring post machining. The careful design of the beam shape and energy distribution can minimize surface melting problems. During closure of a circular hardening pass or during overlapping of several passes, back-tempering can occur in the overlapped regions. If this is detrimental, then the overlapped area should be located in a position of low service wear.

For example, the interior walls of internal combustion engines can be hardened by spiral laser passes, which results in some unhardened regions without causing harmful effects. It has been claimed that the use of a laser-hardened spiral on a shaft that runs in a bearing results in one-third of the wear that is experienced when using a uniformly induction-hardened shaft, because, in the former case, the wear debris becomes safely embedded in the softer interspiral regions [7]. In one line-hardening procedure for plain carbon steels, the laser-heated areas rise above the sample surface and act as hard bearing tracks in a tough material, because of the volume expansion of the martensite [8].

In laser transformation hardening, the thermal energy is generated by absorption of the laser radiation at the surface. The increase in temperature below the surface is by conduction only. If the rate of heat absorption and the thermal properties of the sample are known, then the temperature distribution in the sample can be calculated. By this means it is possible to determine the optimum laser processing conditions.

Heat flow expressions are available in the literature to make these calculations [5]. The simplest expression assumes heat flow to be one-dimensional, normal to the surface. In reality, heat diffuses in all directions and is lost by radiation from the surface. Furthermore, the thermal properties are temperature-dependent, and the workpiece is not always large enough to be considered an infinite heat sink. These factors need to be considered when determining the temperature distributions. However, for high processing speeds, thick samples, and large beam sizes, the one-dimensional model gives good estimates. Using the one-dimensional heat flow model, cooling rates can also be calculated. When used in conjunction with cooling transformation diagrams, the adequacy of the self-quenching for hardening can be predicted.

Examples of Laser-Hardened Surface Microstructures In ferrous alloys, the formation of martensite is essential for hardening and increased wear resistance. In laser-hardened carbon steels, such as 1018 and 1045, the relative amounts of ferrite and martensite depend on the carbon content. In laser-hardened alloy steels, such as 4340, 8620, and 52100, a very fine martensite forms [9]. Courtney and Steen [10] found that in a 0.36% C steel, the microstructure was dependent on sample translation rate. At low sample speeds, areas of high-carbon martensite formed, whereas at high sample speeds, an inhomogeneous martensite mixed with austenite formed.

In cast irons, laser heat-treated surfaces appear in light contrast, as shown in Fig. 6.1a for gray iron and in Fig. 6.1b for ductile iron [11]. The microstructure of laser-hardened gray cast iron is generally fine martensite that contains flake graphite, as shown in Fig. 6.2a [11]. Acicular bainite can sometimes occur in the martensite matrix [12]. The microstructure of laser-hardened ductile iron is also fine martensite, but contains "bull's eye" graphite nodules, as shown in Fig. 6.2b [11]. A closer

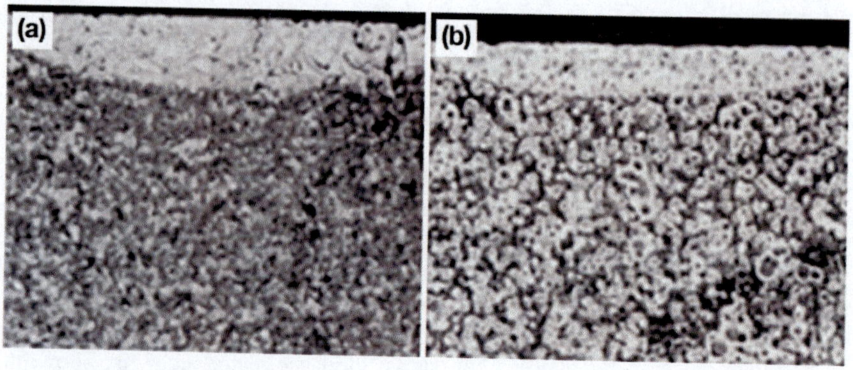

Fig. 6.1 Cross section of laser heated-treated surfaces in cost irons. (**a**) Gray iron. (**b**) Ductile iron. (Source: Ref. [11])

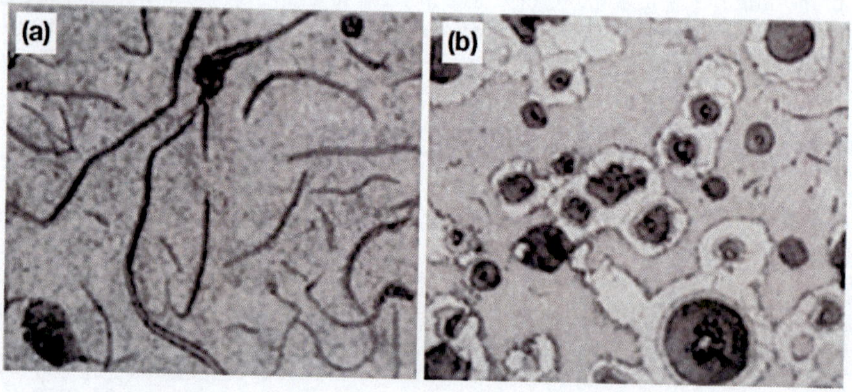

Fig. 6.2 Microstructures of laser-hardened cast irons. (**a**) Gray iron. (**b**) Ductile iron. (Source: Ref [11])

examination of Fig. 6.2b reveals a narrow region of martensite between the graphite nodule and the ferrite ring, which indicates the extent of carbon diffusion occurring during processing.

According to Molian and Baldwin [13], the extensive case that developed during laser hardening of gray and ductile irons was made up of sequential regions with varying degrees of microstructural modification. The region just below a very thin melt zone contained plate martensite, retained austenite, and graphite, whereas the region just above the base metal contained a mixture of refined martensite and untransformed pearlite.

Wear Behavior of Laser-Hardened Surfaces Gnanamuthu [14] reported pin-on-disk wear test results on laser-hardened AISI 1045 steel specimens. After testing for 10 h, the weight loss that was due to wear in the laser-treated samples ranged from 0.6 to

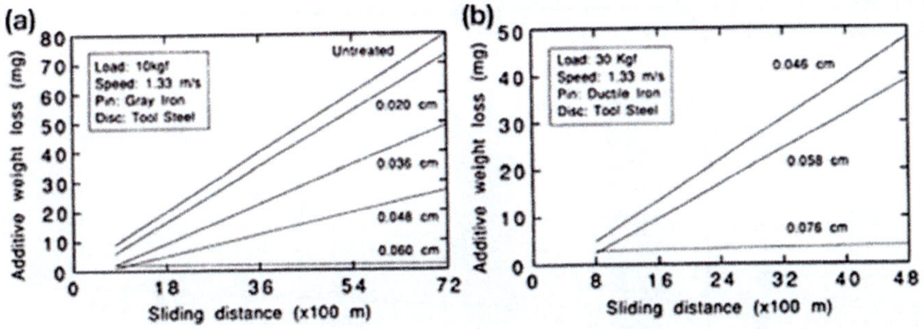

Fig. 6.3 Sliding wear behavior of laser-hardened cast irons as a function of case depth. (**a**) Gray iron. (**b**) Ductile iron. (Source: Ref [11])

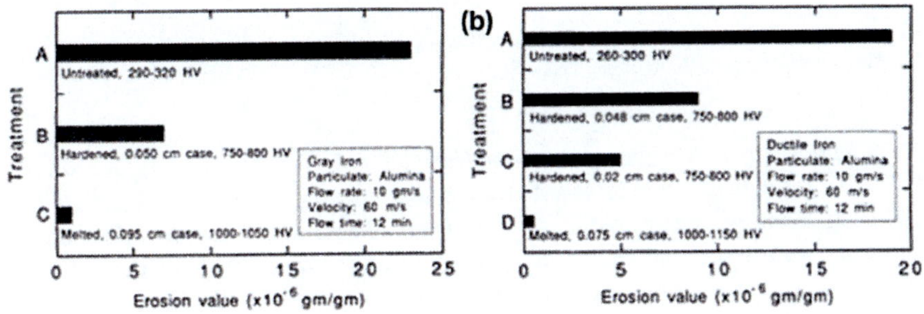

Fig. 6.4 Erosive wear behavior of laser-hardened cast irons as a function of surface hardness and case depth. (**a**) Gray iron. (**b**) Ductile iron. (Source: Ref [15])

1.4 mg (2.1 to 4.9 $\times$ 10^{-5} oz), whereas the weight loss in the untreated samples was 4.8 mg (1.7 $\times$ 10^{-4} oz). Using a two-disk machine, Lu and Zhang [15] found that laser hardening significantly improved the wear resistance in normalized or austempered ductile iron by the formation of strain-induced martensite. From their study of pin-on-disk wear behavior of laser-hardened gray and ductile cast irons, Molian and Baldwin [13] found an improvement in scuffing and sliding wear resistance with an increase in case depth as shown in Fig. 6.3. On the same samples, erosive wear test results showed that the erosion rate depended on the surface hardness and the case depth, as shown in Fig. 6–4, and increased as the matrix microstructure varied from ledeburite to tempered martensite to pearlite [16]. (Fig. 6.4)

Tomlinson et al. [17] observed that besides reducing wear, laser hardening of flake graphite iron resulted in very little plastic deformation, no graphite deposition on the opposing wheel, and no adhesive damage. Trafford et al. [18, 19] used a reciprocating pad-on-plate technique to evaluate the wear resistance of laser-processed gray iron and found that fully martensitic and surface-melted ledeburite microstructures had the lowest wear rates (Fig. 6.5).

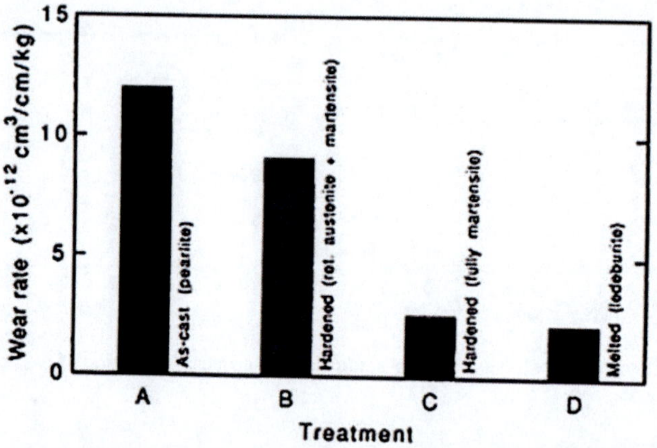

Fig. 6.5 Abrasive wear rates in gray iron as a function of microstructure arising from various laser-hardening treatments. (Source: Ref [18, 19])

Applications of Laser Transformation Hardening Wineman and Miller (Ref 16) describe laser hardening of selected areas on an automotive power steering gear housing made of ferritic malleable cast iron. The bands of highest wear on the housing were laser hardened to 60 HRC by transforming a ferritic–pearlitic-tempered carbon microstructure to temper carbon surrounded by martensite. No distortion was measured, and wear was reduced by 90%. Previously, the part was nitrided, requiring a costly annealing operation to maintain concentricity. Yessik [20] describes laser hardening of a 4150 steel gear spline using a long depth of field and a soft focus to ensure simultaneous hardening of tip and root.

Sandven [5] describes a cast iron cam-shaft lobe that was laser hardened by rotating under a rectangular beam. The cam-shaped lobe was hardened to a constant case depth of 1 mm (0.04 in.) by varying the angular speed from 12.5 mm/s (0.5 in./s) at the cylindrical portion to 3 mm/s (0.12 in./s) at the flat portion. The laser-hardened surface required no post machining, but induction hardening had been used previously, causing considerable distortion that required post machining.

Other laser hardening examples given by Sandven [5] are an SAE 1045 steel gear and a 25 mm (1 in.) diameter 4140 steel cylinder. The gear was hardened to 60 HRC by splitting the laser beam using a prism and directing each half to the gear groove in the form of a 12.5 mm (0.5 in.) square shape. Using a toric mirror, a 5-mm (0.2 in.)-wide band of laser energy was formed to harden the conical top of the cylinder. The entire cone was treated by translating the workpiece axially. Because the focusing distance varied along the conical top, the laser power was increased from 3.5 kW at the tip to 7.6 kW at the cylindrical portion.

By properly shaping the laser beam, Seaman and Gnanamuthu [21] describe hardening of only the sides of the grooves in a gray iron piston ring. Melting of the sharp corner was avoided by precisely directing the beam. This is an inaccessible area that would have been difficult to harden by conventional means. The authors also describe laser hardening of the faces of a malleable cast iron spacer to a hardness of 58 HRC (hardness Rockwell C). In this case, the laser beam power was ramped down by an aperture to minimize tempering where the circle closed out.

La Rocca [22] reports laser hardening of gear and rack teeth at an Italian automotive manufacturer. An oscillating beam and beam splitting mirrors were used to harden the two faces of each tooth. Hardened cases of 50 to 55 HRC were obtained on C-45 steel steering racks. At the electromotive division of an American automotive manufacturer, cylinder liners in high-performance turbocharged diesel engines were made more scuff-resistant by using a 10 mm (0.4 in.) diameter laser beam [23]. The entire surface was treated by continuously rotating the component and overlapping several paths. No heat treating had been performed previously, and laser hardening improved the service life.

6.3 Laser Melting

Processing Laser melting requires higher power densities than the levels used for laser transformation hardening. The workpiece is often made absorptive either by using coatings similar to those used for laser heating or by increasing surface roughness, for example, by sand blasting. Laser melting can harden alloys that cannot be hardened by laser transformation hardening. In ferritic malleable gray iron, melting enhances the diffusion of carbon, and the ensuing rapid quench produces a hardened region.

One problem with laser melting is that once the melt forms, the absorptive coating is removed, and the melt pool becomes reflective. Water-cooled, optical feedback systems described by Steen [24] can be used to increase the coupling efficiency. An oscillating beam can be employed to melt larger areas, which makes the mode structure of the beam less important.

According to Bergmann [25], the production of a crack-free, oxidation-free, and porosity-free layer requires special arrangements for the shielding gas, which allow for changes in the gas flow rate, angle of impingement, and shape of the gas jet. Proper direction of the gas jet can prevent massive plasma formation, which is often responsible for the formation of cracks, as well as oxidation. Processing conditions for laser melting are typically a power density from 10 to 3000 MW/m2 (6.5 to 1935 kW/in.2) and an interaction time from 0.01 to 1 s. Inert gas shielding is used to prevent oxidation of the surface.

Examples of Laser-Melted Surface Microstructures Microstructural changes with laser melting are in the forms of grain refinement, solid solutions, and fine dispersions of precipitates. All of these can contribute to the hardening and strengthening of the surface. Laser-melted surfaces of cast irons appear dendritic, as shown in Fig. 6.6a for gray iron and in Fig. 6.6b for ductile iron [11]. Below the melt zone is the heat-affected zone, which appears in lighter contrast in Fig. 6.6.

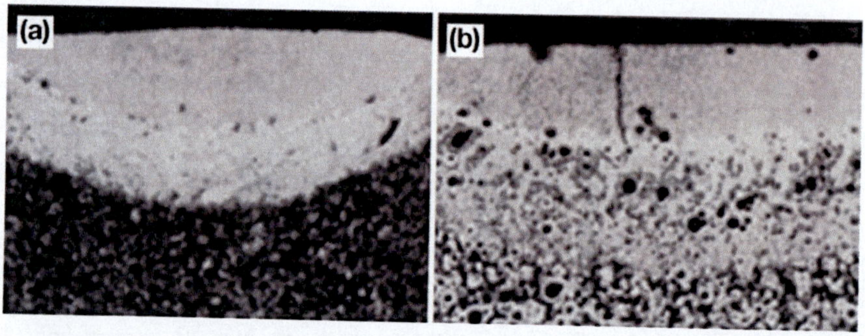

Fig. 6.6 Cross sections of laser-melted cast iron surfaces. (**a**) Gray iron. (**b**) Ductile iron. (Source: Ref [11])

In the solidified melt in cast irons, a ledeburite (mixture of austenite and cementite) structure generally forms [24, 25]. Hardening is caused by graphite dissolution to form cementite and austenite transformation to martensite. Molian and Baldwin [13] described the formation of predominantly dendritic ledeburite with small amounts of plate-like, high-carbon martensite and retained austenite in the melt zone in gray and ductile irons.

Chen et al. [26] determined that microstructures in laser-melted ductile iron depended on the solidification rate. Dendritic retained austenite with a continuous interdendritic carbide, and a microhardness from 400 to 650 HV, formed at high solidification rates. A lamellar mixture of ferrite and cementite plates, with a microhardness from 1000 to 1250 HV, formed at low solidification rates.

Bamberger et al. [27] found that in laser-melted gray cast iron, full dissolution of the graphite occurred, leading to the formation of hot tears on the surface. On the other hand, in laser-melted nodular cast iron, partial dissolution of the graphite occurred, resulting in increased ductility and no hot tearing. Steen [24] found that martensitic structures were formed in some stainless steels, whereas fine carbide dispersions were formed in tool steels.

In laser-melted titanium alloys, highly dislocated, fine, plate-like martensitic structures were reported by Steen [24]. In Ti-15Mo, the crystal growth was epitaxial from the β grains in the substrate in a direction normal to the interface [28]. In Ti-6Al-4 V, β martensite transformation in the solid state occurred upon rapid cooling. Plates of a twinned substructure were common. In Ti-10 V-2Fe-3Al, martensite with some formed upon solidification.

Wear Behavior of Laser-Melted Surfaces Bergmann [25] has summarized the wear properties of laser-melted cast irons. For laser-melted gray iron containing flake graphite in a pearlitic matrix, the dry pin-on-disk test demonstrated that the wear behavior improved by an order of magnitude and was better than a fully martensitic structure. In a test involving rolls that ran against one another with a fixed relative slip, it was found that the wear resistance of laser-melted ductile iron was superior to that of 0.6% C steel, case-hardened 16MnCr5 steel, nitrided or carburized 16MnCr5 steel, and a gas tungsten arc welded melted surface.

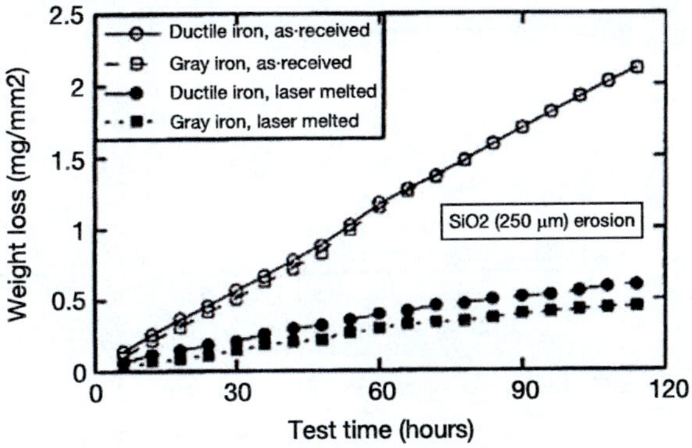

Fig. 6.7 Erosive wear behavior of as-received and laser-melted gray and ductile irons. (Source: Ref [29])

Ju et al. [29] reported significant improvement in erosive wear behavior in laser-melted ductile and gray irons, as shown in Fig. 6.7. This improvement was due to the presence of mechanically metastable austenite that transforms to martensite when plastic deformation of the near-surface region occurs during wear.

Tomlinson et al. [30] reported that laser melting reduced the amount of cavitation erosion of gray cast iron in distilled water by a factor of 0.3 and in 3% saltwater by a factor of 0.57.

Kusinski and Thomas [31] found that by laser melting, the wear resistance of Fe–Cr–Mn–C steels was improved by a factor of 1.4 to 1.6 over that of conventionally hardened steels only when the melt zone was shallow. This improvement was due to the formation of a stress-induced martensite with a highly dislocated lath and internally twinned structure.

De Beurs and De Hosson [32] found stress-induced hardening in laser-melted chromium steel. During wear, the austenite structure with segregated M_3C carbides, which had a microhardness of 550 HV, transformed to martensite, which had a microhardness of 900 HV. In laser-melted steels, the improvement in wear resistance was best when the dendritic growth of the austenite was perpendicular to the surface, because in this orientation the concentration of carbides was the highest [33].

Hsu and Molian [34] reported that the tool life of laser-melted M2 steel tool bits was from 200 to 500% higher than if they were conventionally hardened using catastrophic failure criterion (Fig. 6.8a). For laser-melted M35 steel tool bits, the tool life was from 20 to 125% higher than if they were conventionally hardened using flank wear failure criterion (Fig. 6.8b). High-alloy martensite, fine austenite grain size, and finely dispersed carbides all contributed to high hardness, good toughness, and low coefficient of friction.

Coquerelle and Fachinetti [35] laser melted Al–Si alloys and found that the resulting reduction in dendrite size and Si particle size improved wear resistance. Using a pin-on-disk wear measurement technique, the wear volume was reduced tenfold for a hypoeutectic alloy. The wear rate decreased with increasing silicon content as long as the silicon particle size was less than 2 m (80 in.).

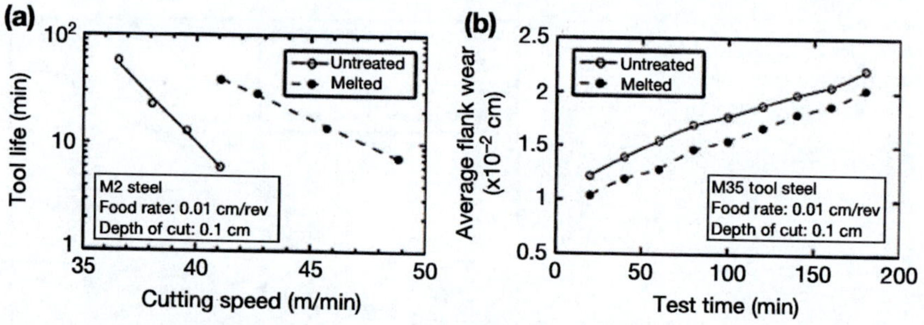

Fig. 6.8 Tool life of conventionally heat-treated and laser-melted tool bits. (**a**) M2 tool steel. (**b**) M35 tool steel. (Source: Ref [34])

Applications of Laser Melting The benefits of laser melting cast irons are achieving very high hardness and wear resistance on a relatively cheap material that is tough in the bulk, in the case of ductile iron, and easy to shape. Tool steels can also be laser surface melted to give very high hardness and wear resistance properties. Although several studies on specific alloys have been performed, there are very few commercial applications for laser surface melting to improve wear resistance. However, during laser transformation hardening, a very thin surface layer is sometimes melted, and the tribological benefits of laser melting are realized in the laser-hardening examples.

Bergmann [25] describes the application of laser melting to a cast iron rocker arm and camshaft. The laser-melted components achieved a lifetime that was four times better than was achieved by melting using gas tungsten arc welding. A German automotive manufacturer routinely uses laser melting to harden camshafts.

6.4 Laser Alloying

Processing A technique of localized alloy formation is laser surface melting with the simultaneous, controlled addition of alloying elements. These alloying elements diffuse rapidly into the melt pool, and the desired depth of alloying can be obtained in a short period of time. By this means, a desired alloy chemistry and microstructure can be generated on the sample surface; the degree of microstructural refinement will depend on the solidification rate. The surface of a low-cost alloy, such as mild steel, can be selectively alloyed to enhance properties, such as resistance to wear, in such a way that only the locally modified surface possesses properties typical of tribological alloys. This results in substantial cost savings and reduces the dependence on strategic materials. Typical processing parameters for laser alloying are a power density from 10 to 3000 MW/m2 (6.5 to 1935 kW/in.2) and an interaction time from 0.01 to 1 s. An inert shielding gas is normally used.

One method of alloying is to apply appropriate mixtures of powders on the sample surface, either by spraying the powder mixture suspended in alcohol to form a loosely packed coating or by coating a slurry suspended in organic binders [36]. The use of metal powders in laser alloying is the least expensive, but, with appropriate process modifications, alloys in the form of rods, wires, ribbons, and sheets can also be added. Because of inconsistency in coating application and possible loss during processing, the composition of the surface alloy may not reflect that of the applied alloying elements.

Powders that are added in controlled quantities using powder feeders with electronic metering can reduce this problem. Powders usually range from 10 to 50 m (0.4 to 2 mils); the finer size facilities dissolution and alloying. Typically, the powder flow rate is from 0.005 to 0.1 cm3/s (0.0003 to 0.006 in. 3/s), particle velocity is from 1 to 5 m/s (3.3 to 16.4 ft./s), and carrier gas velocity is from 3 to 10 m/s (10 to 33 ft./s). Whether preplaced or fed, the powder increases the coupling coefficient of the surface, thereby avoiding the need for special absorptive coatings.

During surface alloying, temperature gradients form in the melt pool, which, along with the addition of alloying elements, influences the surface tension. Convection currents are established as a result of the surface tension gradients and can cause variations in concentration. If the laser beam is oscillated, the melt spreads out over a wider region and, because the beam sweeps the same area several times, a potentially beneficial mixing action can occur. Any alloy concentration gradients that may have formed initially are thus diminished. If the melt pass is made with a very rapid sample translation rate that is greater than 50 mm/s (2 in./s), then inhomogeneities in the microstructure can result, but the high quench rate ensure minimal segregation. The rapid quench also facilitates alloying with hard-to-alloy elements, such as iron, chromium, carbon, and manganese. Examples of laser alloying are cast irons with chromium, silicon, and carbon; steels with chromium, molybdenum, boron, and nickel; stainless steels with carbon; aluminum with silicon, carbon, nitrogen, and iron; and superalloys with chromium.

Another method of laser alloying is gas reaction, for which a shielding gas of appropriate composition is chosen [37]. For the nitriding of titanium alloys, a dilute (10 to 20%) mixture of N_2 in argon is used as the reactive gas. The extent of nitriding depends on the partial pressure of N_2 in the atmosphere, which is determined by the law of mass action, and the temperature, which depends on the laser power density. Alloys of titanium can be carburized by using a methane/argon mixture and carbonitrided by using a methane/N_2 mixture.

Folkes et al. (Ref 25) nitrided titanium alloys by blowing N2 with a flow rate of 100 to 250 cm3/s (6 to 15 in. 3/s) onto the melted surface. Influenced by the mixing effects derived from surface tension forces and from the impingement of the N_2 jet on the surface, nitrogen was uniformly distributed, and TiN formed throughout the melt zone. The authors also carbonitrided titanium alloys by applying a carbon-based paint over the surface and then laser melting in the presence of a N_2 gas jet.

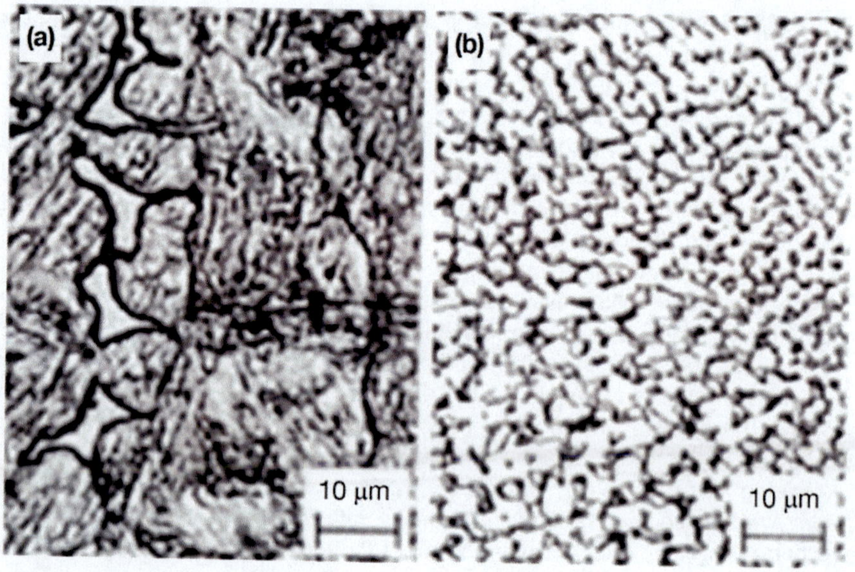

Fig. 6.9 Microstructures of laser-alloyed 1018 steel. (**a**) Addition of 16% Cr. (**b**) Addition of 43% Cr and 4.4% C. (Source: Ref [36])

Examples of Laser-Alloyed Surface Microstructures With laser alloying, a microstructure that strengthens the surface is produced to develop resistance to wear. Surface alloying of AISI 1018 steel with carbon and chromium produces stable carbides, such as M7C3 and M3C, in austenitic, pearlitic, or martensitic matrices [36]. The microhardness of these carbides is from 1100 to 1200 HK and, when uniformly dispersed in a pearlitic or martensitic matrix, results in an improved resistance to abrasive and adhesive wear. With a 16% addition of chromium, the microstructure of laser-alloyed 1018 steel is martensitic, with small islands of ferrite, as shown in Fig. 6.9a. With additions of 43% Cr and 4.4% C, the microstructure consists of hexagonal-shaped M_7C_3 carbides, as shown in Fig. 6.9b.

Marsden et al. [38] found that laser melting of graphite-coated stainless steels produced dendritic austenite and interdendritic carbides with a microhardness of 600 HV, whereas carbon-powder-injected microstructures contained martensitic cells, with segregated carbon and chromium forming a small amount of eutectic in the intercellular regions. With increasing carbon content, the hardness was increased to 1500 HV, because of the increase in carbide volume.

The reactivity of titanium makes its alloys ideal targets for laser alloying by gas reaction. TiC, TiN, and carbonitrides form as dendrites in a titanium matrix throughout the melted area. The matrix is α in Ti-6Al-4 V [37] and β in Ti-15Mo [28]. Laser gas alloying can also be used to increase the carbon content of low-carbon steels and to produce ledeburite structures on steels.

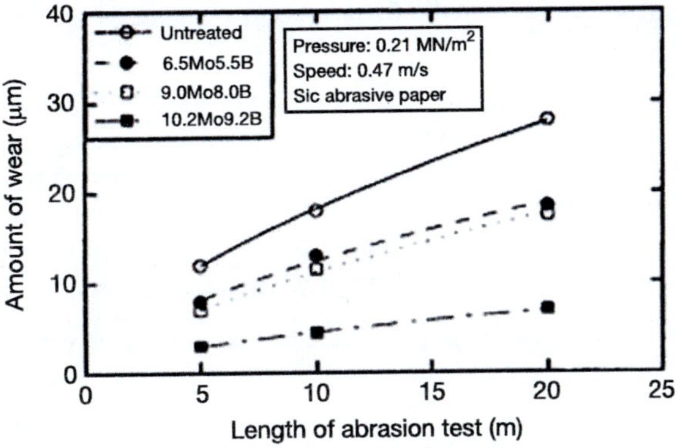

Fig. 6.10 Wear behavior of untreated and laser-alloyed ferritic stainless steel. (Source: Ref [40])

Wear Behavior of Laser-Alloyed Surfaces Yu and Quan [39] laser alloyed steel with carbon, nitrogen, and boron and found that the abrasive wear resistance improved 13 times, compared to that of unalloyed steel. This improvement was due to the hardening caused by a uniform dispersion of nitrides, carbides, and borides in a fine dendritic ferritic matrix. Rieker et al. [40] laser alloyed a free-machining ferritic stainless steel with various combinations of molybdenum and boron by applying the elements as paint. For chemical and structural homogeneity, a second re-melting was done. As shown in Fig. 6.10, the wear behavior improved under all alloying conditions. A microcrystalline ferritic structure with a dispersion of boride particles was responsible for the improved abrasive wear resistance.

Using a dry sand–rubber wheel abrasion test, Tan and Doong [41] found that when AISI 1018 steel was layer alloyed with chromium, the abrasion resistance improved because of the formation of a martensite–ferrite structure. Xiaoping et al. [42] laser alloyed low-carbon steel with two Co–Cr–W combinations. With 17Co7Cr3W (815 HV), abrasive wear resistance was better, whereas with 52Co22Cr9W (625 HV), corrosive wear resistance was better. In the former, martensite formed, whereas in the latter, a mixture of γ -Co and $Cr_{23}C_6$ formed.

Huntz et al. [43] improved the pin-on-disk wear behavior of 70Ni30Cr by laser melting the alloy that was precoated with carbon. The formation of a variety of chromium-rich carbides in a-phase solid solution contributed to the improvement in wear resistance.

Mordike [37] reported improved rolling and adhesive wear resistance of titanium by laser gas alloying. The increased surface hardness improved resistance to penetration and, thus, the resistance to plowing wear and erosion.

Applications of Laser Alloying Laser alloying has been primarily applied to improved corrosion resistance. A very common technique is to alloy steels with chromium. The only examples of laser alloying to improve wear resistance are exhaust valves fabricated by Fiat Research Laboratory.

6.5 Laser Cladding

Processing In laser cladding, a necessarily different material is fused to the substrate surface with minimal dilution. The cladding has a microstructure and properties that are entirely different from those of the substrate. Lasers enable the cladding of high-melting-point alloys over low-melting-point workpieces [37]. Cobalt-, nickel-, and iron-based hard-facing alloys are the usual cladding materials used for metal-to-metal contact, impact, erosion, and abrasion wear resistance.

Cladding material selection depends on its metallurgical compatibility with the substrate material. Alloys that form brittle intermetallic phases with the substrate are undesirable. For example, cobalt forms harmful intermetallics with titanium, and cobalt-base cladding alloys would be unsuitable on titanium alloy substrates.

The cladding alloy is melted by a laser beam and allowed to spread freely and freeze over the substrate. The beam also melts a very thin layer of the substrate, which combines with the liquid cladding to the least extent necessary and solidifies to form a strong metallurgical bond. A good fusion bond can be achieved with a dilution zone that is only from 10 to 20 m (0.4 to 0.8 mil) thick.

The clad alloy can be in several forms, examples of which are a pre-alloyed powder that is applied to the sample surface with or without a binder, a self-fluxing powder that is flame sprayed, a hard-facing alloy that is plasma sprayed, or a chip that is preplaced. Laser consolidation of these coatings results in densification and smoothening, eliminates channels to the substrate, improves the bonding between coating and substrate, and reduces porosity, all of which contribute to the strength and integrity of the clad layer. Processing parameters for laser cladding are a power density that ranges from 10 to 1000 MW/m^2 (6.5 to 645 kW/in.^{2}) and an interaction time from 0.1 to 1 s. The shielding gas could be any of the inert gases or a combination of gases, such as He/Ar and H_2/Ar.

It is sometimes difficult to keep preplaced powder in place as it is melted by the laser beam. The external addition of powder avoids this problem. The clad material can be gravity-fed just ahead of the beam spot, so that as the sample translates, the laser beam melts and fuses the powder to the workpiece.

With the help of a suitable carrier gas, the powder can also be blown into the laser-melted pool. In this case, the melt pool is mostly clad material, with the leading edge incorporating a small portion of the substrate surface. Particle size usually ranges from 45 to 150 m (1.8 to 6 mils), with the larger size helping to minimize alloying.

The amount of dilution and clad thickness are controlled by the interaction time and the powder flow flux. Desirable dilutions are from 3 to 8%. Clad thickness can be varied from 0.05 to 4 mm (2 to 155 mils). The optimum powder flow flux is one that gives minimum dilution and maximum deposition [4]. The lowest particle velocity needed to uniformly convey the powder to the melt pool is desirable to minimize losses from the molten pool. The powder delivery nozzle is positioned in such a manner as to flood the entire melt pool with powder. The feeding angle is generally from 35° to 45° to the horizontal, and the feed tube, which typically has a

3 mm (0.12 in.) diameter, is positioned from 10 to 12 mm (0.4 to 0.5 in.) from the substrate. Typically, the powder flow rate is from 0.005 to 0.1 cm^3/s (0.0003 to 0.006 in^3/s), particle velocity is from 1 to 2 m/s (3.3 to 6.6 ft./s), and carrier gas velocity is from 3 to 7 m/s (10 to 23 ft./s). Uniform feeding of the powder ensures uniform clad layers. Unlike plasma-sprayed coatings, porosity and unmelted powder particles are almost never observed within a laser-clad layer.

The shape of the cladding can be controlled within certain limits, the most preferred shape being one with an obtuse angle to the substrate. Multiple clad tracks produced by overlapping such a clad layer will have a good surface finish and be porosity-free. Cladding is commonly achieved by relatively large "doughnut-shaped" beams, or by somewhat focused beams that are oscillated. These ensure a desirable clad profile. Clad material can also be added in rod, wire, or sheet form, but special procedures are needed because of reflectivity problems.

Thermal stresses in the cladding can cause harmful cracking, but this can be eliminated by an appropriate preheating practice. Low power densities, large beam diameters, and slow sample translation rates tend to produce crack-free clad layers.

Ayers and Gnanamuthu (Ref 33) give several examples of laser cladding for improved wear resistance, such as Tribaloy T-800 on powder-coated ASTM A 387 steel; Haynes Stellite alloy No. 1 cast rods placed on AISI 4815 steel; and a mixture of WC and Fe powder applied without binder on AISI 1018 steel.

In addition to stainless steel on mild steel, Steen [4] has reported successful laser cladding of nickel, bronze, Fe–B, and Stellite SF6 on mild steel, chromium on titanium, stainless on aluminum, mild steel on stainless, and stellite on brass. Other examples of laser cladding are AISI 1016 steel clad with a mixture of chromium, manganese, and carbon powders [45] and gray cast iron clad with a powder-fed mixture of molybdenum, Cr_2C_3, chromium, nickel, and silicon [45].

Examples of Laser-Clad Layer Microstructures The key to improved wear resistance is the presence of hard phase distributions in the laser-clad microstructures. The wear resistance of Tribaloy is attributed to the presence of hard Laves intermetallic phases surrounded by solid solution and a eutectic matrix [36]. In T-800 alloy, the Laves phases make up nearly 50 vol% of the microstructure (Fig. 6.11a), and the cladding has an average microhardness of 690 HK. Stellite alloy No. 1 contains M7C3 and M6C carbides finely dispersed in an austenitic cobalt matrix (Fig. 6.11b) and has a microhardness of 730 HV. Microstructural analysis of the WC–Fe mixture revealed a distinct region on the WC particles, which formed because of interaction with the liquid iron. Microhardness of the WC particles was 1100 HK, and that of the surrounding matrix was 870 HK.

The Fe–Cr–Mn–C cladding of Mazumder and Singh [44] had a microstructural consisting of M_7C_3 and M_6C type carbides in a ferritic matrix, a solid solubility extension of chromium in ferrite by 50%, and a microhardness of 550 HV. The molybdenum-clad coating on gray iron of Belmondo and Castagna [45] consisted of molybdenum dendrites surrounded by a Cr–Ni matrix containing Cr_2C_3, which had a microhardness of 900 HV.

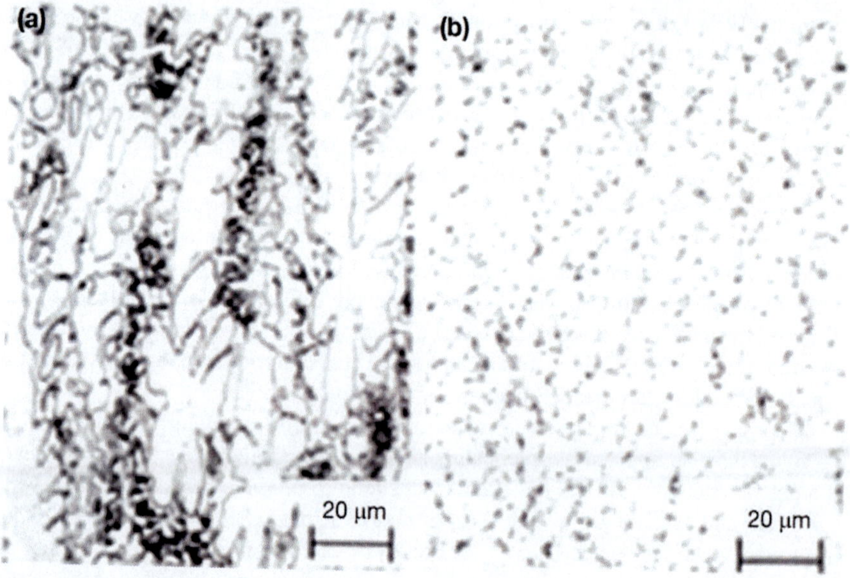

Fig. 6.11 Examples of laser-clad microstructures. (**a**) Tribaloy T-800 alloy on ASTM A 387 steel. (**b**) Haynes Stellite alloy No. 1 on AISI 4815 steel. (Source: Ref [36])

Wear Behavior of Laser-Clad Layers Mazumder and Singh [44] found that the Fe–Cr–Mn–C cladding on AISI 1016 steel resulted in tribological properties that were superior to Stellite 6. The width of the wear scar was reduced from 3.5 mm (0.14 in.) in the base alloy to about 0.6 mm (0.02 in.) in the laser cladding. Under the same test conditions, Stellite 6, a common hard-facing alloy, developed a wear scar which of 1.4 mm (0.06 in.).

Abbas et al. [46] laser clad a mild steel with Stellite 6, Alloy 4815, and their composites with SiC by using pneumatic powder delivery. Wear tests, conducted by grinding the samples against a revolving alumina disk, showed that the composite clad samples had the best abrasive wear resistance (Fig. 6.12).

Belmondo and Castagna [45] performed wear tests on the Mo–Cr_2C_3 clad cast iron samples using a reciprocating motion testing machine. When compared with plasma-sprayed coatings of similar composition, the performance of laser-clad coatings was far superior under all testing conditions, especially at the highest pressures. With a load of 25 MPa (3.6 ksi) and slider roughness of 0.3 m (12 in.), the wear loss in the laser sample was 0.4 mg (1.4 × 10–5 oz), compared to 1.3 mg (4.6 × 10–5 oz) in the plasma-sprayed sample. In addition, the plasma-sprayed coating developed cracks at the interface, which seriously impaired its integrity.

Molian and Hualun [47] performed pin-on-block reciprocating sliding wear tests on Ti-6Al-4 V alloy substrate laser clad with BN, both with and without a NiCrCoA1Y addition, and found that the wear resistance improved from 10 to 200 times more than that of age-hardened and laser-melted samples (Fig. 6.13).

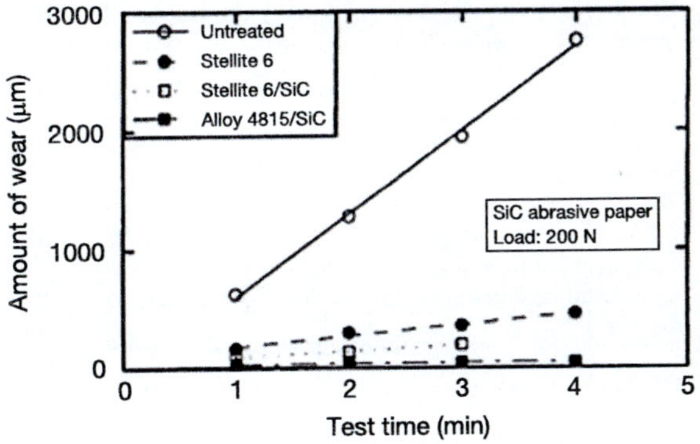

Fig. 6.12 Abrasive wear behavior of mild steel laser clad with Stellite 6, Stellite 6/SiC, and Alloy 4815/SiC. (Source: Ref [46])

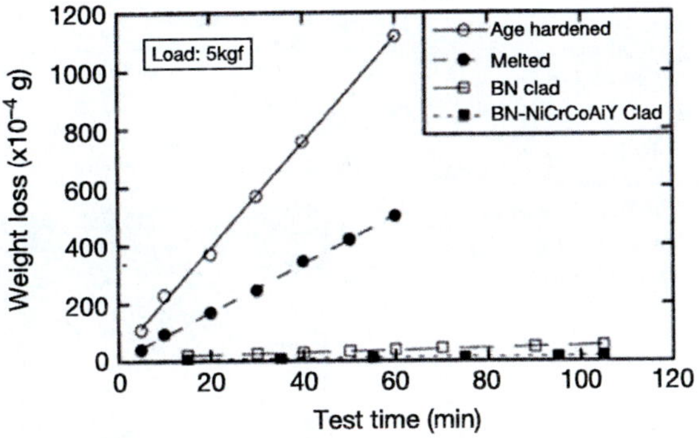

Fig. 6.13 Sliding wear behavior of laser-clad Ti-6Al-4 V alloy. (Source: Ref [47])

The presence of solidification reaction products, TiN and TiB2, resulted in a cladding microhardness of 1600 HV, which led to improved wear resistance by preventing ploughing and reducing friction.

Boas and Bamberger [48] determined the abrasive wear characteristics of plasma-sprayed and laser-melted plasma-sprayed coatings of Tribaloy T-400 on AISI 4130 steel using a block-on-cylinder test. The plasma-sprayed coating was prone to rapid wear by delamination, whereas laser consolidation of the plasma coating removed flaws, such as porosity and micro-fissures and generated an adherent wear-resistant layer. Figure 6-14 demonstrates the improvement in wear after laser consolidation of the plasma coating. For comparison, results on a hard D2 steel are also given.

properties of components, several applications have been found for the technique. Some of them are in the exploratory stage, others are in the pilot-plant stage, and a few have reached the production stage.

Bruck [49] demonstrated cladding within a confined space by coating the inside surface of small-bore pipes (inside diameter of 50 to 100 mm, or 2 to 4 in., and length of 0.3 to 1.2 m, or 1 to 4 ft), as shown in Fig. 6.15. An oscillating mirror placed in the pipe formed a 12.5-mm (0.5 in.)-wide melt pass, into which the clad alloy powder was fed. The pipe was rotated and translated to cover the entire inner surface.

Previously, chrome plating was used, but the laser technique appreciably improved galling resistance. Using suitable cooling techniques, the pipe temperature was maintained below 500 °C (930 °F) to avoid deterioration of the core properties, as well as distortion (Fig. 6.14).

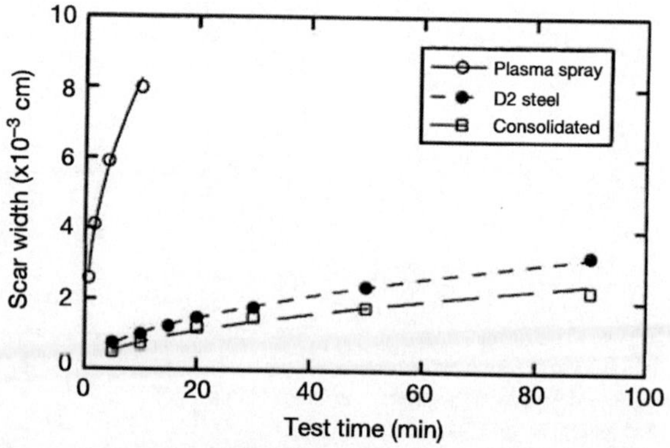

Fig. 6.14 Wear scar growth curves of plasma-coated 4130 steel before and after laser consolidation. (Source: Ref [48])

Fig. 6.15 Laser cladding of small-bore pipes. (Source: Ref [49])

La Rocca [22] describes laser cladding of exhaust valves with Stellite at an Italian automotive manufacturer. The Stellite powder was fed into the laser melt pool. The laser-clad material was superior to gas tungsten arc welded material in terms of thickness uniformity and uniform microstructure and elemental distributions and had better adhesion. Powder utilization also was better (30% that of gas tungsten arc welding), and over-metal removal was reduced by 10–15%.

One British automotive manufacturer uses production-stage laser cladding of a nickel alloy turbine blade shroud interlock by powder feeding either Tribaloy or Nimonics [50]. Previously, a manual arc melting technique was used. Laser cladding reduced cladding time from 14 min to 75 s, improved productivity and quality, reduced cost by 85%, and reduced powder consumption by 50%. Table 6.2 lists components that are laser clad for commercial applications.

Table 6.2 Component, cladding alloy, and cladding technique

Component	Clad alloy	Cladding technique	Company	References
Gate and seat of steel valves for oil-field, geothermal, and nuclear energy production and chemical processing	Stellite 6	Powder feed	W-K-M Div. of Joy Mfg., Houston, TX	[51]
Diesel engine valve	Stellite SF6	Preplaced powder paste	Cabot Corp., Kokomo, IN	[52]
Gate valve seat	Deloro 60	Preplaced powder paste	Cabot Corp., Kokomo, IN	[52]
Gas turbine blade "Z" notch	Co-based alloy	Powder feed	Quantum Laser, Edison, NJ	[53]
Leading edge of turbine blades	Co-based alloy	Powder feed	Quantum Laser, Edison, NJ	[53]
Pump bushings and impellers	Co-based alloy	Powder feed	Quantum Laser, Edison, NJ	[54]
Engine valve bases	Ni–Cr alloy	Powder feed	Quantum Laser, Edison, NJ	[54]
Oil-field valve gates	WC and Co alloy	Powder feed	Quantum Laser, Edison, NJ	[54]
Tractor bushings	Stellite 6	Powder feed	Quantum Laser, Edison, NJ	[54]
Turbine blades, pump valve seats	Stellite, Colmoloy,	Preplaced beds,	Westinghouse, Pittsburgh, PA	[55]

(continued)

Table 6.2 (continued)

Component	Clad alloy	Cladding technique	Company	References
	blended powder	gravity feed		
Valve stem, valve seat, aluminum block	CrC2, Cr, Ni, Mo/cast Fe	Preplaced powder	Fiat, Turin, Italy	[55]
Off-shore drilling and production parts, valve components, boiler firewall	Stellite, Colmonoy, alloys/ carbides	Powder feed	Combustion Eng., OH	[55]
Aerospace components	Stellite, Tribaloy	Powder feed	Rockwell Int., CA	[55]
Turbine blade shroud interlocks	PWA 694, Nimonics	Preplaced chip	Pratt & Whitney, West Palm Beach, FL	[55]
Turbine blade "Z" notch	Hardfacing	Powder feed	United Technol., E. Hartford, CT	[56]
Swivel joints, petroleum valves	Hardfacing	Powder feed	Spectra physics, San Jose, CA	[56]
Gas engine turbine blades	Hardfacing	Powder feed	General Electric, SC	[57]

6.6 Laser Melt/Particle Injection

Processing Laser melt/particle injection produces an in situ, metal matrix/particulate composite surface layer by mixing, but not melting, the second phase with the substrate. The particulate material is injected with sufficient velocity as a spray into the melt pool formed by the laser beam. If the second phase is hard, such as a carbide, the injected layer can be made to resist wear.

Figure 6.16 depicts the laser injection process. An oscillating CW CO2 laser beam produces a shallow melt pool on a substrate, into which are injected powder particles via a nearby nozzle. As the sample translates, an injected layer forms on the surface. By varying the amplitude of oscillation, melt widths ranging from 3 to 20 mm (0.12 to 0.8 in.) are possible [58]. Processing conditions for laser melt/particle injection are a power density from 10 to 3000 MW/m^2 (6.45 to 1935 kW/in.^{2}) and an interaction time from 0.1 to 1 s. Any inert shielding gas is normally used.

The particles are propelled by pressurized helium gas. The pressure depends on the particle size and the relative densities of the powder and the molten substrate. Larger and heavier particles require lower gas pressures. Typically, powder flow rates are from 0.1 to 0.5 cm3/s (0.006 to 0.03 in.3/s) and particle velocities are a few

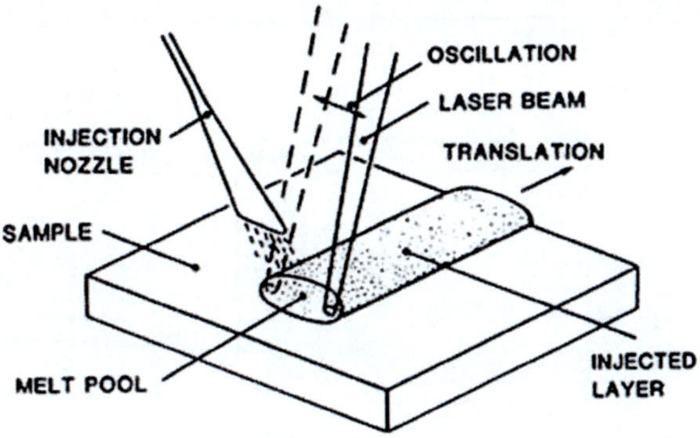

Fig. 6.16 Laser melt/particle injection process

m/s. Carrier gas pressures typically range from 50 to 120 kPa (7.3 to 17.4 psi). The carrier gas also serves as a cover gas and keeps the particles relatively cool, preventing them from melting or fusing. Particle sizes range from 45 to 150 m (1.8 to 6 mils). Finer particles tend to either fuse or dissolve in the melt, whereas coarser particles do not flow easily.

The injection nozzle, usually made of copper, is inclined at 60° to the horizontal and is positioned from 10 to 20 mm (0.4 to 0.8 in.) away from the sample. Its slotted opening is designed to produce a rectangular spray that is of the same size as the melt pool, to ensure that most of the powder is incorporated into the melt.

Although minimum interaction between the carbide phase and the melt is desired, some dissolution does occur. It was found that the degree of carbide dissolution, as well as other physical characteristics of the injected layer, such as penetration depth, mounding, and carbide volume, could be reasonably controlled. Variations in these characteristics, as well as hardness, with laser power and powder feed rate, are determined for Ti-6Al-4 V alloy injected with TiC (Fig. 6.17) [59]. The carbide volume can be varied from 15% to a limit of 60%. For improved wear resistance, higher carbide volumes are desirable, but the conditions that produce increased carbide volume also increase dissolution. An estimation of the degree of carbide dissolution is given by matrix microhardness. As the trends in Fig. 6.17 show, matrix microhardness is higher when the carbide volume is greater.

Initial work on laser melt/particle injection was done on low-density/high-strength aluminum and titanium alloys, for which few surface-hardening methods exist. Some of the alloy/particulate systems that have been investigated are 5052 Al/TiC, Al bronze/TiC, Ti-6Al-4 V/TiC, Ti-6Al-4 V/WC, 304 stainless/TiC, 4340 tool/TiC, Inconel 625/TiC, and Inconel 625/WC [60].

Microstructures of Laser Melt/Particle-Injected Layers The injected layer is a composite of particles surrounded by a metal matrix. It is desirable that the only surface modification be the presence of the particulate phase and that the surrounding metal should remain unchanged and retain most of the substrate properties, such as corrosion resistance and toughness. Figure 6.18 shows the cross-sectional view of a WC-injected layer on Inconel 625 alloy substrate. The top surface is rough but is

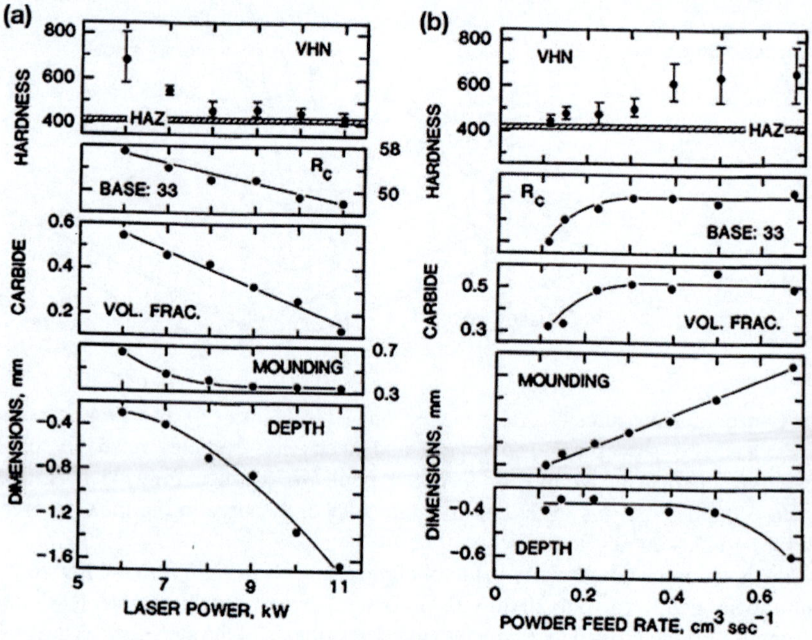

Fig. 6.17 Effect of processing parameters on injected layer characteristics in Ti-6Al-4 V injected with TiC. (**a**) Laser power. (**b**) Powder feed rate. (Source: Ref [59])

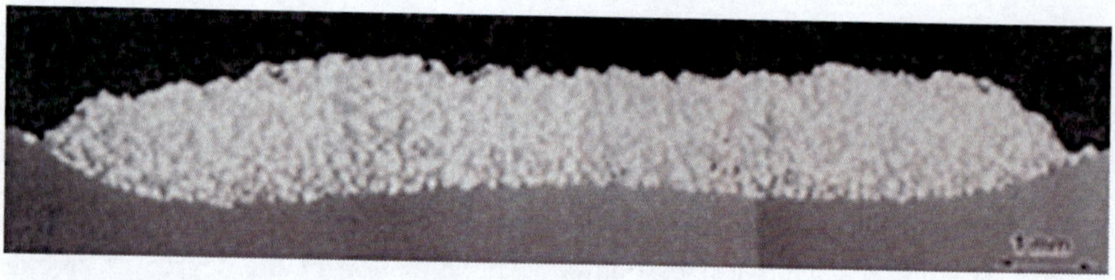

Fig. 6.18 Cross section of Inconel 625 alloy injected with WC

free of deep channels. The light grains are the WC particles, which are surrounded by the dark Inconel. The carbides are uniformly distributed throughout the injected layer and occupy about 50% of the total volume. Because the matrix phase is essentially made of the same material as the substrate, there is chemical continuity across the interface and a strong metallurgical bond.

Carbide dissolution products formed during solidification can influence the matrix microstructure. The carbide particles sometimes develop a perturbed or scalloped, highly alloyed interface with the metal matrix. Eutectic and dendritic carbides are some of the dissolution products that appear within the matrix [61]. Figure 6.19 shows examples of some of the dissolution products. Besides

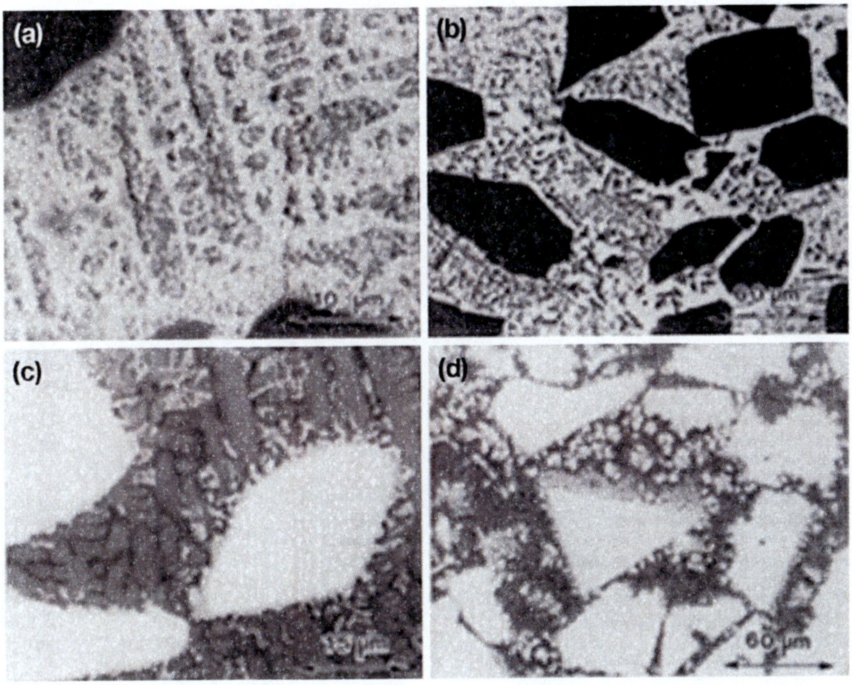

Fig. 6.19 Carbide morphological changes and matrix re-solidification products as a result of carbide phase dissolution in Inconel 625 alloy. (**a**) Injected with TiC, eutectic carbides. (**b**) Injected with TiC, dendritic carbides. (**c**) Injected with WC, eutectic carbides. (**d**) Injected with WC, dendritic carbides

influencing properties such as microhardness and friction wear, these re-solidification products can cause matrix embrittlement and microcracking, although with suitable preheating, cracking can be eliminated [62].

In order to retain some of the substrate properties in the modified surface, it is important to keep carbide dissolution to a minimum. Some of the matrix microstructural changes are inevitable. Depending on the base alloy, melting and rapid solidification can result in microstructural refinement, formation of solid solutions, phase transformations, and precipitation. These microstructural modifications, along with the dissolution products, can harden the matrix by various mechanisms. Carbide dissolution products harden the matrix by a dispersion-hardening mechanism. This effect was observed in stainless steel, tool steel, Inconel, and titanium alloys. Transformation hardening comes into play when martensite forms in steels and titanium alloys. In tool steels, precipitation hardening also plays a role, and in aluminum bronze, microstructural refinement results in modest hardening [60].

Wear Behavior of Particle-Injected Surface Layers On a macroscopic scale, hardening of particle-injected alloys is due to plastic flow inhibition by the injected phase. The relative increase in macro-hardness of the composite surface is from 1.1

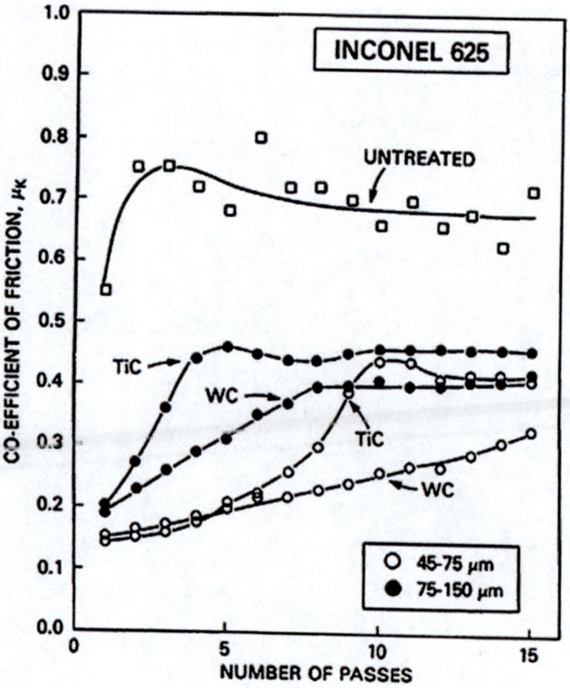

Fig. 6.20 Coefficient of friction as a function of number of slides in untreated and particle-injected Inconel 625. (Source: Ref [63])

to 2.6 times, depending on the alloy [60]. On a microscopic scale, hardening of the metal matrix varies with the type of alloy and has been found to affect friction wear behavior [63].

Friction wear was evaluated using a balt-on-surface test, which measures the coefficient of kinematic friction, μ_k, between a hard steel ball and a polished sample surface. A pressure of 1 N (0.22 lbf) and a slide velocity of 0.0001 m/s (0.02 ft./min) were used. This test is sensitive to minute changes in microstructure. For untreated Inconel 625, μ_k was about 0.7. Particle-injected Inconel 625 had a k that ranged from 0.15 to 0.20 after the first slide, which increased to a range of 0.3 to 0.4 for WC and 0.4 to 0.45 for TiC after several slides (Fig. 6.20). The wear scar was hardly evident in the optical microscope, and there was very little abrasive damage. Although μ_k increased with number of slides, it was not due to continuous wear of the sample surface. Rather, it was due to formation of wear debris.

The wear mechanism was different for the two carbides. With WC injection, the carbide and the harder (600 to 650 HV) metal matrix were strong enough to cause the slider to wear, and the wear debris was mostly that from the steel ball. With TiC injection, the wear debris was a mixture of material from the steel ball and from the softer (400 to 500 HV) metal matrix [63].

From dry sand–rubber wheel tests done on aluminum- and titanium-based alloys injected with TiC, it was found that the wear volume decreased rapidly with modest

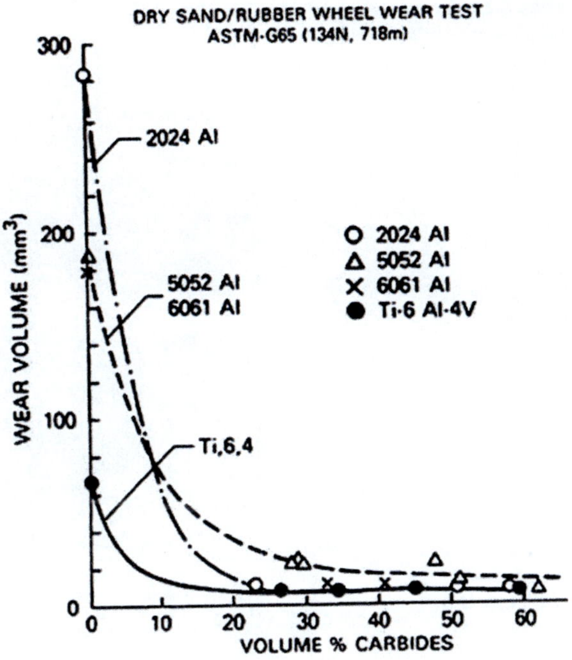

Fig. 6.21 Reduction in abrasive wear rate with increase in vol% carbide in particle-injected aluminum and titanium alloys. (Source: Ref [64])

volumes of the carbide phase. For aluminum, the reduction is from 0.18 to 0.28 cm3 (0.01 to 0.02 in.3) to 0.02 to 0.025 cm3 (0.0012 to 0.0015 in.^{3}). For titanium, it is from 0.06 to 0.065 cm^3 (0.0037 to 0.0040 in.3) to 0.01 cm^3 (0.0006 in.3) (Fig. 6.21) [64]. Examination of the worn surface showed that the softer metal matrix eroded but that the hard carbide particles prevented further erosion of the surface. The abrasive action did not uproot the carbides from the matrix, but the normally angular carbides appeared slightly rounded, which helped reduce further erosion.

Ayers and Bolster [65] found that abrasive wear with 3 and 30 μm (120 and 1200 in.) diamond particles of aluminum and titanium alloy samples was reduced by introducing WC or TiC into their surfaces. The wear resistance of aluminum alloys improved by a factor of 30 in samples containing TiC, whereas for Ti-6Al-4 V containing TiC, the improvement was less dramatic (factor of 4).

Applications of Laser Melt/Particle Injection Process Most of the laser melt/particle injection processing work has involved process optimization, microstructure evaluation, and test sample preparation. One application involved fabrication of wear-resistant surfaces on Inconel alloy shaft seal test rings, such as the one shown in Fig. 6.22 [66]. The injectant is WC and the circular pass is 127 mm (5 in.) in diameter and 10 mm (0.4 in.) wide. The coverage rate was 130 mm^2/s (0.2 in.^{2}/s). Before testing, the rough surface was ground flat. The seal ring was successfully tested against a graphite mating surface using water as the lubricant.

Fig. 6.22 Inconel alloy shaft seal test ring particle-injected with WC. (Source: Ref [66])

References

1. ASM Handbook Volume 1 Properties and Selection Irons, Steels, and High-Performance Alloys (06181) by Rudolf Steiner, American Society for Metals
2. Meister, D.: *Principles of Atoric Lens Design*, in: *Lens Talk*, Vol. 27, No. 3 (Jan. 1998)
3. Volk, D.: *Aspheric Lenses* Archived 2012-03-12 at the Wayback Machine, in *Duane's Ophthalmology*, chapter 50 (Lippincott, Wilkins & Williams/Wolters-Kluwer Health, Chicago, USA)
4. W.M. Steen, *Proceedings of the NATO Advanced Study Institute on Laser Surface Treatment of Metals,* C.W. Draper and P. Mazzoldi, Ed., Martinus Njihoff, The Netherlands, 1986, p 369
5. Sandven, *Heat Treating,* Vol 4, 9th ed., *Metals Handbook,* American Society for Metals, 1981, p 507
6. F.D. Seaman, *The Industrial Laser, Annual Handbook,* D. Belforte and M. Levitt, Ed., PennWell Books, Tulsa 1986, p 147
7. J. Taylor, *Metalwork. Prod.,* Sept 1979, p 138
8. J. Meijer, M. Seegers, P.H. Vroegop, and G.J.W. Wes, *Proceedings of the International Conference on Applications of Lasers and Electro-optics,* C. Albright, Ed., IFS Ltd., United Kingdom, 1986, p 229
9. Yessik, M., Scherer, R.P.: In: Metzbower, E. (ed.) *Sourcebook on Applications of the Laser in Metalworking*, p. 219. American Society for Metals (1981)
10. Courtney, C., Steen, W.M.: In: Metzbower, E. (ed.) *Sourcebook on Applications of the Laser in Metalworking*, p. 195. American Society for Metals (1981)
11. P.A. Molian, Iowa State University, Private Communication, 1992
12. Engle, S.L.: In: Metzbower, E. (ed.) *Sourcebook on Applications of the Laser in Metalworking*, p. 149. American Society for Metals (1981)
13. Molian, P.A., Baldwin, M.: *J. Tribol.* **108**, 326 (1986)
14. D.S. Gnanamuthu, *Sourcebook on Applications of the Laser in Metalworking,* E. Metzbower, Ed., American Society for Metals, 1981, p 324
15. G. Lu and H. Zhang, *Wear,* Vol 138, 1990, p 1
16. P.A. Molian and M. Baldwin, *J. Tribol.,* Vol 110, 1988, p 462

17. W.J. Tomlinson, R.F. O'Connor, and T.A. Spedding, *Tribol. Int.*, Vol 21 (No. 6), 1988, p 302
18. Trafford, D.N.H., Bell, T., Megaw, J.H.P.C., Bransden, A.S.: Heat Treating, Proceedings of the International Conference, p. 198. Metallurgical Society (1983)
19. J.A. Wineman J.E. Miller, *Sourcebook on Applications of the Laser in Metalworking,* E. Metzbower, Ed., American Society for Metals, 1981, p 209
20. M.J. Yessik, *Optical Eng.,* Vol 17 (No. 3), 1978, p 202
21. F.D. Seaman and D.S. Gnanamuthu, *Sourcebook on Applications of the Laser in Metalworking,* E. Metzbower, Ed., American Society for Metals, 1981, p 179
22. A.V. La Rocca, *Proceedings of the NATO Advanced Study Institute on Laser Surface Treatment of Metals,* C.W. Draper and P. Mazzoldi, Ed., Martinus Nijhoff, The Netherlands, 1986, p 521
23. General Motors, *Sourcebook on Applications of the Laser in Metalworking,* E. Metzbower, Ed., American Society for Metals, 1981, p 227
24. W.M. Steen, *Industrial Laser, Annual Handbook,* D. Belforte and M. Levitt, Ed., PennWell Books, 1986, p 158
25. H.W. Bergmann, *Proceedings of the NATO Advanced Study Institute on Laser Surface Treatment of Metals,* C.W. Draper and P. Mazzoldi, Ed., Martinus Nijhoff, The Netherlands, 1986, p 351
26. C.H. Chen, C.P. Ju, and J.M. Rigsbee, *Mater. Sci. Technol.,* Vol 4 (No. 2), 1988, p 161
27. M. Bamberger, M. Boas, and O. Akin, *Z. Mettallkde.,* Vol 79, 1988, p 806
28. Folkes, J., West, D.R.F., Steen, W.M.: In: Draper, C.W., Mazzoldi, P. (eds.) *Proceedings of the NATO Advanced Study Institute on Laser Surface Treatment of Metals*, p. 451. Martinus Nijhoff, The Netherlands (1986)
29. C.P. Ju, C.H. Chen, and J.M. Rigsbee, *Mater. Sci. Technol.,* Vol 4 (No. 2), 1988, p 167
30. W.J. Tomlinson, J.H.P.C. Megaw, and A.S. Bransden, *Wear,* Vol 116, 1987, p 249
31. Kusinski, J., Thomas, G.: *Proc. SPIE-Int. Soc. Opt. Eng*, vol. 668, p. 150. Society of Photo-optical Instrumentation Engineers (1986)
32. H. De Beurs and J. Th. M. De Hosson, *Scr. Met.,* Vol 21, 1987, p 627
33. H. De Beurs, G. Minholts, and J.Th. M. De Hosson, *Wear,* Vol 132, 1989, p 59
34. M. Hsu and P.A. Molian, *Wear,* Vol 127, 1988, p 253
35. G. Coquerelle and J.L. Fachinetti, *Proceedings of the 5th International Congress on Applications of Lasers and Electro-optics,* C.M. Banas and G.L. Whitney, Ed., IFS Ltd., United Kingdom, 1987, p 19
36. J.D. Ayers and D.S. Gnanamuthu, *Metals Handbook,* 9th ed., Vol 6, *Welding,* American Society for Metals, 1983, p 793
37. B.L. Mordike, *Proceedings of the NATO Advanced Study Institute on Laser Surface Treatment of Metals,* C.W. Draper and P. Mazzoldi, Ed., Martinus Nijhoff, The Netherlands, 1986, p 389
38. C. Marsden, D.R.F. West, and W.M. Steen, *Proceedings of the NATO Advanced Study Institute on Laser Surface Treatment of Metals,* C.W. Draper and P. Mazzoldi, Ed., Martinus Nijhoff, The Netherlands, 1986, p 461
39. Z.S. Yu and Z.K. Quan, *Mater. Chem. Phys.,* Vol 25, 1990, p 277
40. C. Rieker, D.G. Morris, and J. Steffen, *Mater. Sci. Technol.,* Vol 5, 1989, p 590
41. Y. Tan and J. Doong, *Wear,* Vol 132, 1989, p 9
42. J. Xiaoping, H. Zhuangqi, G. Yunlong, J. Ming, and S. Changxu, *Mater. Res. Soc. Symp. Proc.,* Vol 80, Materials Research Society, 1987, p 331
43. A.M. Huntz, T. Puig, L. Confignal, F. Charpentier, and M. Condat, *Mater. Sci. Eng.,* Vol A121, 1989, p 555
44. J. Mazumder and J. Singh, *Proceedings of the NATO Advanced Study Institute on Laser Surface Treatment of Metals,* C.W. Draper and P. Mazzoldi, Ed., Martinus Nijhoff, The Netherlands, 1986, p 297
45. A. Belmondo and M. Castagna, *Sourcebook on Applications of the Laser in Metal-working,* E. Metzbower, Ed., American Society for Metals, 1981, p 310
46. G. Abbas, D.R.F. West, and W.M. Steen, *Proceedings of the Materials Processing Conference,* Vol 69, Laser Institute of America, 1989, p 116

47. P.A. Molian and L. Hualun, *Wear,* Vol 130, 1989, p 337
48. M. Boas and M. Bamberger, *Wear,* Vol 126, 1988, p 197
49. G.J. Bruck, *J. Met.,* Vol 39 (No. 2), 1987, p 10
50. R.M. Macintyre, *Proceedings of the NATO Advanced Study Institute on Laser Surface Treatment of Metals,* C.W. Draper and P. Mazzoldi, Ed., Martinus Nijhoff, The Netherlands, 1986, p 545
51. P. Koshy, *Proc. SPIE-Int. Soc. Opt. Eng.,* Vol 527, R.R. Jacobs, Ed., Society of Photo-optical Instrumentation Engineers, 1985, p 80
52. S.J. Matthews, *Lasers in Materials Processing,* E. Metzbower, Ed., American Society for Metals, 1983, p 138
53. "Quantum Laser Technical Bulletin No. 1–86," Quantum Laser Corp., 1986
54. B.F. Kuvin, *Weld. Des. Fabr.,* May 1987, p 35
55. G.M. Eboo and A.E. Lindemanis, *Proc. SPIE-Int. Soc. Opt. Eng.,* Vol 527, R.R. Jacobs, Ed., Society of Photo-optical Instrumentation Engineers, 1985, p 86
56. D.A. Belforte, Belforte Associates, private communication, 1987
57. P. Mehta, General Electric, private communication, 1987
58. Cooper, K.P., Beigel, R., Slebodnick, P.: In: Banas, C.M., Whitney, G.L. (eds.) *Proceedings of the 5th International Congress on Applications of Lasers and Electro-optics*, p. 169. IFS Ltd, United Kingdom (1987)
59. K.P. Cooper, *International Conference on Surface Modifications and Coatings,* R.D. Sisson, Jr., Ed., American Society for Metals, 1986, p 409
60. K.P. Cooper, *Proc. SPIE-Int. Soc. Opt. Eng.,* Vol 957, G. Sepold, Ed., Society of Photo-optical Instrumentation Engineers, 1988, p 42
61. K.P. Cooper and J.D. Ayers, *Proceedings of the 1988 Conference on Laser Surface Modification,* American Welding Society, 1989, p 115
62. K.P. Cooper and P. Slebodnick, *J. Laser Appl.,* Vol 1 (No. 4), 1989, p 21
63. K.P. Cooper, J. Vac. *Sci. Technol.,* Vol A4 (No. 6), 1986, p 2857
64. J.D. Ayers, *Wear,* Vol 97, 1984, p 249
65. J.D. Ayers and R.N. Bolster, *Wear,* Vol 93, 1984, p 193
66. K.P. Cooper and J.D. Ayers, *Surf. Eng.,* Vol 1 (No. 4), 1985, p 263

Chapter 7
High-Power Laser-Matter Interaction and Related Computer Codes

7.1 Introduction

The Strategic Defense Initiative (SDI)—also known as "Star Wars"—wanted to have a defensive umbrella over the United States to shoot down incoming missiles from space. This program started during President Ronald Reagan's administration against incoming threat from adversaries, especially the Soviet Union around 1985. Soviets could not afford to keep up with this kind of science due to the total cost of ownership of such a complex system as well as being able to afford the cost of Research and Development (R&D). A holistic depiction of such system is presented in Fig. 7.1.

The arms race would not fizzle out until Mikhail Gorbachev came to power in the USSR in 1985.

However, as part of the Cold War Diplomacy and being able to negotiate with the USSR and forcing them to break down the Berlin Wall and unification of East Germany with the West, President Reagan launch such huge and complex program in the US industry, and as a side effect of economy and jobs creation, such a movement was very beneficial during the 1980s and 1990s and was a driven factor for economy enhancement. See Fig. 7.2.

During this era, President Regan stated that:

> *The defense policy of the United States is based on a simple premise: The United States does not start fights. We will never be an aggressor. We maintain our strength in order to deter and defend against aggression -- to preserve freedom and peace.* – Ronald Reagan

Reagan viewed the SDI as a defensive program, while Gorbachev believed it would escalate to an arms race in space.

The Strategic Defense Initiative (SDI), a former US government program, was responsible for the research and development of a space-based system to defend the nation from attack by strategic ballistic missiles. The program is now administered by the Missile Defense Agency (MDA) (originally the Strategic Defense Initiative Office), a separate agency in the US Dept. of Defense (DOD).

B. Zohuri, *Thermal Effects of High Power Laser Energy on Materials*,
https://doi.org/10.1007/978-3-030-63064-5_7

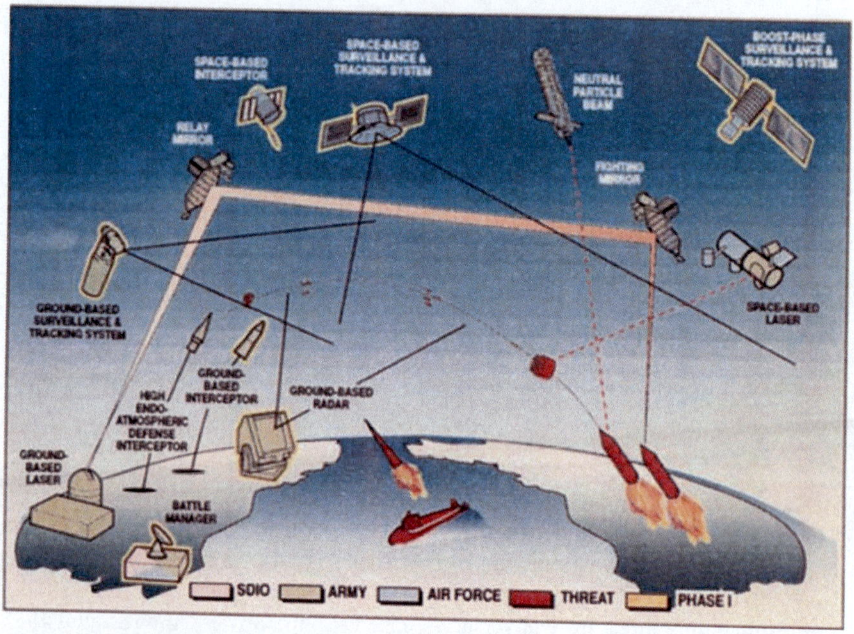

Fig. 7.1 Artistic depiction of Star Wars system and sub-systems

Fig. 7.2 Era of Reagan and Gorbachev

The main and primary goals of the Strategic Defense Initiative or now known as Missile Defense of Missile Offensive can be sated as follows:

- One of the most controversial issues discussed throughout these summits was the United States' Strategic Defense Initiative (SDI).
- In 1962, the Soviet Union had deployed missiles to Cuba, a close proximity to the United States. Although the missiles were removed, the United States realized the potential threat of incoming missiles.
- Reagan's refusal to eliminate SDI forced Soviets to compromise in the fear of potential consequences.

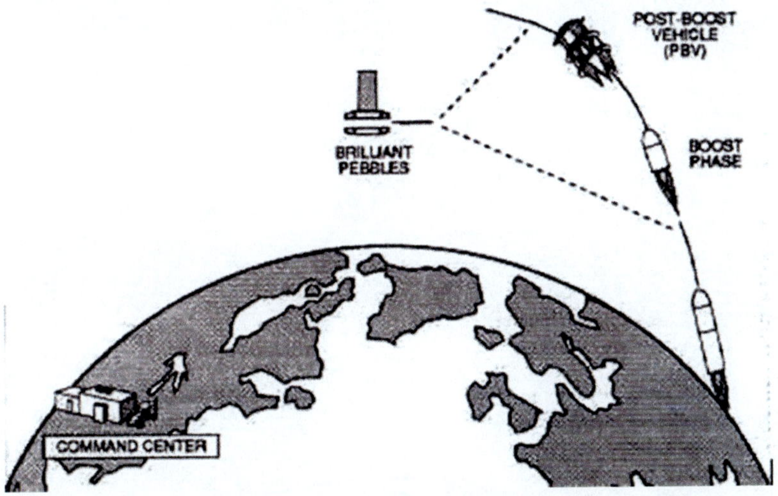

Fig. 7.3 GPALS architecture: space-based protection against ballistic missiles with a range greater than 600 kilometers. (Source: Strategic Defense Initiative Organization, *1991 Report to the Congress on the Strategic Defense Initiative* (May 1991), 2–5)

Furthermore, the Soviet Union had their own missile defense system that was kept highly confidential; however one of President Gorbachev's primary goals throughout the summit meeting was to terminate the American SDI Program, and as he stated:

> *We want to accomplish three things: ... Three, we want the Americans to stop proceeding with Strategic Defense Initiative.* – Mikhail Gorbachev

As part of the US strategy under Strategic Defense Initiative Office (SDIO) of past time (1980s–1990s) and present Missile Defense Agency (MDA), managing to architect a Global Protection Against Limited Strikes (GPALS) program, given recent experience with theater ballistic missile threats in the Persian Gulf Was and the fact that these threats will become more sophisticated in the future, the United States pursued development and deployment of advanced theater defenses by the mid-1990s as an urgent priority. These ground-based theater defense interceptor systems would provide midcourse and terminal defense capability. Interceptors could be based in-theater continuously (or moved to hotspots as needed), on ships, or on aircraft.

Figure 7.3 shows an artistic version of GPALS with sophisticated intercepting systems, and these interceptor systems would be cued and supported by space-, air-, and ground-based sensors.

Based in GPALS' architecture planning, the US Department of Defense under the supervision of the MDA organization has developed a US ground-based defense as illustrated in Fig. 7.4.

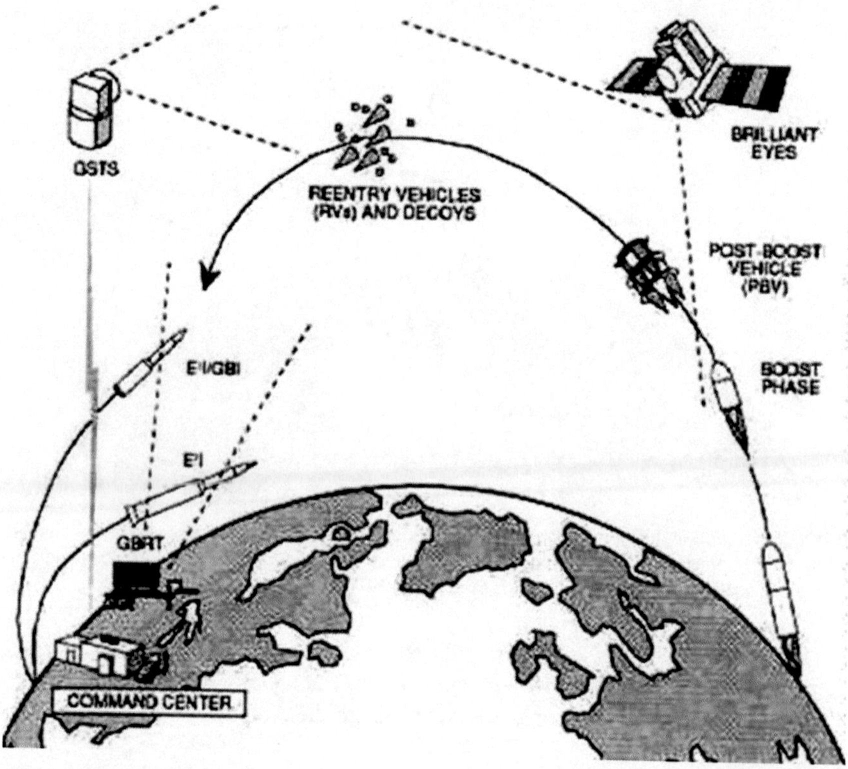

Fig. 7.4 GPALS architecture: ground-based protection against strategic ballistic missiles. (Source: Strategic Defense Initiative Organization, *1991 Report to the Congress on the Strategic Defense Initiative* (May 1991), 2–7)

Figure 7.4 depicts the architecture for a ground-based defense against strategic ballistic missiles. The ground-based architecture consists of a command center and a combination of Brilliant Eyes satellites, terminal phase ground-based radar trackers (GBRT), and high-speed accurate interceptors. These interceptors would be terminal phase endo-exoatmospheric interceptors (E^2I) and/or midcourse phase exoatmospheric ground-based interceptors (GBI).

An option also exists to add the Ground-based Surveillance and Tracking System (GSTS) to the architecture. Brilliant Pebbles (BE) satellites are derivatives of the previously planned space-based surveillance and tracking system satellites. The E^2I is a derivative of the previous high-endoatmospheric defense interceptor.

There are more to such an ambitious program as we have describe above, with details beyond the scope of this book; thus we refer our readers to References [1, 2], as well as what can be found on the Internet that are open to the public as part of unclassified aspect these programs are concerned.

7.2 High-Power Laser Energy-Induced Directed-Energy Weapon

With Star Wars Missile Defense being back on the radar of defense department, defense companies and contractors are again paying attention to high-energy laser (HEL) as a means of directed-energy weapon (DEW) and as a source of the Missile Defense System (MDS) as depicted in Fig. 7.5 [1].

The classification of this type of laser as weapon application creates a new scenario in the traditional battlefield by taking old-fashioned war engagement to a new era of modern stage, where directed-energy weapons will be considered as a new age in defense. Tomorrow's battles will be fought at the speed of light rather than the speed of a bullet.

Beam weapons that fall under the category of directed-energy weapon (EW) systems include lasers, linear particle accelerators or particle-based weapon, microwave, and plasma-based weaponry. The author and other scientists are also suggesting a feasibility of a new beam weapon as longitudinal scalar wave (LSW) and as an additional means of defense system in particular, when it comes to dueling with incoming threats such as hyper-velocity moving objects that travel from a launching source to the aimed target at a speed beyond 5–15 times the speed of sound [2, 3].

Fig. 7.5 Artistic illustration of the Star Wars Missile Defense System

Particle beams involve the acceleration of charged or neutral particles in a stream toward a target at extremely high velocities, the impact of which creates a reaction causing immense damage. Most of these weapons are theoretical or impractical to implement currently, aside from lasers which are starting to be used in terrestrial warfare. That said, directed-energy weapons are more practical and more effective in a vacuum (i.e., space) than in the earth's atmosphere, as in the atmosphere, the particles of air interfere with and disperse the directed energy. A brief description of each of these suggested technologies can be found in Reference [1] by Zohuri, or readers can do their own research on the Internet based on available unclassified sites describing them.

High-power lasers (HPLs) as a weapon can be deployed into theater of battle in three forms and modes [1, 2].

In summary, an air defense HEL weapon designed to shoot down airplanes, helicopters, and missiles successfully must have the ability to keep a very powerful beam at one point on the target for a long enough time to deliver at least 5000 joules per square centimeter. This requires a laser in the megawatt range. If the shot is to be successful, it must be directed to a certain part of the target that is limited in size and very sensitive and then kept there until the desired effect is reached. Thus, the laser beam must track and follow a target if any great length of time is needed to achieve the desired effect.

Many parts of an aircraft or helicopter are highly resistant to a HEL weapon, but there are still enough thin-skin parts and sensitive areas to produce a devastating effect or destruction if hit precisely. On the other hand, it is obvious that at battlefield ranges even an extremely high-energy laser weapon cannot penetrate the heavy armor on a tank or other armored vehicles, and thus a HEL weapon is of no use for destroying resistant ground targets in the battlefield. However, sensors, optics, and related devices are still valid targets wherever they appear on the battlefield, even in a tank.

7.2.1 Airborne-Based Laser (ABL)

The Boeing YAL-1 Airborne Laser Testbed (formerly Airborne Laser) weapon system, as illustrated in Fig. 7.6, was a megawatt-class chemical oxygen–iodine laser (COIL) mounted inside a modified Boeing 747-400F. It was primarily designed as a missile defense system to destroy tactical ballistic missiles (TBMs) while in boost phase. The aircraft was designated YAL-1A in 2004 by the US Department of Defense [4].

The YAL-1 with a low-power laser was test-fired in flight at an airborne target in 2007 [5]. A high-energy laser was used to intercept a test target in January 2010 [6] and in the following month successfully destroyed two test missiles [7]. Funding for

Fig. 7.6 Airborne-based laser (ABL) antiballistic missile weapon system

the program was cut in 2010, and the program was canceled in December 2011 [8]. It made its final flight on February 14, 2012, to Davis–Monthan Air Force Base in Tucson, Arizona, to be kept in storage at the "Boneyard" by the 309th Aerospace Maintenance and Regeneration Group. It was ultimately scrapped in September 2014 after all usable parts were removed [1].

7.2.2 *Space-Based Laser (SBL)*

Figure 7.7 is an artistic illustration of light amplification by stimulated emission of radiation (LASER) pump energy into molecules, creating an electronic state that releases energy in the form of photons. The photons pass by other molecules, spreading energy, making more photons. To make an actual laser, a beam has to pass through mass quantities of laser medium by bouncing back and forth between mirrors placed at opposing ends. Then the light beam exits through one of the mirrors which is more transparent than the other. Making a functional laser requires the electrons to not only reach their excited state but also be reliant on the time it takes for them to get excited and also the time for the energy created to reach new electrons. The efficiency of the laser relies on the amount of heat that exits. In terms of lasers, the power of the laser far outweighs the chemical efficiency. Of course, the trajectory of the laser matters as well as its ability to hit the target it is aimed at, but when lasers are placed in space, diffraction can cause interference [9].

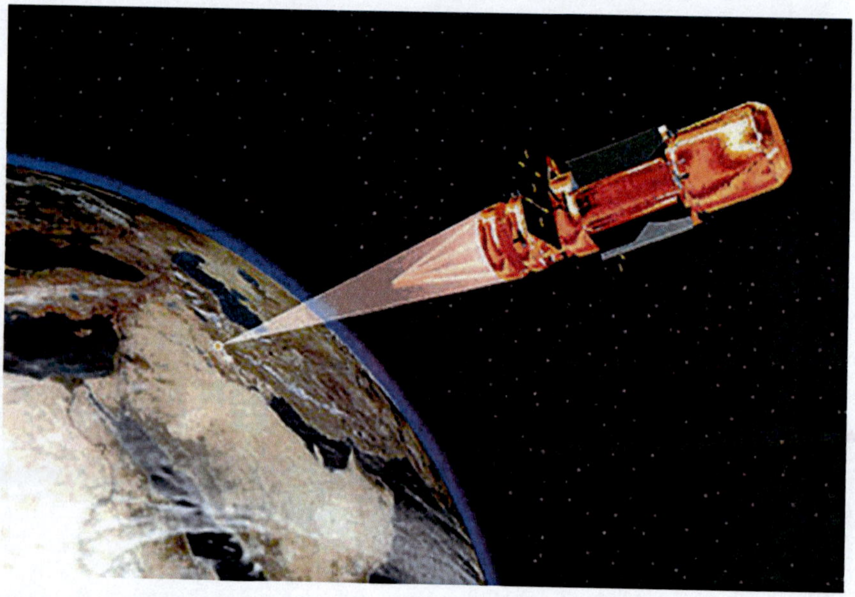

Fig. 7.7 Space-based laser (SBL) system, artistic illustration

"Waiting until an adversary is in midcourse [phase of flight] is giving the adversary a free pass to launch," Michael Griffin, the undersecretary of defense for research and engineering, told reporters during a media roundtable at the Space and Missile Defense Symposium.

7.2.3 *Ground-Based Laser (SBL)*

As illustrated in Fig. 7.8, a second major alternative to destroying theater ballistic missiles with laser weapons is to place the laser on the ground and relay the beam to the missile with large mirrors in space.

The distinct advantage of this architecture is that the high-energy laser is kept on the ground, which eliminates the need to fit a laser platform onto an existing launch vehicle and the need to refuel the laser weapon's chemicals in space.

In addition, the complex and maintenance-intensive equipment, i.e., the laser, fuels, and pumping systems, are left on the ground. If problems develop with the ground laser systems, the equipment is readily accessible without the need for planning, funding, and recovering satellites from orbit. A further benefit is that the ground laser and beam director are not as constrained by diameter, weight, or volume as is the case for a space platform that must fit within a launch vehicle.

One of the issues associated with high-energy laser (HEL)-driven directed-energy weapon (DEW) is that the beam needs to propagate through the earth's atmosphere

Fig. 7.8 Ground-based laser (GBL) system, artistic illustration

no matter which class of these three laser types we are using as an application of weaponry. The only exception is if we are engaging and dwelling with a target above the earth's atmosphere with spaced-based laser (SBL), where the beam is not required to travel through atmospheric environment, however if this class of laser has to go after a target within earth atmosphere.

Problems that these laser weapons need to overcome are:

1. **Atmospheric propagation problem**
2. **Thermal blooming effects**
3. **Laser beam jittering**

Note: The beam is affected by the stability of the optical platform (called jitter||), power fluctuations, impurities in the lasing medium, and atmospheric conditions between the source and the target. The ability to direct large amounts of energy accurately, near-instantaneously to a small spot from extremely long distances opens up a new range of possibilities for military use [1, 11, 12].

With atmospheric propagation and thermal blooming comes the issue of light scattering and intensity loss of laser energy at very-high-power stage in particular. The details of the physics are were defined by Zohuri [1], where reader can refer to it for more granular information and suggested solutions no matter if we engage our target of opportunity in a tactical or strategical way within theater of battlefield, where these lasers are used as a new weapons and it does not matter even of the target is within line of site or out of it beyond the horizon of operator and directed beam observation of the weapon.

However, one of the solution that could be offered and to be investigated is Relay Mirror technology and associated technical with these mirrors and that is why we need codes such as PUFF-TFT, ASTHMA series by way of versioning, XRTH, MOM, etc. All these codes are described in more detail in the following sub-sections.

7.3 Relay Mirrors-Driven High-Power Laser Energy Beyond Horizon

Directed-energy weapons represent a potential revolution in military capabilities for tomorrow's warfighters. Lasers are already used for making precise measurements, cutting, communications, and power transmission. At higher power levels, lasers have the potential to be formidable weapons, both lethal and non-lethal. In recent years, the United States has made several advances in high-powered laser technology, producing a megawatt-class laser, with several other countries not far behind [10].

Combining lasers with space-based relay mirrors could offer the capability of transmitting a ground-based laser toward any point the earth, an orbiting satellite, or a ballistic missile in flight. By the year 2035, several nations will likely have developed this capability, which will significantly affect US national security. This paper examines the threat posed to the United States by a laser/relay mirror system and addresses the issue of how the United States should deter hostile nations from employing such a system.

As illustrated in Fig. 7.9, a general and high-level description of a space-based relay mirror system that supports ground-based laser (GBL) follows, along with a description of technical challenges for implementation and threats posed by such a system against US ground, air, and space systems.

Fig. 7.9 Artist's impression of ground-based SDI laser (Photograph by the US Department of Energy)

As we have learned from our college physics, in basic terms, a laser is created when coherent light waves interact in such a way as to form excited-state atoms creating a beam of electromagnetic radiation (Fig. 7.10) [11]. The beam is formed when a group of excited atoms is energized by an outside power source through creation of a chemical reaction (chemical oxygen–iodine laser), an electric discharge, or flashlamp (ruby laser) [11]. The distinguishing characteristic of a laser beam is the way the beam remains relatively intact, in a tight cylinder (i.e., collimated and coherent), for extremely long distances.

Below are the types of lasers that can sustain high-power energy, beside chemical lasers that are currently in use which may have military applications and also could be used as matter of interaction with matter:

1. *Free electron lasers*—these types of lasers are classified as electric lasers and can be designed to illuminate any wavelength or frequency, from gamma rays to microwaves. Each free electron laser is also tunable, able to change frequencies within a limited range centered on the frequency for which the laser was designed [11]. Free electron lasers produce radiation by sending fast electrons through a magnetic field, typically a series of alternating magnets.
2. *Fiber lasers*—these types of lasers are produced by injecting light into a superthin optical fiber containing an amplifying medium. Since there is a limit to the power produced within each fiber strand, these strands can be bundled to produce a laser with a higher combined power output [11].
3. *Solid-state lasers*—these types of lasers excite molecules inside a crystal solid until they can be induced to return to a lower-energy state in unison. They tend to generate a great deal of heat that must be dissipated, which is a current limitation for this type of laser [13].

The lethality effect of lasers with high-power energy is primarily due to the effect driven by laser heating the target surface, where it is radiating upon it. This is caused by the target surface absorbing the beam's energy [1, 12]. A laser's damage potential is measured in terms of fluence, or energy per unit area (joules/cm^2). For high-energy laser weapon applications, such as anti-ballistic missile defense or anti-tank applications, a fluence of 10,000 joules/cm^2 is considered the minimum required. This corresponds to a laser with a power of approximately 1 megawatt [11]. If the laser dwells on a target long enough, it can cause melting and vaporization of metal. For example, the US Air Force's Advanced Tactical Laser program is essentially a scaled-up version of a laser cutting tool. In October 2009, this C-130 fired a laser at a truck on a test range, burning a hole in the vehicle's front hood and into the engine [12]. If the laser power were scaled up further, the result could be a formidable battlefield weapon.

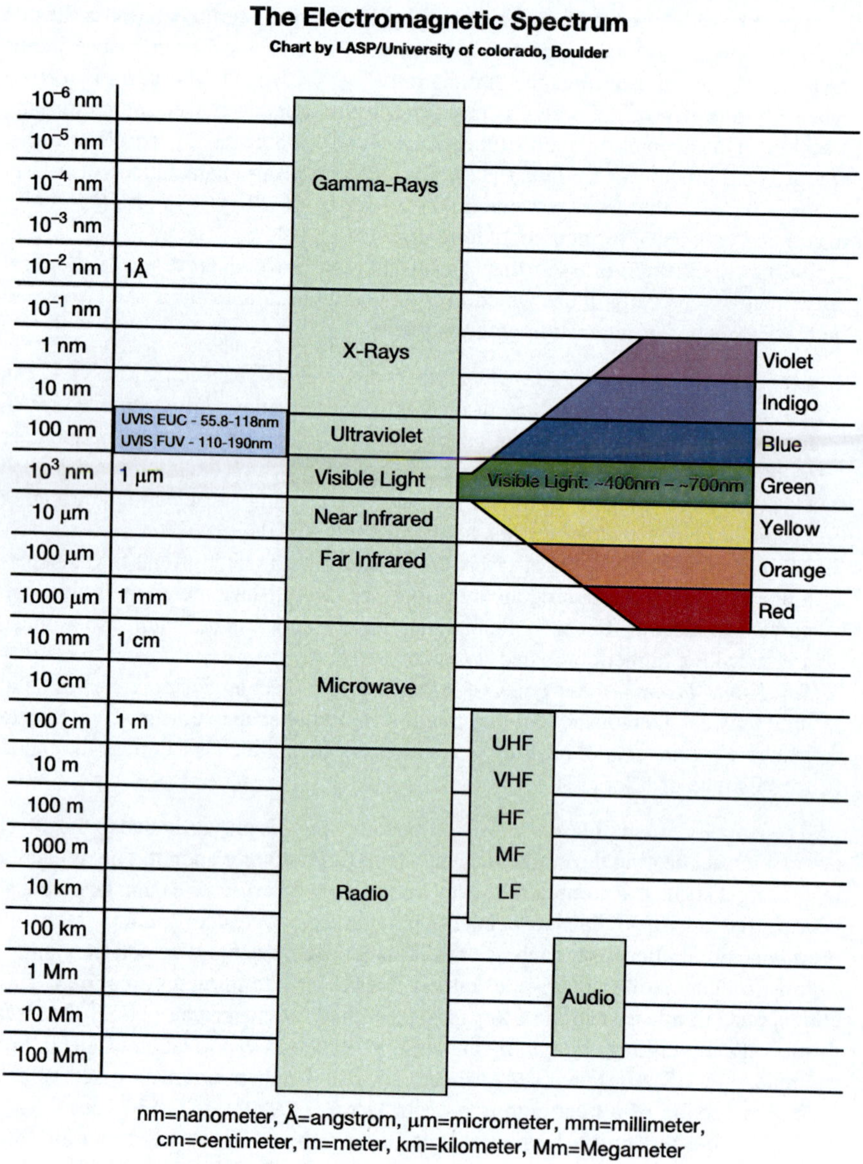

Fig. 7.10 Electromagnetic spectrum (Ref. [12])

In addition to thermal and vaporization effects, lasers can have a devastating effect on human beings. A threshold of approximately 15 joules/cm^2 is all that is required to burn human skin, and less than 1 joule/cm^2 causes permanent eye damage [14]. A high-energy laser shot would severely burn or vaporize a human being, but even reflected laser energy can still blind a person and cause corneal burns or retinal damage [14]. A high-energy laser pulse can be reflected in the vicinity of any people in the target area, often without their immediate knowledge. High-energy military lasers are often not in the visible portion of the electromagnetic spectrum, which makes it extremely difficult for people to actively avoid collateral eye damage from laser beams [15].

The demand for overcoming physics complexity of problem of atmospheric propagation laser of high power energy and consequently, thermal blooming effects and finally jittering of laser beam was demanding a new innovative approach such as Space-Based Rely Mirror System (SBRMS) or Aerial Relay Mirror System (ARMS) or existing Air Force program knows it as Aerospace relay Mirror System and they are described as follow.

The laser's beam of concentrated light is designed to produce 2.2 million watts of energy, making it the nation's most powerful military laser, experts outside the Government say. Officially, the power of the beam is secret, with contractors saying only that it is hot enough to melt metal and that the energy intensity at the core of the laser is several times that of the surface of the sun. However, the main challenge of such innovative approach is to turn a chemical laser technology that had been proven on the ground into a device light enough to be launched into space. Thus, the laser is largely aluminum, since among the laser's heavier components are its mirror, which has to have ponderous systems of cooling water to prevent extreme heat from shattering them.

One item developed over the past few years and tested during the August 1994 firing was coatings that make mirrors so reflective they need no cooling. Computer codes such as PUFF-TFT is geared to achieve such a mission as this code is described later on in this chapter.

Currently, the laser is being linked to a system of mirrors known as Large Advanced Mirror Program (LAMP). Its biggest circular mirror is 13 feet in diameter, and the LAMP apparatus is housed in a separate vacuum chamber at the San Juan Capistrano site. LAMP is tasked to take the raw Alpha beam and simulate how it could be focused and directed across space to hit enemy missiles. Note that this program also in the community of scientists and engineers involved with AFRL program of missile defense is known as Alpha-type weapon and would be effective only in space and be able to penetrate into the earth's atmosphere no deeper than 5 or 6 miles above the ground, if it is operating at a wavelength of 2.7 microns, which is in the infrared part of the electromagnetic spectrum and invisible to the human eye as depicted in Fig. 7.10.

A laser/space-based relay mirror system greatly enhances the range and lethality of laser weapons. In the case of ground-based laser (GBL), the primary driver behind the lethality of this system is a high-powered, ground-based laser pointed toward the orbiting relay mirrors. Although there are several types of high-powered lasers currently in development, all of them face the challenge of being able to produce a beam with power in the megawatt range. Using a ground-based laser instead of a space-based or airplane-carried laser greatly simplifies logistics and power production, which is a significant advantage.

Theoretically, only one ground laser is required, but multiple sites can be built to ensure redundancy for the nation during a conflict.

The defining characteristics of this type of system are beam power and coherence. In order to achieve the minimum 10,000 joules/cm^2 on a ground target, studies indicate that a minimum power output of approximately 1.4 megawatts would be required for an effective missile defense system, based on an assumption of 25 percent cumulative losses due to atmospheric effects to and from the relay mirrors, absorption, and jitter [16, 17]

Historically, as part of the initial "Star Wars" program, as far as part of Research and Development around SDI is concerned, there existed a Rely Mirror Experiment (RME) at around June 26, 1990. This program is described here.

Relay Mirror Experiment (RME)

Relay Mirror Experiment (RME) is known as *USA 52* and *Losat-R*. RME was launched as a dual payload with *LACE* (Low-power Atmospheric Compensation Experiment). Both satellites carried defense experiments to help validate the technology needed for a space-based antimissile laser system. RME was designed to show stabilization, tracking, and pointing technologies working at performance levels needed for Strategic Defense Initiative (SDI) missions. It 61-centimeter mirror tested laser pointing technology by deflecting ground-based beams back to Earth. A problem occurred immediately after launch when RME's attitude control malfunctioned, shutting off a reaction wheel. The Maui Optical Observatory on top of Mount Haleakala established the first relay on June 26, 1990, with Kihei, Hawaii. RME's payload also included the Wideband Angular Vibration Experiment (WAVE) which measured low-level angular vibrations affecting performance of acquisition, tracking, and pointing (ATP) systems.

(continued)

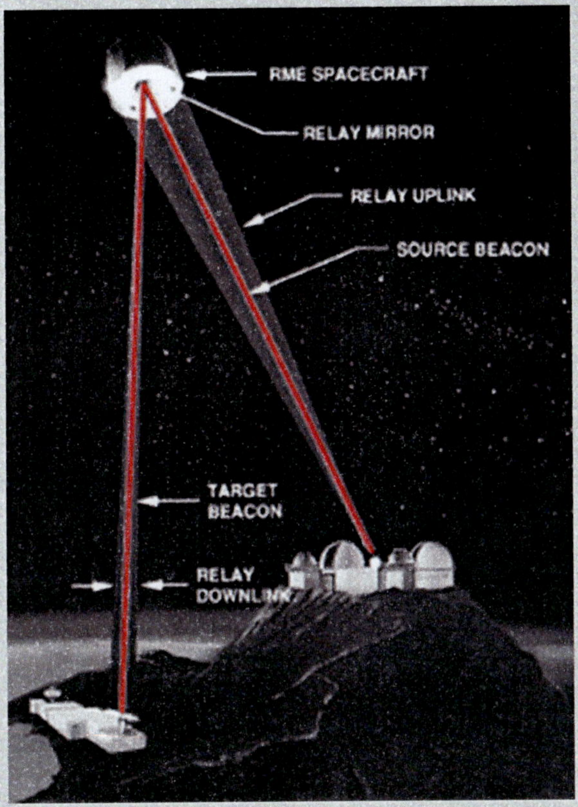

Launch date	Feb 14, 1990
Launch vehicle	Delta 6925
Launch site	Cape Canaveral
Orbit	261 × 281 km × 43°
Mass	1040 kg

Low-power Atmospheric Compensation Experiment (LACE)

LACE (Low-power Atmospheric Compensation Experiment) was part of a dual payload with *RME* (Relay Mirror Experiment) carrying laser defense experiments. LACE, also known as Losat-L, was built by the Naval Research Laboratory as a target for ground lasers, to investigate atmospheric distortion and compensation methods. Low-power lasers were beamed from the Air Force Maui Optical Station and picked up by onboard infrared and phased detectors. A laser then locked onto a reflector mounted on a 46-meter boom to

(continued)

acquire the satellite, and the sensor array returned data on the laser coverage, allowing the laser's adaptive optics to be adjusted to compensate for atmospheric distortion. The payload included ABE, the Army Background Experiment, an instrument that monitored background levels of neutron radiation in order to be able to discriminate between warheads and decoys, and the UPI, Ultraviolet Plume Instrument, to track rocket plumes.

Figure 2: LACE in the acoustic reverberation chamber

Launch date	Feb 14, 1990
Launch vehicle	Delta 6925
Launch site	**Cape Canaveral**
Orbit	463 × 480 × 43.1°
Size	2.4 × 1.4 m
Mass	1430 kg

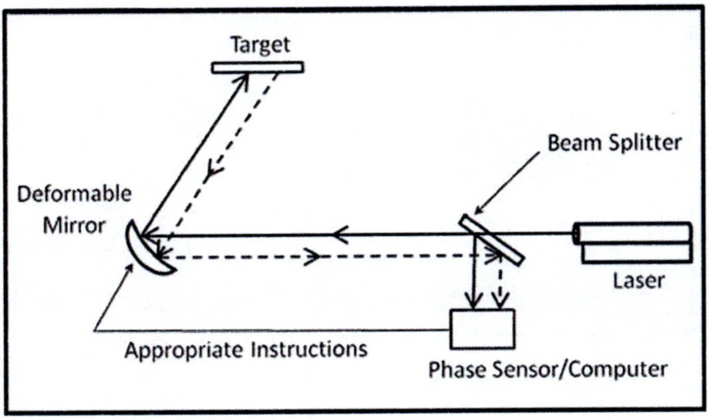

Fig. 7.11 Schematic of an adaptive optics system [1]

However, Air Force Research Laboratory (AFRL) plans include a goal of 4 to 10 megawatts for a US system. As a laser beam travels from the ground to outer space, it encounters moisture, pressure, and density changes in the atmosphere, all of which contribute toward unrousing the laser beam by the time it reaches the satellite relay mirror. An adaptive optics system is a feedback mechanism that allows the initially distorted beam to be fed back to the laser source, where a computer algorithm can determine how to compensate for the distortions. To correct these distortions, a computer makes thousands of small deformations in the laser's optical mirror in such a way that the laser beam is deliberately distorted prior to being shot through the atmosphere. The atmosphere then acts as a lens, refocusing the beam to a coherent point at the relay mirror. Figure 7.11 shows a schematic of an adaptive optics system.

Note: The principle behind adaptive optics is to use a deformable mirror to compensate for the distortion caused by the atmosphere. The system first sends out an artificial "star" created by a low power laser. When that laser beam is scattered by the atmosphere, the scattering radiation is reflected back and measured so that the system knows just how much the atmosphere is distorting the laser. By feeding this information into a complex control system, the deformable mirror, with its hundreds of small actuators positioned behind the mirror, alters the surface of the mirror to compensate for atmospheric distortion. Thus, a high-energy laser can be "pre-distorted," so it will regain its coherence as it passes through the atmosphere [1].

7.4 Relay Mirror Satellites

The relay mirror satellite system is the enabling part of the system that allows the power of the laser to be extended worldwide. The satellite itself consists of a standard satellite bus, including small rocket motors for attitude/position control, a

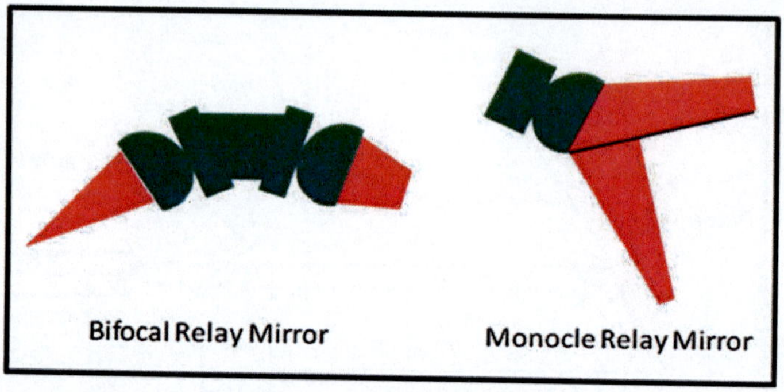

Fig. 7.12 Bifocal versus monocle relay mirror

cooling system, attitude gimbals, and a central computer. The relay mirrors would be mounted on the outside of the satellite and would likely be between 3 and 8 meters in diameter, depending on design. For worldwide coverage, at least 4 satellites would be required if placed in geosynchronous orbit (40,000 kilometers above the surface of the earth), or 20 if placed in low or medium orbit (4000 kilometers above the surface of the earth) [17, 18].

The key to the laser/relay mirror system functioning properly is the design of the mirrors. There are two primary types of mirrors to be considered: a monocle mirror and a bifocal mirror (Fig. 7.12). A monocle mirror receives a laser beam and refocuses it toward a target with the same mirror. A bifocal mirror, or binocular system, has two lens systems, with the laser beam passing through the first aperture to be—conditioned to correct atmospheric distortions. The beam is then amplified with a chemical or solid-state amplifier and then refocused on to the target through the second aperture [19]. The binocular system is more complex but is a more accurate system and is better at refocusing the beam to create maximum fluence on the target.

Figure 7.13 shows a schematic of a laser/relay mirror system with a laser beam originating from the ground and being re-directed from one relay mirror to another and then to a ground target on the other side of the world from the originating beam. The capability to instantaneously hit a target anywhere on the planet with a potentially lethal laser beam would be a dramatic leap in technology. With relay mirror satellites in orbit in appropriate locations, a beam could be directed to any ground point on the planet [20].

Figure 7.14 shows a schematic of a laser/relay mirror system with a laser beam originating from the ground and being re-directed from a relay mirror toward an enemy satellite.

With the United States' heavy reliance on satellites for a variety of uses, this is also a significant vulnerability. Instead of pointing the beam at another orbiting satellite, the beam could also be directed at a missile or airplane in flight.

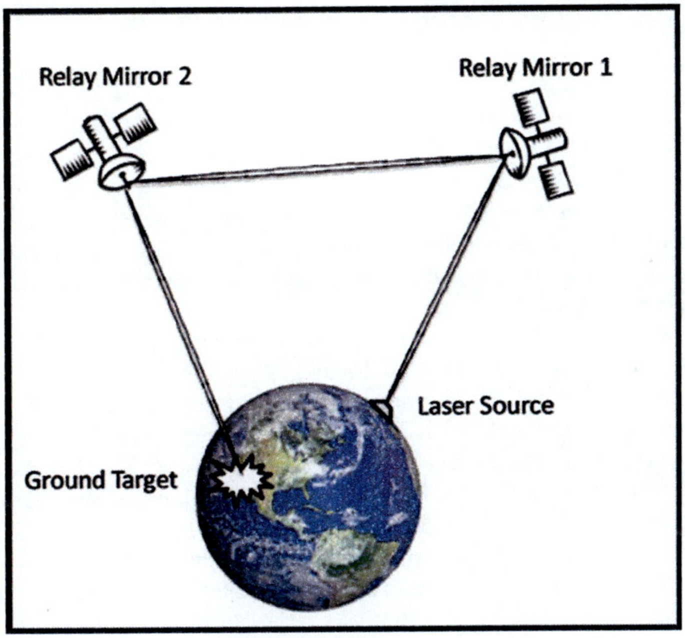

Fig. 7.13 Laser/relay mirror system engaging ground target

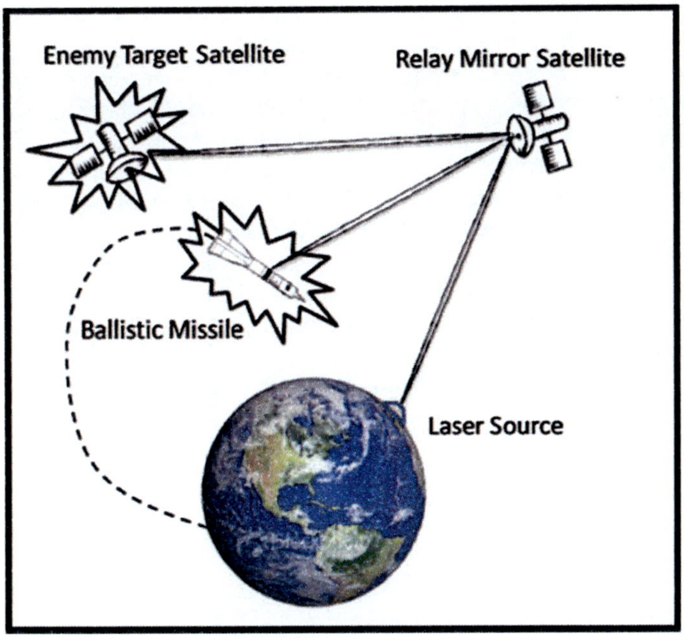

Fig. 7.14 Laser/relay mirror system engaging space targets

7.5 Relay Mirror Satellite Technical Challenges

The original premise of this paper was addressing the question of how the United States should posture itself to deter a potential adversary using a laser/relay mirror system in 2035. The areas listed below outline the primary areas of emphasis [21].

Ground-Based High-Powered Laser—There is still much research and testing to be accomplished in order to build a reliable, repeatable, possibly mobile, megawatt-class ground laser. Technology areas requiring further breakthroughs include optical lens and mirror materials, optical alignment, power generation and transmission, and thermal management [22].

Satellite Vehicle/Relay Mirror Design—A relay mirror satellite will be different from other satellites in that its ability to stay pointed in a particular orientation is critical. Not only does the satellite itself have to be able to remain orientated properly, but the relay mirror has to be able to make microscopic precision movements in order to redirect the laser beam in the proper direction. In order to accommodate the high-powered laser beams without melting, advances are required in optical coating materials for the relay mirrors [22].

Vibration and thermal management, attitude and jitter control, and slewing all require additional research. Vibration must be controlled because the laser must point precisely at the target. Vibration and material expansion due to heat buildup need to be addressed in satellite design. Controlling the satellites' attitude and motion is needed to ensure exact orientation toward the target and the incoming beam, which is necessary for relaying the laser toward its aim point. Lastly, slewing and momentum control present challenges, as once a satellite rotates to orient itself toward a target, the satellite must be able to stop its angular momentum [22].

Acquisition/Tracking/Targeting Algorithms—This is an area of critical importance that brings the entire system together. First, the ground laser must be able to detect and track the first relay mirror satellite. If the target is not within the line-of-sight of the first relay mirror, the beam must be directed to a second relay mirror. At the end of the line, the target, which itself may be moving (missile or airplane), must be detected, tracked, and targeted. All of this detection, tracking, and targeting must be accomplished simultaneously. The 1991 Gulf War saw the transformation of reducing the targeting kill chain from hours to minutes. This system will further reduce the timeline from minutes to seconds from target identification to target engagement [22].

Further information on the topic of relay mirror satellite can be found either in Reference [18] or on the Internet for readers who are interested in the subject.

7.6 Relay Mirror Satellite Limitation

For very obvious and directly related reason supported with science of physics, any high-power laser energy with short wavelength such as krypton fluoride (KrF) laser with an energy of 20,000 joules (20,000 watts per second of delivered energy) and

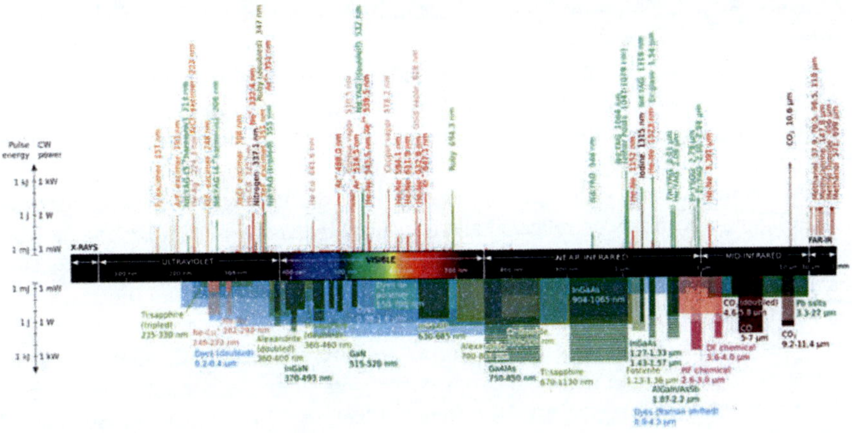

Fig. 7.15 Wavelengths of commercially available lasers [39]

laser light wavelength of 0.25 μm (micron, or micrometer), neodymium-doped yttrium aluminum garnet (Nd:Yag) or $Y_3Al_5O_{12}$ laser with an energy of around 19,000 joules (19,000 watts per second) and wavelength of 1.06 μm, and carbon dioxide (CO^2) laser with high-energy power and wavelength of 10.6 μm is a very good candidate as part of directed-energy beam weapons, which transmit well through the atmosphere. In fact, any high energy laser with a characteristic of "short-wavelength," "high-frequency"-type lasers are highly desirable for their application as high-energy laser (HEL) and as directed-energy weapon (DEW), when it comes to ground-based laser (GBL) or airborne laser (ABL).

Shorter-wavelength and higher-frequency laser light (see Fig. 7.15) is qualitatively superior in "coupling" its light energy onto either a fusion-driven fuel of deuterium (D)–tritium (T) target [40] or a military target. In crude terms, the percentage of the incident light energy that gets converted into a shock wave propagation through the target is greater. Less energy is used merely to heat the target. So, the disabling for a target such as a missile occurs by a fast punch, rather than a slower burn.

However, above a certain limit, the frequency is too high to be refocused by any mirror materials. This takes place when radiation gets into this high ultraviolet or X-ray range, and all mirrors become transparent. This limit is at about 0.1 μm (micron) wavelength. Thus, the X-ray popup laser, at a wavelength of only 0.04 μm (micron), will use no mirror, nor need any at all.

Although the X-ray laser will be superefficient in terms of target destruction and kill, it cannot propagate through the atmosphere at all and must be fired from space; thus such laser is an excellent candidate for space-based laser (SBL) to engage an incoming target or threat through its mid-course phase.

On the other hand, the KrF type of laser will be a crucial complement to X-ray lasers in overall defense systems.

Note: Laser types with distinct laser lines are shown above the wavelength bar, while below are shown lasers that can emit in a wavelength range. The height of the lines and bars gives an indication of the maximal power/pulse energy commercially available, while the color codifies the type of laser material (see the figure description for details). Most of the data comes from Weber's book *Handbook of Laser Wavelengths* [39], with newer data in particular for the semiconductor lasers.

At about 0.25 μm (micron), this laser is near the very shortest wavelengths of laser light that can still be reflected and refocused by a relay mirror. By comparison, the large chemical lasers that are under development for antimissile systems generated laser wavelengths from 2.7 μm (micron) down to about 1.0 μm (micron).

Furthermore, comparing the use of these high-power lasers (HPLs) with refocusing mirrors, the size of the mirror required to *catch* and reflect the beam decreases rapidly as the wavelength of the light gets shorter, while a 2.5 μm (micron) laser beam may require a mirror 10 meters (33 feet) in diameter (e.g., see Section 5.3 of this book for a calculation of the mirror diameter). For example, a krypton fluoride (KrF) laser with a 0.25 μm wavelength will need a mirror only 1 meter in diameter.

Since the refocusing mirrors must be incredibly smooth and evenly curved and must be put into orbit in space, this difference is extremely important in building a laser antimissile system, and computer codes such as PUFF-TFT and other related codes are able to provide the basic infrastructure parameters to build such mirrors.

Although not as powerful as their chemical counterparts, solid-state lasers are suited to defense missions such as destroying and illuminating targets for air defense, mine destruction, ship protection, and optoelectronic warfare, Northrop Grumman officials say.

"Solid-state and fiber lasers offer the potential of being more compact, thus suitable to a wider range of applications and platforms," says Jackie Gish, director of Directed Energy Technology at Northrop Grumman Space Technology. "With solid-state lasers, the challenge will be to make them more powerful and compact, and to dispose of and/or store waste heat compactly."

Solid-state lasers require only electrical energy to run, which makes them easier to support than their chemical laser counterparts.

Electric lasers are the future, experts believe. Solid-state and fiber lasers offer the potential of being compact and suitable to a wide range of applications and platforms. Bulk solid-state lasers are more mature than fiber lasers, but fiber lasers offer the promise of higher efficiency.

The funding for solid-state laser technology does not touch that of the airborne laser, and fielding is further away, but researchers are making progress.

The US Army is funding the Joint High Power Solid-State Laser (JHPSSL) program to develop "military-grade," solid-state laser technology that is expected to pave the way for the US military to incorporate high-energy laser systems across all services, including ships, manned and unmanned aircraft, and ground vehicles.

Increasing the average power of solid-state lasers are a major focus of high-energy laser (HEL) technology.

The HEL-JTO, the US Army (USA), and US Air Force (USAF) have been funding the JHPSSL program, with the goal of producing a 100 kW solid-state laser. The Defense Advanced Research Program Agency (DARPA) has been funding the High-Energy Liquid-Laser Area Defense System with a similar goal. Under the JHPSSL program, a 1.06 μm (micron) Nd:YAG laser has been developed and demonstrated the following characteristics: 19 kW average power, beam quality of 1.7 μm (micron), and a run time of larger than 5 min.

A significant materials development affecting solid-state lasers has been the introduction of Nd:YAG ceramics. Replacing crystalline material with ceramics has the potential to considerably improve the manufacturing ability of solid-state lasers.

Under the current phase, the program's goal is for a laser system to reach 100 kilowatts, setting the stage for a variety of force protection and strike missions such as shipboard defense against cruise missiles; wide-area and ground-based defense against rockets, artillery, and mortars; and precision strike missions for combat aircraft. These are still laboratory lasers and fielding of an actual weapon is still years away.

"These systems promise to provide the warfighter with enhanced capability in terms of shorter engagement timelines, high-precision targeting combined with long reach, while inflicting low collateral damage," says Jim Stamboni, senior vice president of the Textron Systems Advanced Solutions Center. These solid-state laser concepts also need to provide the warfighter with compact, affordable, and rugged directed-energy weapon systems capable of withstanding the most severe operational and environmental battlefield conditions.

7.7 Available Public Domain Computer Codes

There exist few available public domain computer codes, which are a production of "Star Wars" era, and they may be found for purchase or be obtained from government facilities and laboratories such as AFRL or NASA as well as national laboratories such as Sandia National Laboratory or Lawrence Livermore National Laboratory. Interested readers should be able to find resources about them on the Internet. These codes are listed here, that this author is familiar with them and to best of his knowledge although they are considered unclassified, yet they are very sensitive computer codes.

Normally, distribution of these codes is authorized to US Government agencies and their contractors; critical technology. Other request for these codes and their documents must be referred to AFRL or associated national labs. Mainly, their user manuals and any other related documents fall under the restriction rule by the Am8 Export Control Act (T i t l e *22,* U.S.C., Sec 2751, et seq.) or The Export Administration Act of *1979*, as amended. Title *50,* U.S.C., App. 2401, et seq. penalties.

7.7.1 Thin Film Transport (PUFF-TFT) Computer Code

The PUFF-TFT codes was developed around the physics rationale for the Thin Film Transport code. The code was written in Fortran language for VAX/VMS, CRAY, and presently its PC version is available from Galaxy Advanced Engineering (www.gaeinc.com) and runs on Windows PC. The author of this book took the effort of porting the code to the lateral computing platform, fixing existing bugs and errors that were in the original code on VAX/VMS, as well as developing online graphics capable of producing most of the calculation results sample by plotting them during the code run on Windows PC.

The code originally was written for the Air Force Weapons Laboratory (AFWL) primarily to allow evaluation of thin-layer stack response of materials to X-ray deposition resulting in one-dimensional (1-D) strain stress responses. The code takes into account the X-ray generation of secondary cascade particles such as photoelectrons, Auger electrons, and fluorescent photons, using a cascade routine, and incorporates a thermal condition routine allowing the effects of rapid thermal diffusivity to be included. The source of heat may be a high-power laser energy of interest as condition of problem with given boundary and initial conditions of user's choice.

The output of the X-ray/cascade/thermal routines is used as input to an updated version of the previous version of this code known as PUFF74 hydrodynamic code, which includes hydrodynamic, elastic-plastic, porous, and dispersive material responses in a fully coupled manner, and also accounts for simple phase changes during thermal and heat exposure of the target surface. This code has an update that does include the Laser-Damage Modeling as well.

The Air Force Weapons Laboratory (AFWL) is presently investigating the potential use of high-power laser weapons in space. An essential component of such systems is the mirror needed to guide the laser beam to its target. Such mirrors usually employ a metallic reflecting layer overlaid on a suitable substrate. The metallic layer is itself overlaid by a dielectric package to enhance overall reflectivity. Alternatively, an all-dielectric stack with no metal can also be used. The individual component layers (dielectrics and metal) are generally very thin (typically <10,000 A = 1.0 pm each), and there may be as many as 10–100 total.

An essential problem to be assessed is the vulnerability of such mirrors to exoatmospheric attack by a nuclear weapon. rapid energy deposition due to weapon X rays. It is known, and will be shown later, that the energy deposition in such thin layers is not the same as that obtained for thicker layers, e.g., 25 pm = 0.001 in, typical of most other vulnerability problems. The reason for this difference lies in the fact that X-ray absorption results in the production of secondary particles (electrons and photons) which have ranges of deposition comparable to the thin mirror layer thicknesses. Consequently, a degree of energy sharing between layers occurs, thereby modifying the expected depositions based on normal X-ray interactions.

Additionally, thermal conduction can be very significant for thin layer packages since the thermal time scales for conductive energy transfer are frequently less than, or comparable to, the X-ray deposition times, again modifying the expected depositions based on normal X-ray interactions.

Similar problems can also occur for circumstances involving very steep X-ray or other deposition profiles in single layers. In particular, thermal conduction can seriously modify the true loading conditions, reducing the expected peak loadings near a material surface and transporting energy further in depth. This effect can drastically alter the expected thermomechanical responses.

Although the initial Air Force Weapons Lab (AFWL) rotary interest was in the ability to model damage in the thin layer optical stacks, recent interest has also concentrated on the more general response of either thin layer systems or single layers with steep loading profiles, when driven strongly. Thus, in addition to the ability to determine incipient damage, there also exists a need to predict such effects as the recoil stress response and impulse generation of packages driven strongly into vaporization, which could be important for passive decoy analysis.

Since significant geometry changes can themselves interactively affect the thermal response (due to diffusivity changes and adiabatic energy feedback), it is also necessary to couple the thermal and stress responses in a consistent manner.

Initially the AFWL had no computational capability for assessing these problems. Existing codes which could have been used were SANDYL, TIGERP, and XRT. However, SANDYL is a very complex 3-D code which is expensive to use, as is TIGERP. XRT is a simpler 1-D code, cheap to run, and easier to use.

However, SANDYL and TIGERP are available to the AFWL, while XRT is presently unavailable because it was a proprietary code of S-CUBED.[1] Equally to the point, these codes can handle only the energy deposition aspects of the problem. For *a* full thermomechanical analysis of a mirror system, the data from these codes would need to be used as input to a hydrodynamic code such as PUFF in an interactive manner.

The AFWL thus had a need for a suitable code capable of determining true energy deposition from X-rays in a mirror assembly and predicting the true thermomechanical response of the mirror. This code was to be available to the AFWL for their own use and ideally should be simple to use, relatively inexpensive, and easily understood and implemented.

Two papers, Part I and Part II, have been written around this code with its new version, namely, PUFF-TFT, where Part I describes the overall problem, physics rationale, and development of a computer code to handle these aspects. Part II describes the coding the code itself.

The eventual aim was to present the AFWL with a code which formed subroutines in the existing old version of this code, namely, PUFF74 code, with

[1]S-CUBED, 3398 Carmel Mountain Road, San Diego, CA 92121. Unfortunately, this company is dissolved as of today, and the new version of the code XRT, namely, XRTH, can be obtained from Galaxy Advanced Engineering at www.gaeinc.com.

switchable routines that need only be employed when deemed necessary, e.g., on a thin film package and off for most normal vulnerability problems involving thicker layers. By editing PUFF at the cascade deposition stage (no hydrodynamic or thermal calculations), the resulting data would be similar to that presently obtainable from SANDYL or XRT/XRTH.

By allowing full PUFF computation, the resulting data would be the overall thermomechanical response of the system being assessed, including the consequences of modified energy deposition in thin layers. The new routines, which were developed and named Thin Film Transport (TFT), use the existing routine within PUFF that computes X-ray deposition. They reprocess these data to handle thin film energy sharing and, thus, redefine the X-ray depositions. This new deposition information is then automatically used as input for the hydrodynamic calculations in PUFF, in a fully coupled manner.

Part II of the PUFF-TFT manual contains the descriptions of the subroutines that have been added to the PUFF74 code to produce the PUFF-TFT code. As a result of the modifications and additions, the input stream and commons have been modified. Part II describes the updated and new commons as well as the new input stream.

These modifications and additions were written to conform with the FORTRAN 77 standards. The code is double precision for implementation on VAX family computers and currently occupies 10.2 megabytes of storage in a non-overlaid form.

Today finding this code at source and executable level is out of question, and it seems AFRL in its today formation no longer has this code not KTECH company is in existence at Albuquerque, New Mexico. However, this code in both forms PUFF74 and PUFF-TFT running on Window PC can be obtained from Galaxy Advanced Engineering (www.gaeinc.com), and this code has gone through some modification and enhancement that allow the output result of calculation to be plotted while code is running its course, and it can be obtained for a nominal purchase fee.

It is assumed that the user is familiar with the information contained in "PUFF 74 -- A MATERIAL RESPONSE COMPUTER CODE," Volumes I and II. Due to changes in existing commons and the addition of new variables and commons to accommodate the TFT and transmitted fluence model, the input stream and variable descriptions are reprinted from Volume I.

The code in its present form can handle an X-ray input, compute the consequential secondary cascade energy redistribution, compute the consequential thermal conduction, and yield stress responses for 1-D strain problems. The code is fully coupled so that the interactive nature of thermal and stress responses is incorporated.

7.7.1.1 The Thermal Conduction Routine (General Points)

Thermal conduction can be handled to high numerical precision by finite difference techniques. However, there are many uncertainties in the thermal properties of thin layers. These uncertainties are:

1. Such layers rarely have solid density. Typically, $0.8 \le \rho/\rho_0$ solid≤ 1.0.
2. Whereas in bulk form a layer material, e.g., a metal, would be crystalline, the very thin layers are frequently semicrystalline or even amorphous and may also be porous.
3. For very thin layers, the diffusion time across a layer can approach, or even decrease below, the sound wave transit time when computed using bulk data.
4. Interface thermal contact resistances are generally unknown.

Accordingly, there are significant uncertainties in all relevant thermal parameters. Specific heat can markedly change with crystal structure and even be layer thickness-dependent for very thin layers (due to restricted lattice phonon wavelengths). It is physically impossible to have diffusion time less than sound transit time, since the former is a random walk process of phonons and the latter is a linear motion. Again, this results from restricted phonon wavelengths for thin layers, which, in turn, implies a redefinition of conductivity. Conductivity is also a function of material porosity, crystal structure, and temperature. Diffusivity, which is the most important parameter for transient heat transfer, is conductivity/(density x specific heat). Thus, diffusivity has the greatest uncertainty. To compound the problem further, crystals generally have high diffusivity at low temperature, decreasing with increasing temperature, whereas amorphous materials generally have low diffusivity at low temperatures, increasing with increasing temperatures. Finally, for very thin layers, thermal contact resistance can dominate heat flow. Even the melting temperature and enthalpy can be uncertain. Recent work on silicon [23] has shown thin layers of amorphous silicon to have grossly differing thermal properties compared to bulk crystalline silicon, e.g., T_{melt} amorphous is $= 0.7 \times T_{melt}$ crystalline, and thermal conductivity amorphous is $= 10^{-2} \times$ conductivity crystalline.

Nevertheless, a finite difference routine has been written for TFT.

There are two basic finite difference schemes that can be used to compute one-dimensional heat flow (appropriate since the optical packages have very large diameter-to-thickness ratios). These schemes are the explicit and implicit techniques. The explicit technique marches forward in time, while the implicit technique marches backward. The implicit approach has several advantages. It can be easily made unconditionally stable, gives good accuracy, and has the virtue of readily allowing updates in the time-steps employed. As time progresses and thermal gradients decrease, ever larger time-steps can be used. Consequently, the total computational run time can be reduced. The one drawback to the implicit approach is that at each time-step, a matrix of simultaneous equations must be solved for each mesh point in the problem. For large numbers of mesh points, this again produces a slow code.

If the combined X-ray/cascade/thermal routines are intended to be a standalone code, which does not feed into a hydrodynamic routine, then the implicit approach is frequently the preferred route. Thus XRT (initially such a standalone code) uses the implicit approach.

TFT, however, was intended from the outset to feed data into the PUFF hydrodynamic code, and it was necessary to ensure compatibility between the thermal steps and the associated hydrodynamic time-steps. Since the latter are frequently smaller than the former, for typical problems, there was no virtue in allowing the thermal time-steps to significantly increment for a combined thermal-stress response. However, for a thermal-only calculation, larger time-steps can frequently be used.

An early version of TFT used the explicit approach. This was done on the premise of being easier to code, and the method avoids the need to solve a matrix of boundary solutions (for each spatial zone). However, it was difficult to use both compatible spatial zoning and time-steps simultaneously when in the coupled thermal-stress mode for problems involving significant expansions (e.g., melt and vapor). Accordingly, all subsequent editions of TFT use the implicit approach.

This was chosen because:

(i) It is still simple to code.
(ii) Can easily be made stable with identical zoning and time-steps as used in the stress routines.
(iii) Can be made to give high accuracy.
(iv) Can easily conserve energy.
(v) Easily handles phase changes (solid to melt], etc.

7.7.1.2 The Finite Difference Equations

The finite difference equations that are developed around this routine are as follows.

TFT operates on the temperatures at material mid-zones, rather than on interfaces. Using the following nomenclature:

Let time, t, equal $p\Delta t$ where Δt is a time-step and p is a running integer, increasing by unity per time step.

Let $m - 1$, m, and $m + 1$ represent three consecutive material zones, running left to right. The problem is always scanned left to right at any time.

Then T^p_{m-1}, T^p_m, and T^p_{m+1} indicate the temperatures of these three zones at time p (i.e., $t = p\Delta t$), while T^{p+1}_m indicates the temperature of zone m at one time-step later, etc.

For one-dimensional ‘neat flow, the following generalized equation applies:

$$\begin{aligned} T^{p+1}_m - T^p_m = u\Big[A\Big(T^{p+1}_{m-1} + T^p_{m-1}\Big) + B\Big(T^{p+1}_{m+1} + T^p_{m+1}\Big)\Big] \\ -u(A+B)\big(T^{p+1}_m + T^p_m\big) \end{aligned} \tag{7.1}$$

where:

- $u = \alpha_m \Delta t/(\Delta x_m)^2$ with $\alpha_m = k_m/\rho_m c_{pm}$.
- $k =$ conductivity.
- $\Delta t =$ time-step.
- $\rho =$ density.
- $c_p =$ specific heat at constant pressure.

And
$A = \frac{1}{(\gamma_1+1)}$, with $\gamma_1 = \left(\frac{k}{\Delta x}\right)_{m-1} / \left(\frac{k}{\Delta x}\right)_m$
$b = \frac{\gamma_2}{(\gamma_2+1)}$, with $\gamma_2 = \left(\frac{k}{\Delta x}\right)_m / \left(\frac{k}{\Delta x}\right)_{m+1}$
The terms A and B (containing γ_1, and γ_2) take into account the situation where the thermal properties vary on a zone basis.
For the simple case of uniform properties, $\gamma_1 = \gamma_2 = 1$; hence $A = B = 1/2$.
Thus Eq. 7.1 reduces to the following form as Eq. 7.2:

$$T_m^{p+1} - T_m^p = u\left[\frac{T_{m-1}^{p+1} + T_{m-1}^p}{2} + \frac{T_{m+1}^{p+1} + T_{m+1}^p}{2}\right] - u\left(T_m^{p+1} + T_m^p\right) \tag{7.2}$$

For vacuum boundaries, two options are set.
For *Left* side vacuum:
$\gamma_1 = 0 \quad \therefore\ A = 1$ since $k_{m-1} = 0$
For *Right* side vacuum:
$\gamma_2 = \infty \quad \therefore\ B = 1$ since $k_{m+1} = 0$

7.7.1.3 Spatial Zoning

In case of taking spatial zoning under consideration, the preceding equations allow solutions for all zones and time-steps. Consequently, the TFT code uses the same spatial zones as required by the PUFF stress zoning routine. Likewise, for a coupled thermal-stress problem, the code uses the same time-step for both the thermal conduction routine and the stress routine. This procedure guarantees that all zonal properties can be properly tracked throughout the calculations. Thus, for each zone, internal energy (cal/gm), temperature (°C), density (gm/cm^3), pressure (dynes/cm^2), particle motion (cm/s), etc. can be tracked.

The spatial zoning is chosen automatically in PUFF on the basis of the initial X-ray deposition profile (without cascade). The code chooses fine zoning for steep deposition profiles and slowly coarsens the zoning as the deposition profile becomes flatter. This automated system guarantees good spatial resolution for deposition and simultaneously satisfies the hydrodynamic stability criteria. Thus, good spatial and time resolution is obtained.

However, it should be noted that this good (and necessary) resolution can result in large numbers of zones and, consequently, small time-steps. The user is warned that this can sometimes result in slow calculations. This can only be avoided by using hand zoning and deliberately degrading the problem resolution by choosing larger (hence fewer) zones and correspondingly larger time-steps. This is not generally recommended, since the user can lose necessary data.

7.7.1.4 Time-Steps

In case of time-steps, the coupled thermal-stress mode, the code uses the hydrodynamic choice of time-step, which is automated within the stress routine. This hydro time-step is always sufficiently small to guarantee good, accurate, behavior for the thermal routine.

For an implicit routine, it is well-known that the problem is unconditionally stable, regardless of time-step. However, it is equally well-known that the accuracy can be poor if time-steps are too large; set the time-step according to:

$$\Delta T_T = \frac{\Delta x^2}{\alpha} \quad \text{minimum} \tag{7.3}$$

where Δx is a zone width, a is diffusivity, and the minimum for the entire problem is selected. This choice is conservative for thermal-only calculations.

The time-step for hydro is set to:

$$\Delta \tau_H = \frac{0.8\Delta x}{c} \quad \text{minimum} \tag{7.4}$$

where Δx is a zone width and c is the sound. Except for exceedingly thin zones:

$$\nabla \tau_H < \Delta \tau_T \tag{7.5}$$

Hence, use of the hydro time-step always errs on the side of conservatism and guarantees very stable thermal and hydro results with good accuracy.

Thin zones for which $\Delta \tau_H > \Delta \tau_T$ are violating the continuum-physics assumptions of the code since thermal waves cannot travel faster than sound waves. While this situation almost never occurs for automated zoning, the user must be very careful not to violate this rule if using hand-zoning. The result could be "numerical gibberish."

7.7.1.5 Phase Change: Melt

There is only one phase change of significant interest for typical space-based laser (SBL) mirror assessments: the solid/melt transition. However, the liquid/vapor transition is also considered in TFT. A solid component optical system is guaranteed damage even before vaporization occurs and would disrupt violently during vaporization. Indeed, damage is assumed to be inevitable if any zone significantly enters the pure liquid phase, since then its rate of cooling becomes very slow, and in all probability, such a zone will see an attempted tension. Since the zone has negligible strength in pure melt, damage can occur.

A rigorous modeling of the phase change is very difficult. First, the specific heat has a delta function behavior (a step jump in enthalpy equals the latent heat) which is difficult to incorporate. Second, there are practically no data defining the thermal conductivity and diffusivity within the latent heat region of any materials used for SBL mirrors. Data for these parameters do exist for some of the materials (in bulk from only) when in pure melt phase.

To allow modeling of the solid/melt transition in TFT, the following approach is used. First the latent heat is assumed to occur over a small finite temperature range, δT_m. This can be set arbitrarily small. At present $\delta T_m = 1$ °C, which represents a small fractional width in melt temperature for all SBL mirror materials of current interest (typically $\delta T_m/T_m = 10^{-2}$).

Then an effective specific heat for fusion ($c*$) is defined as:

$$c* = \frac{L_H}{\delta T_m} \tag{7.6}$$

where L_H is the latent heat.

The diffusivity is arbitrarily kept constant, and Eq. 7.1 to the solid value, for the latent heat region. To do so, conductivity is redefined as:

$$k* = \frac{c*}{c} \cdot k \tag{7.7}$$

Then $\frac{\alpha*}{\alpha} = \frac{k*}{\rho c*} \cdot \frac{\rho c}{k} = 1$ assuming $\rho =$ constant.

Since u is constant for the problem and Δx is essentially constant while crossing t, i.e., transition (TFT does not recognize step-jumps in expansion at melt), this procedure also keeps the time-step Δt constant.

Once the temperature of a zone crosses the latent heat and enters pure melt (i.e., $T > T_{melt} + \delta T_m$), the thermal parameters are set to those for liquid. If these are not known (which is sometimes the case), the code defaults to the solid parameters.

This procedure works successfully in practice. Instabilities do not occur, and energy is conserved to very high accuracy (generally << 1 percent). The method also successfully tracks the moving solid/melt boundaries. At each time-step, the code checks whether $T_m^p \geq T_{melt}$ are reset as appropriate and computation continues until $T_m^p > T_{melt} + \delta T_m$, If true, the parameters whereupon they are reset again. For all practical, realistic problems, crossing the latent heat region always takes many thermal time-steps. Consequently, there is no danger of the transition occurring within a single time-step, and such phase transitions have always been traversed in a very smooth manner.

There is a lot into this code including the vapor transition and choice of parameters to set up an example or solve a problem with utilization of this code, which are beyond the scope of this book to fit the two volumes of PUFF-TFT code manuals here; however we recommend that readers get hold of the two volumes by reaching out to Galaxy Advanced Engineering at www.gaeinc.com to purchase them.

7.8 Review of Existing Computer Codes

Ideally an existing code would compute:

- (1) X-ray interactions to give X-ray dose
- (2) Electron/photon cascade energy redistribution
- (3) Thermal conductivity energy redistribution
- (4) Thermomechanical stress-wave response incorporating (1), (2), and (3)

No previous single existing code handled all four aspects simultaneously. handles topics (1) and (4) only, while SANDYL Ref. [24] and TIGERP Ref. [25] handle (1) and (2), and XRT or its new version of it, XRTH Ref. [26, 27] handles (1), (2) and (3). Other codes and techniques also exist, such as POEM (similar to SANDYL) and the analytic diffusion approach of J.C. Garth et al. [28, 29], but these only handle topics (1) and (2).

The most promising codes for adaptation or incorporation are the Monte Carlo SANDYL or TIGERP and the diffusion-analytical XRT (i.e., new version XRTH, which is a combination of XRT and hydro in one code).

7.9 SANDYL Computer Code

This code is Monte Carlo-based and can handle 1-D, 2-D, and 3-D geometries. The physics included are rigorous, allowing for correct X-ray interactions of all types (photoelectric, Compton scattering, pair production and annihilation, fluorescence, and Bremsstrahlung). It handles X-rays in the energy range 0.1 keV to 100 MeV. It handles X rays in the energy range 0.1 keV to 100 MeV. The code fully predicts secondary interactions such as photoelectron production, fluorescence, Auger electron production, and Compton electron excitation, including all relative probabilities and energies. It correctly incorporates true directionality of photoelectrons relative to photon path, i.e., the Fischer and Sauter formulations, and includes Coster-Kronig autoionization effects. The code accounts for electron energy deposition using condensed Monte Carlo techniques (as in the ELTRAN code, Ref. 8) and employs suitably modified Bethe energy-loss formulations. The code easily handles mixed layer geometries allowing for the interface electron/photon cascade energy redistribution effects. The code clearly handles relativistic effects.

The main theoretical limitation concerns low-energy electron behavior for E < 10 keV. The Rohrlich-Carlson formulation of the H. Bethe [30] stopping-power theory is known to give progressively inaccurate answers below 10 keV, yielding unrealistically small electron ranges and correspondingly high stopping powers. These inaccuracies become very significant below 1 keV. Suitable modifications could be developed to correct this deficiency. However, it is probable that no great need exists for such a modification because very few X-ray sources are both low energy and of very short deposition time. As a general rule, a cold X-ray source, whether blackbody or machine-produced, e.g., PITHON[2] or Blackjack-5[3] (BJ-5), tends to have a moderate to long deposition time, e.g., 30 to ISO ns. With the thin optical packages, thermal diffusivity will generally dominate with these time scales, making the necessity for accurate electron behavior prediction less relevant.

The operational problems associated with SANDYL are those common to most Monte Carlo codes: (1) very complex problem definition and setup, (2) generally long run times and high costs, and (3) the use of user-defined parameters.

The user-defined parameters include *N*, the number of input particle histories to be tracked; E_c, the low-energy cutoff below which particles are no longer tracked; and *dz*, the initial zone size for the problem. An increase in *N*, or decrease in E_c and *dz*, will yield answers of increasing accuracy. The penalty, of course, is increased run time and cost.

Ideally, the code user should first perform a parametric study of a typical problem to determine the appropriate values of *N*, E_c, and *dz* necessary to yield an answer of specified accuracy. These values can then be used for other similar problems. However, this is rarely done. Instead, the user either uses rules of thumb or inspired guesses for such parameters or makes use of code-generated quotes of variances. Consequently, it is rarely possible to know the accuracy of a given run, though usually it will be good to a few percent. Thus, although in principle SANDYL can give very precise answers, in practice, the accuracy is frequently undetermined.

It is necessary to distinguish between precision and accuracy. By increasing N, the code will quote ever-smaller variances, with variance proportional to $N^{-1/2}$. However, this is a measure of precision (repeatability), not absolute accuracy. Accuracy can only be determined by comparison with unambiguous veil-defined experimental data. Unfortunately, such data rarely exist either. However, it must be admitted that the physics in SANDYL are remarkably rigorous; one can therefore hope that small variance implies good accuracy.

To compound the problem, consider the parameter E_c. This is a single value for the entire code run. Whereas a value E – 1 keV would give an accurate c description of a 100 keV electron behavior, it clearly would fail to do so for a low-energy electron, e.g., 1 to 10 keV. An electron born with energy <1 keV would be instantly terminated and its energy deposited locally without tracking. When the X-ray input is a blackbody one containing photons from 1 keV to 100 keV, this clearly gives variable accuracy for differing parts of the electron/photon cascade. However, since SANDYL is inaccurate for low-energy electrons, there is no benefit gained in choosing a very low value of E_c. At low energies, the code is also inconsistent in that it can handle photons down to 0.1 keV but cannot correctly handle the resulting lower-energy electrons.

[2]PITHON: A Defense Nuclear Agency-owned electron beam facility operated by Physics International Co., San Leandro, CA. The machine can be used as a plasma radiation source.

[3]Blackjack-5: A Defense Nuclear Agency-owned electron beam facility operated by Maxwell Laboratories, San Diego, CA. The machine can be used as a plasma radiation source.

It should be noted that most of the experimental data against which SANDYL is tested concerns high-energy input photons (>100 keV). The primary use of such a Monte Carlo code is the determination of charge-related problems in semiconductors. Such devices are usually well shielded, resulting in attack by hard X-rays only. Likewise, Aboveground Test (ACT) simulations usually make use of Bremsstrahlung generation, which again is mostly >100 keV. The result is that very little direct experimental data exists to confirm the code accuracy for X-rays in the 1–100 keV range, appropriate for SBL mirror analysis. The use of cold X-ray Prompt Radiation Sources (such as Blackjack-5) fail to provide the necessary data. Such experiments are usually strongly dominated by thermal conduction, thereby washing out the details of direct X-ray/cascade interactions.

The use of cold X-ray Prompt Radiation Sources (such as Blackjack-5) fail to provide the necessary data. Such experiments are usually strongly dominated by thermal conduction, thereby washing out the details of direct X-ray/cascade interactions.

In conclusion, SANDYL has the capability of yielding very precise results in complex geometries for input photons greater than 10 keV and moderately accurate results for photons of 1 to 10 keV. The code's drawbacks are complexity, run time, and cost.

7.10 TIGERP Computer Code

Many of the comments concerning SANDYL also apply to TIGERP. The latter code is used more frequently and has many of the features of SANDYL. Concerning the drawbacks of Monte Carlo codes, it should be noted that TIGERP has been used to model simple thin layer packages with monoenergetic X-ray inputs of order 20–50 keV. Depressingly, despite run times of 100 CPU hours on a micro-VAX, the resulting doses still had quoted variances of up to S percent, using 500,000 input histories. This problem arises because for very thin layers, the interactions per input photon per layer per pass are small. Consequently, large number histories are required to reduce variances. Since variance $\approx N^{-1/2}$ the problem rapidly becomes onerous if small variances are required and is even worse if a full input spectrum (say 50 energy bins) is being modeled. Unless a user has access to a free CRAY computer, it is difficult to justify an attempt to use a Monte Carlo code as the cascade solver for a combined X-ray/cascade/thermal/stress response code.

7.11 XRT and XRTH Computer Codes

The XRT code is a mixed analytical-diffusion code, while its latest version known as XRTH is exactly what XRT code handles, yet the hydro problem has been combined with it, and that is why it is called XRTH. It handles only 1-D problems, appropriate to the mirror. It computes the initial X-ray interactions (photoelectric absorption, Compton scattering, and coherent scattering) for a multilayer package using the FSCATT routine. FSCATT (i.e., available from www.gaeinc.com) has a logic somewhat similar to that of TRIDEP (in PUFF). However, it is more accurate in that it uses the discrete ordinates logic and defines fluence in terms of discrete angular bins which are resolved for 1-D calculations. It therefore more accurately calculates scattering effects. For thin layers, however, such scattering effects are usually minor since the layers are only a small fraction of the scattering mean free paths. For thick materials FSCATT is somewhat more accurate than the TRIDEP routine, but only for strong scattering. The FSCATT routine first computes total interactions per volume throughout the mesh. The interactions are apportioned according to true photoelectric absorption or scattering. The scattered portion is treated as a series of internal sources, and sequential forward and reverse sweeps through the mesh are performed on this scattered fluence. On each sweep the interactions are again apportioned to photoelectric or scattered events. Total photoelectric absorption per zone is summed over all sweeps until the value is asymptotic within a required accuracy. The technique gives accurate values for final deposition throughout the multilayer mesh. The scattered portion also includes energy deposition due to Compton collisions and accounts for the downgrading of the scattered photon's energy. Hence, the total deposition is a mixture of true photoelectric effects and Compton effects.

For each photoelectric absorption, XRT/XRT computes the ensuing electron/photon cascade, generating relative probabilities and energies for each secondary. For Compton electrons, XRT assumes local deposition, i.e., no range. For all other particles, XRT tracks the spatial redistribution. Fluorescence is treated similarly to the initial photon input, but the isotropic nature of emission from the absorbing atom which captures the original X-ray photon is taken into account.

Electron deposition behavior is treated in a similar manner to that used by L.V. Spencer [31], i.e., method of moments, and embodies the M.J. Berger [32] version of the H. Bethe stopping powers for $E > 10$ keV and modifications fitting the ORNL data used by E.A. Burke [33] for $E < 10$ keV. This gives more realistic ranges and stopping powers at low energies. In this respect, XRT is better than SANDYL or TIGERP. However, XRT is not intended to handle high-energy events accurately, e.g., >100 keV, and, consequently, does a poor job of computing Compton events and cannot handle relativistic effects, pair production, or Bremsstrahlung. These limitations are of little, or no, concern for most problems concerning the mirrors.

The Spencer approach to computation of electron deposition is essentially a diffusion formulation. However, whereas Spencer used high-order moments in his calculations, XRT restricts itself to only the first two terms of the Legendre polynomial, i.e., the P(1) approximation. However, a warning must be given. It is well-known that the P(l) approximation can give erroneous results for interactions which occur within a mean free path of a boundary. Unfortunately, for thin layers and high-energy particles (which have large ranges, or mfps), the entire package can be within this regime. It is not easily determined what error arises from this polynomial truncation (a computational comparison is required to evaluate this), but it is probably not large. XRT treats all electrons similarly as though isotopically emitted from the X-ray absorbing atoms. For Auger and Coster-Kronig electrons, this is true but not for photoelectrons, i.e., the Fischer/Sauter formulations are not included. Except for the combination of high X-ray energy and low Z material, this will not give large errors. The same comment applies to the failure of XRT to donate a range to Compton excited electrons.

XRT/XRTH computes the electron/photon redistribution per input photon and sums over the input spectrum to yield a realistic, and reasonably accurate, redistribution dose profile within the multilayer package. This profile will be less accurate (on average) than a SANDYL computation for high-energy particles but probably more accurate than SANDYL for low-energy particles, i.e., <5 to 10 keV.

XRT/XRTH further has the capability for computing thermal conduction effects by using the Backward Euler algorithm with a finite spatial difference mesh. It is capable of incorporating temperature-dependent specific heats. The thermal conduction mode applies both during and after the X-ray deposition time and, thus, includes initial heating and late-term equilibrium effects.

Operationally XRT is less complex than SANDYL or TIGERP and will run faster. Its accuracy is not dependent on user-input parameters in the same way as a Monte Carlo code (since it involves no statistics). It has the advantage of including thermal conduction effects. Its overall credibility should be good. However, it is not accessible for the general user, being proprietary to S-CUBED; however this company is no longer in existence and has been dissolved over a decade ago. To the best the author's knowledge, the code could be obtained from Galaxy Advanced Engineering at www.gaeinc.com.

7.12 ELTRAN Computer Code

ELTRAN is a one-dimensional Monte Carlo electron transport code which employs a quasi-Monte Carlo computational scheme to compute one-dimensional, electron deposition, in semi-infinite composite slabs. It has the capability of computing either the energy deposition profile distribution of stopped electrons or the spectral properties of electrons transmitted by a thin slab, which may be solid, liquid, or gas. The incident electrons may be mono-energetic, or they may form a spectrum; they may be incident perpendicular or at an angle to the slab.

It employs a quasi-Monte Carlo scheme which involves calculation of the trajectory of an individual electron. This calculation is repeated until a sufficient number of electrons or case histories have been accumulated so that the user obtains the desired degree of statistical precision. Instead of considering each of the tens of thousands of individual collisions, the electron is given an incremental displacement, and the small energy loss and small effective scattering angle are computed. The electron is given additional incremental displacements until the electron is absorbed (stopped), backscattered, or transmitted.

Again, this incrementation is repeated for additional electrons until the desired number of case histories is run. The energy loss is computed using the continuous slowing-down approximation, with the effects of density, range straggling, and bremsstrahlung included. The scattering angle is computed by means of the Moliere multiple scattering angle distribution.

7.13 FSCATT Computer Code

FSCATT is an energy deposition and transport code for slab geometry which includes the effects of photoelectric absorption, fluorescence, and Compton scattering. The intensity of the secondary radiation field (scattered or fluorescent radiation) is resolved using discrete angular and energy groups. The mesh (one or more material layers) is swept alternately in the forward and backward direction until the energy remaining to be distributed is less than some specified fraction of the initial energy incident on the mesh. The motivation for using this type of numerical technique is to overcome the expense and uncertainty involved when using Monte Carlo codes. Observations indicate that results at least as accurate as those obtained with Monte Carlo codes can be obtained with substantially less computer time using the FSCATT/ code. The preceding statement is not intended to imply that all FSCATT calculations are inexpensive. While the majority of calculations run rapidly, increasing the number of angular groups and energy groups used to resolve the scattered and fluoresced energy increases the running time in a nonlinear fashion.

FSCATT is designed to allow the experienced user a great deal of latitude in problem definition while at the same time being a fairly automatic and simple code for the inexperienced user to operate. All of the physical data needed for 40 elements are contained in the code as are the data for a blackbody (Planck) spectrum.

FSCATT allows for the following radiation source options:

1. A blackbody spectrum.
2. A mixture of 1–10 different temperature blackbody spectra, either:
   (a) Distributed over 90 energy groups, each containing equal amounts of energy
   (b) Distributed over an energy group structure supplied by the user
3. A spectrum read from file.
4. Data supplied by the user consisting of points describing the spectrum (energy, intensity) and the number of energy groups desired between adjacent points.
5. The angle of incidence of the radiation on the slab can be specified for any of the spectrum options.

The geometry of the code allows for the specification of ten material layers each consisting of up to five individual constituents (elements). In addition, the various physical processes accounted for in FSCATT (i.e., photoelectric absorption, fluorescence, and scattering) can be suppressed by the user in individual layers, thus decreasing the computer time required to complete a calculation.

Again, note that most of these above codes can be found and purchased for a nominal fee from Galaxy Advanced Engineering, on their website www.gaeinc.com.

7.14 Other Recommended Computer Codes

There are few other codes that are recommended for purpose that are useful for transient heating material ablations, and they are listed below.

7.15 ASTHMA Computer Code

ASTHMA stands for Aerotherm Axisymmetric Transient Heating and Material Ablation program. This code deals with problem of heat transient conduction for two-dimensional axisymmetric bodies. Multiple non-charring, anisotropic materials may be studied with this code.

The surface boundary condition has three options, including an unusually general thermochemical erosion or ablation condition as well as simplified radiation and specified temperature options. It accounts for two specific kinetically controlled carbon oxidation reactions, one specific kinetically controlled carbon reduction reaction, one kinetically controlled water gas shift reaction, and any number of gas-phase equilibrium reactions for any combination of ablating materials and environments.

Chief applications for the computer program are rocket nozzles and entry vehicles. The following sketch serves to clarify various physical aspects of the ablation problem treated. The table below cites the

The sketch in Fig. 7.16 serves to clarify various physical aspects of the ablation problem treated.

Also, Table 7.1 below cites the aspects an 3 indicates which descriptions of the analysis are to be found in the present report and which are to be found elsewhere. Note that the ASTHMA Program described in here is usually used in conjunction with another program which handles the chemical state computations required in the heated surface boundary condition definition. Any one of several thermochemistry computer programs is suitable for this purpose, including EST [34], EST2 [35], EST3 [36], ACE [37], and GASKET [38]. These codes all differ to various degrees in details of the thermochemical solution. The GASKET version is the newest and contains a simplified chemical kinetics treatment of carbon oxidation particularly pertinent to rocket nozzles.

Any one of several thermochemistry computer programs is suitable for this purpose, including EST [34], EST2 [35], EST3 [36], ACE [37], and GASKET [38]. These codes all differ to various degrees in details of the thermochemical solution. The GASKET version is the newest and contains a simplified chemical kinetics treatment of carbon oxidation particularly pertinent to rocket nozzles.

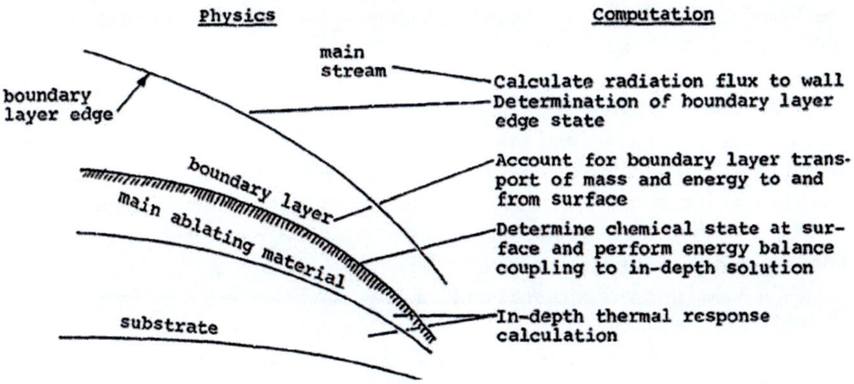

Fig. 7.16 Various physical conditions of the ablation problem

Table 7.1 Summary of computational task

Summary of computational tasks		
Task	Program	Where described
1. Determine radiation flux from free stream to wall	Not computed but accounted for by ASTHMA. User must provide as input	
2. Determine boundary layer edge state (Chemical)	GASKRT, etc.	Ref. [33–38]
3. Account for boundary layer transport of:		
a. Mass	GASKRT, etc.	Ref. [33–38]
b. Energy	ASTHMA, but user must provide convective transfer coefficients as input	
4. Surface thermochemical state	GASKRT, etc.	Ref. [33–38]
5. In-depth temperature	ASTHMA	See below

Consequently, when it comes to ASTHMA code and surface state solution, it deals with GASKET as a companion code for ASTHMA.

7.16 Mass Offgasing Model (MOM) Computer Code

MOM computer code is a general thermo-chemical solver code used to compute the surface conditions of ablating materials under low pressure and high heat flux rates. It employs a rigorous thermo-chemical treatment of solid-vapor equilibrium. This code is intended to model the chemical and physical processes occurring when a solid rapidly ablates into a low-pressure environment due to intense, localized heating from a laser beam. The ablating vapor may contain a number of individual gas-phase species coupled to a surface composed of a number of distinct surface species.

In MOM code arbitrary solid materials may be considered, provided the elemental composition is known and suitable thermodynamic data is available. Materials which "char" or decompose in-depth also may be treated unlike many previous models, which treat only materials where all reactions occur directly at the material surface. In addition, environmental conditions (such as pressures) are not limited allowing nearly any external conditions to be treated. Hence, the MOM code is able to treat a wide variety of materials under a large number of external environments.

While the MOM code is a self-contained entity and may be considered as a "standalone" module, many of the concepts used in the formulation of this code are drawn from other sources. Thermochemical ablation has been studied in great detail in earlier works, and the principles for treating such systems have been well defined. The MOM code is an extension of these efforts as it builds upon the concepts by adding refinements to existing theories. The MOM code does not supersede them. Therefore, users unfamiliar with the general principles of material response modeling should consult additional sources of information. References 1 and 2 present additional information we consider useful.

Again, most of these above codes can be found and purchased for a nominal fee from Galaxy Advanced Engineering on their website www.gaeinc.com.

References

1. Zohuri, B.: Directed energy weapons, physics of high energy lasers (HEL), 1st edn. Springer Publishing Company, Cham (2016)
2. Zohuri, B.: Directed energy beam weapons, 1st edn. Springer Publishing Company, Cham (2019)
3. Zohuri, B.: Scalar wave driven energy applications, 1st edn. Springer Publishing Company, Cham (2019)
4. DoD 4120.15-L, Model Designation of Military Aerospace Vehicles (PDF). US Department of Defense. May 12, 2004
5. Airborne Laser returns for more testing. Air Force. Archived from the original on March 8, 2007
6. Archived January 28, 2010, at the Wayback Machine
7. Wolf, J.: U.S. successfully tests airborne laser on missile. (February 12, 2010). reuters.com
8. Boeing YAL-1 Airborne Laser impacted by Pentagon spending priorities. Flight Image of the Day. Archived from the original on October 20, 2013
9. Mowthorpe, M.: The militarization and weaponization of space, pp. 141–142. Lexington Books, New York (2004) ISBN: 0-7391-0712-5
10. Lambakis, S.: On the edge of Earth: the future of American Space Power, pp. 124–125. The University Press of Kentucky, Lexington, KY (2001) Also gained knowledge of major —near-peer|| capabilities during briefings to Air War College students during tours of Air Force Research Laboratory facilities at Kirtland AFB, NM from 24–28 Aug 2009 and National Air and Space Intelligence Center briefings at Wright-Patterson AFB, OH from 23–24 Sep 2009

11. Beason, D.: The E-bomb: how America's new directed energy weapons will change the way future wars will be fought, pp. 26–27. Da Capo Press, Cambridge, MA (2005) Figure 1 source: http://lasp.colorado.edu/cassini/images/Electromagnetic%20Spectrum.jpg (Accessed 6 Dec 2009)
12. Nielsen, P.E.: Effects of directed energy weapons, p. 170. Directed Energy Professional Society, Albuquerque (2009)
13. O'Hanlon: Neither star wars nor sanctuary, p. 72. Directed Energy Professional Society, Albuquerque
14. Anderberg, B., Wolbarsht, M.: Laser weapons: the dawn of a new military age. Plenum Press, New York (1992)
15. Edward A. Duff (Long Range Strike Product Line Leader, Air Force Research Laboratory Directed Energy Directorate, Kirtland AFB, NM) in discussions with author, 13 January 2010. Most current high-energy lasers produce beams in the shortwave infrared portion of the electromagnetic spectrum (invisible to naked eye)
16. Fusion Energy Foundation: Several variations are possible for a relay mirror system, including placing the relay mirrors on an unmanned aerial vehicle, a high-altitude airship (—blimp||), or on satellites. In: Beam defense: an alternative to nuclear destruction. Aero Publishers, Inc., Fallbrook (1983)
17. William, H.: Possel, —lasers and missile defense: new concepts for space-based and ground-based laser weapons Occasional paper no. 5, Center for Strategy and Technology, Air War College. Air University Press, Maxwell AFB (1998) Estimates for laser transmission efficiency validated by Mr. Edward Duff of the Air Force Research Laboratory Directed Energy Directorate during discussions on 13 January 2010
18. Stephens, M.J. Lt Col, USAF.: Harnessing light: laser/satellite relay mirror systems and deterrence in 2035. A Research Report Submitted to the Faculty In Partial Fulfillment of Graduation Requirements, 16 February 2010
19. Stephen, G.L.: Laser options for national missile defense Air Command & Staff College, pp. 65–66. Air University Press, Maxwell AFB (1998) While amplification of the beam is not necessary, especially if the ground station generates the 4 to 10 megawatts posited by AFRL, amplification is possible
20. David, M.M.: Directed energy weapon system for ballistic missile defense Air War College, p. 31. Air University Press, Maxwell AFB (2009)
21. Air Force Research Laboratory.: Relay Mirror Technology, U.S. Air Force Fact Sheet, Kirtland Air Force Base, NM, April 2006., http://www.kirtland.af.mil/shared/media/document/AFD-070404-044.pdf (accessed 31 October 2009). Also gained insight during Air War College student tour of Air Force Research Laboratory facilities at Kirtland AFB, NM from 24–28 Aug 2009
22. Duff.: 4A – globally deliver directed energy and non-kinetic effects, Air Force Research Laboratory Directed Energy Directorate briefing on long-term research areas, August 2008. Also gained insight during briefings given to Air War College students during tours of Air Force Research Laboratory briefings, Kirtland AFB, NM, 24–28 Aug 2009
23. Webber, H.C., Cullis, A.G., Chew, N.E.: Computer simulation of high speed melting of amorphous silica. Appl. Phys. Lett. **43**(7), 669 (1983)
24. Colbert, H.M.: SANDYL – a computer program for calculating combined photon-electron transport in complex systems, SLL-74-0012 (1974)
25. Halbleib, J.A., Mehlhorn, T.A.: ITS: the integrated TIGER series of coupled electron Photon Monte Carlo transport codes, SAND84-0573. Sandia National Laboratories, Albuquerque (1984)
26. Triplett, J.R.: The XRT code, SSS-R-83-5829. Systems, Science and Software, La Jolla (1982)
27. Triplett, J.R., Sweet, K., Shkoller, B.: XRTH user's manual and reference guide, SSS-R-89-9891, September 1988, revised November 1988

28. Garth, J.C., Chadsey, W.L., Sheppard Jr., R.L.: Monte Carlo analysis of dose profiles near photon irradiated material interfaces. IEEE Trans. Nuc. Sci. **NS-22**(6), 2562 (1975)
29. Garth, J.C.: Diffusion equation model for Kilovolt electron transport at X-irradiated interfaces. IEEE Trans. Nuc. Sci. **NS-25**(6), 1598 (1978)
30. Bethe, H., Ashkin, J.: Experimental nuclear physics, vol. 1. Wiley, New York (1953)
31. Spencer, L.V.: Energy dissipation by fast electrons N.B.S. monograph, vol. 1. Wiley, New York (1959)
32. Berger, M.J.: Monte Carlo calculation of the penetration and diffusion of fast charged particles Methods in Computational Physics, vol. 1. Academic Press, Inc., New York (1963)
33. Burke, E.A.: Soft X-ray induced electron emission. IEEE Trans. Nuc. Sci. **NS-24**(6), 2505 (1977)
34. McCuen, P.A., Schaefer, J.W., Lundberg, R.E., Kendall, R.M.: A study of solid-propellant rocket motor exposed materials behavior (AFPJL-TR-65-33 (AD-462 331)). Vidya Division, Itek Corp, Palo Alto (1965)
35. Aerotherm Corporation, Palo Alto, Calif: User's manual, Aerotherm equilibrium surface thermochemistry program, version 2. Aerotherm Corporation, Palo Alto (1966)
36. Aerotherm Corporation, Mountain View, California: User's manual, aerotherm equilibrium surface thermochemistry program, version 3 Volume I – program description and sample problems, volume I – FORTRAN variables, flow charts, and program listings, AFRPL-TR-70-93. Aerotherm Corporation, Mountain View (1970)
37. Powars, C.A., Kendall, R.M.: User's manual, aerotherm chemical equilibrium program. Aerotherm Corporation, Mountain View (1969)
38. Aerotherm Corporation, Mountain View, California: User's manual, aerotherm. Aerotherm Corporation, Mountain View
39. Weber, M.J.: Handbook of laser wavelengths. CRC Press, Boca Raton (1999) ISBN 978-0-8493-3508-2
40. Zohuri, B.: Inertial confinement fusion driven thermonuclear energy, 1st edn. Springer, Cham (2017) 2017 edition (January 29, 2017)

Index

B. Zohuri, *Thermal Effects of High Power Laser Energy on Materials*,
https://doi.org/10.1007/978-3-030-63064-5